Lecture Notes in Networks and Systems

Volume 257

The series "Lecture Notes in Networks and Systems" publishes the latest developments in Networks and Systems—quickly, informally and with high quality. Original research reported in proceedings and post-proceedings represents the core of LNNS.

Volumes published in LNNS embrace all aspects and subfields of, as well as new challenges in, Networks and Systems.

The series contains proceedings and edited volumes in systems and networks, spanning the areas of Cyber-Physical Systems, Autonomous Systems, Sensor Networks, Control Systems, Energy Systems, Automotive Systems, Biological Systems, Vehicular Networking and Connected Vehicles, Aerospace Systems, Automation, Manufacturing, Smart Grids, Nonlinear Systems, Power Systems, Robotics, Social Systems, Economic Systems and other. Of particular value to both the contributors and the readership are the short publication timeframe and the world-wide distribution and exposure which enable both a wide and rapid dissemination of research output.

The series covers the theory, applications, and perspectives on the state of the art and future developments relevant to systems and networks, decision making, control, complex processes and related areas, as embedded in the fields of interdisciplinary and applied sciences, engineering, computer science, physics, economics, social, and life sciences, as well as the paradigms and methodologies behind them.

Indexed by SCOPUS, INSPEC, WTI Frankfurt eG, zbMATH, SCImago.

All books published in the series are submitted for consideration in Web of Science.

More information about this series at https://link.springer.com/bookseries/15179

Irena Tušer · Šárka Hošková-Mayerová
Editors

Trends and Future Directions in Security and Emergency Management

 Springer

Editors
Irena Tušer ⓘ
Department of Security and Law
Ambis College
Praha, Czech Republic

Šárka Hošková-Mayerová ⓘ
Faculty of Military Technology
Department of Mathematics and Physics
University of Defence
Brno, Czech Republic

ISSN 2367-3370   ISSN 2367-3389   (electronic)
Lecture Notes in Networks and Systems
ISBN 978-3-030-88909-8   ISBN 978-3-030-88907-4   (eBook)
https://doi.org/10.1007/978-3-030-88907-4

This Springer imprint is published by the registered company Springer Nature Switzerland AG
The registered company address is: Gewerbestrasse 11, 6330 Cham, Switzerland

Preface

The book deals with basic theoretical and methodological approaches in the field of security sciences and security management in connection with its subsystems—the application of emergency management and risk management. It presents the results of research and other scientific activities, with a reference on potential and real applications, of the research team.

The concept of security and vulnerability of society to threats is the basic starting point for the study of social aspects of emergency situations and disasters.

The task of all governments of developed countries is to ensure, to the appropriate extent, the security of the population, the defense of the sovereignty and territorial integrity of the country, and the preservation of the requirements of a democratic state governed by the rule of law. The institutional tool for achieving these objectives is a comprehensive and functional security system that continuously adapts to the current security situation in the country and in the world.

Security is based on the principle of the ability to ensure the safety of an individual, the protection of his life, health, freedom, human dignity, and property. To systematically fulfill this principle, it is essential to support the security of state institutions, including their full functionality, and to develop processes and tools that serve to strengthen the security and protection of the population. The government is primarily responsible for ensuring security, but active cooperation between the country's population, legal and natural persons, and public authorities is also desirable to reduce the risk of threats. In this way, the overall resilience of the society to security threats is strengthened.

For instance, unilateral attempts of some countries to build spheres of influence through a combination of political, economic and military pressure, and intelligence activities can be considered a threat, where these activities and repressions also take place in cyberspace. These implicit tendencies are associated with the violation of political and legal commitments leading to European security.

The growing severity of non-military threats (e.g., mass migration, cyber-attacks, soft target attacks, or abiotic or biotic threats) and the worsening security situation in the territories immediately adjacent to the Member States of the North Atlantic Treaty Organization (NATO) and the European Union (EU) place ever-increasing demands on Europe's ability to react independently. These aspects also highlight the shortcomings of the international community in their military capabilities and preparedness for resistance to security threats.

In addition to positive aspects, the process of globalization also brings negative aspects which, in the absence of an effective international regulation system, include the possibility of abusing the interconnectedness of financial markets, misuse of information and communication technology and infrastructures. Uneven economic development and the easy dissemination of radical political and religious ideas also contribute to instability. In connection with the high mobility and migration of people, the rate of spread of infectious diseases, including diseases with the potential for a pandemic, is increasing. One of the effects of the current development of the world economy may be the weakening of the position of Europe and the USA, the prioritization of national interests at the expense of shared ones, and other tendencies leading to reduced solidarity and efficiency in the international communities NATO and the EU.

Non-state actors, in comparison with countries and international organizations, are able to respond more flexibly and take advantage of the opportunities arising from globalization, especially from the integration of information and communication technology, trade, and transport. This reduces the importance of the position of countries as entities that use a monopoly to utilize their power and regulate information and economic flows. In contrast, the ability of non-state actors to threaten the interests of countries, replace elements of the state system with their own structures, realize their ambitions in a given territory, and threaten the security of the population, stability, and integrity of affected states through violence increases. In this context, demographic change is also an increasing risk in the area of security threats, especially the risks arising from the aging of the population in developed countries and uncontrollable migration.

Among the security threats of a country, we can also include the probability of occurrence of extremism, crime, or local armed conflicts growing in connection with the issues of poverty, long-term social exclusion, and lack of fulfillment of basic needs and services. The emergence of these threats also instigates social and economic backwardness of regions, especially in the developing world. The threat of terrorism, as a method of forcibly pursuing political goals, is also still high. The importance of critical infrastructure protection is becoming more and more intense, especially in the area of transport and logistics of strategic raw materials, the vulnerability of which is exploited by state and non-state actors. The trend of abusing the position of the sole supplier of these raw materials or as a transit country to assert its interests has an impact on meeting the basic needs of the countries concerned and threatens the political cohesion of NATO and the EU. Another example of current threats may be the effects of climate change on human health and the environment. However, fears of this change alone can lead to growing tensions between countries, result in humanitarian crises with implications for local, state, or even international structures, including the potential escalation of local conflicts that may be accompanied by increased migration. The effects of extreme weather can also have an impact on a country's economy or the supply of raw materials. Finally, the spread of infectious diseases (pandemics) also increases the vulnerability of the population and places greater demands on the protection of public health and the provision of health care.

Key international organizations are also responding to current trends and changes in the security environment. In its 2010 Strategic Concept, NATO defines its three main tasks: to ensure the collective defense of its members, manage security crises outside their territory, and build cooperative security with their partners. In addition to the ability to defend member states from attack, it emphasizes the building of relevant skills to address the full spectrum of crises, from preventive tools to achieving stability in post-conflict situations.

The European Union is also working to strengthen its activities in the field of crisis prevention and crisis management. Its aim is to strengthen the Common Security and Defense Policy (CSDP) so that it is a reliable element of external influence. It therefore seeks to continue to develop civilian and military CSDP skills, to take a comprehensive approach to preventive measures and crisis management, and to also strengthen the capacity of third countries and regional organizations to deal with emergency situations through the provision of training or equipment. In accordance with the EU Internal Security Strategy, the EU aims to strengthen long-term forms of cooperation and solidarity among the Member States in order to protect EU citizens more effectively from internal security risks.

Higher frequency of extraordinary events (EE) and emergency situations (ES), associated with threats of anthropogenic and natural nature, places higher demands on ensuring the protection of the population, the environment, material assets, and increasing the resilience of society in general, including adaptation measures. The tools for the preparation and solution of EE/ES are also provided by the state internal security subsystem, i.e., emergency management and risk management. The goal of emergency management of a country is to protect the community by coordinating and integrating all activities necessary to build, maintain, and improve the ability to mitigate, prepare for, respond to, and recover from threats, natural disasters, terrorist acts, or other man-made disasters. Risk management tools are also used for the implementation of these activities. It makes it possible to identify current risks, evaluate their severity, and set priorities for the implementation of countermeasures to minimize their occurrence. The stakeholders involved in the internal security of the state are the armed security forces and integrated rescue system. The topics and tools described, intended for increasing the security of society, related to the above-mentioned issues, are dealt with in the presented chapters of this book.

In relation to the above mentioned, the book deals with basic theoretical and methodological approaches in the field of security sciences and security management, in connection to its subsystems—the application of emergency management and risk management. It presents the results of the research and other scientific activities of the research team.

The mission of this book is to support new scientific knowledge and the organization of security activities. The current direction (dimension) of security practice requires a real and effective transfer of scientific knowledge. The monograph is divided into five consistent sections, which deal with issues in the areas of security, crisis management, threat and risk assessment, cyber risks, preparation and management of emergencies.

The first part—*Safety and Security Science*—presents the issue of forming security sciences as practical sciences. It shows safety as a starting point and an object of research. It presents security sciences in interdisciplinary and multidisciplinary fields.

The second part—*Security and Emergency management*—deals with all phases of crisis management. It focuses on the definition of a conflict from the perspective of international relations as well as on building community resilience in legal norms. It provides with an approach to the protection of soft targets as well as the possibilities of involving municipalities in their protection. It also evaluates the preparation of the country for tackling consequences on society, such as aspects of immigration.

The following section—*Threats and Risks*—focuses on the security threats of the state or the resilience of the territory to anthropogenic threats. There are also predictions of risk phenomena, such as the deployment of the armed forces abroad.

The fourth part—*Cyber Risks*—discusses attacks that take place in cyberspace. The individual chapters deal with different types of attack, such as cyberterrorism or cybercrime. The authors also weigh the effectiveness of risk management in cyberspace.

The final fifth part—*Extraordinary Event, Preparation, and Solutions*—defines certain extraordinary events associated with a particular subject. Subsequently, the preparation of the subject and its crisis solutions are presented. This section also discusses other related disciplines, such as post-traumatic care or infrastructure of logistics, i.e., possible approaches and methods to mitigate supply chain constraints due to an emergency.

The aim of the presented chapters, from the authors' point of view, is to inform about the current empirical results of their work in connecting science, theory, and practice in the individual fields that are related to security and emergency management. The chapters present research work and case studies from international, state as well as regional levels. Methods, suggestions, and recommendations for increasing the resilience of the discussed systems may be applied in practice. The results of these studies may lead other stakeholders, with some modifications, to take timely preventive measures and an approach to mitigating security threats. The results are also interesting for the international community.

It is intended primarily for scientific communities established in security sciences, theorists, and experts working in various positions and levels of security organizations, universities with specializations in security studies, but also for the expert public interested in security issues or entities directly responsible for security and emergency management.

Prague, Czech Republic   Irena Tušer
Brno, Czech Republic   Šárka Hošková-Mayerová

Contents

Extraordinary Events, Preparation and Solutions

Safety and Security Science

Science on Safety and Security Is Interdisciplinary and Multidisciplinary Discipline

Dana Prochazkova and Hana Bartosova

Abstract In the current period, the need for new scientific knowledge and scientific organisation of security activities is significantly increasing. Professional research shows that security activities contain strong elements of social, especially legal, sociological, psychological, natural and technical, but also economic, managerial and informatics. Launched discussions on the status (or even merits) of security sciences require so that the discussion of their theoretical, methodological and social foundations might be preceded by at least a brief sketch of the origins of their constitution, gradual development and systemic integration into the system of existing scientific disciplines. The authors of the chapter present genesis, arguments and the latest findings in favour of shaping safety the science on safety and security.

Keyword Security · Safety · Safety management · Security management · Safety engineering · Interdisciplinary and multidisciplinary scientific field

1   Introduction to the Issue

The main goal of humans is their security and possibility of development. The term security refers to the state (condition) of the system in which humans live, in which the occurrence of damage to protected interests (further assets) has an acceptable probability (i.e. it is almost certain that harm will not arise); antonym to security is a danger. Such state (condition) of the system guarantees humans existence, satisfied life and development. The term safety refers to a well-ordered set of anthropogenic measures and activities by which humans ensure their security [22]. The humans' needs are expressed by the Maslow's Pyramid [20]. To ensure the current needs, it goes on integral safety, i.e. safety, which considers several various protected

D. Prochazkova (✉)
Czech Technical University in Prague, Technická 4, 166 00 Praha, Czech Republic
e-mail: prochdana7@seznam.cz

H. Bartosova
Ambis College, Lindnerova 575/1, 180 00 Praha 8, Czech Republic
e-mail: hana.bartosova@ambis.cz

© The Author(s), under exclusive license to Springer Nature Switzerland AG 2022
I. Tušer and Š. Hošková-Mayerová (eds.), *Trends and Future Directions in Security and Emergency Management*, Lecture Notes in Networks and Systems 257,
https://doi.org/10.1007/978-3-030-88907-4_1

assets at once, which are interconnected by links and flows; antonym to safety is a recklessness/criticality.

The transformations of society and human civilization towards greater complexity, require the systemic solutions and place significantly different requirements on the safety reflection in the form of new scientific knowledge, theory and scientific organization of activities for safety that meet the needs of human civilization in a new stage of its development. This means not only the transformation of society, but also the modification of scientific disciplines and the emergence of new ones, which are derived from societal needs in connection with technological development and current problems of manifestations of diversified security threats. It is mainly about linking the knowledge of scientific disciplines, which is possible only by using the same terms and by mutual efforts of everyone after achieving a common goal, which today includes the term of a safety culture.

The fundamental emphasis on safety is a major phenomenon of recent times, so it needs to be monitored it and to realise, why this has happened and what it means. The concept of integral safety has not been yet widely accepted, as a number of subparts of safety (internal, external, international, fire, nuclear, chemical, food, health, ecological, etc.) are used in practice, and their internal links are overshadowed by the various terms used and the departmental objectives. Only now, the modern concept of safety has been created and elaborated. In doing so, existing contradictions are gradually addressed, e.g. environmental safety objectives, which are very dogmatically addressed, go against:

– the development of technologies,
– the use of nature by humans in recreation,
– the human security in critical situations, etc.

It means that in the new concept, space is primarily defined as the system to which integral safety relates and in which it is ensured. Because its philosophy is built for the needs of humans, such built safety may be referred to as human safety and a system in which it is sought as a human system.

In addition to national safety directed to item "the State", the safety in new concept is associated with natural, social and technological systems. For example, in this context, safety refers to the functioning a particular technological system and it depends on system reliability in carrying out the defined operations, i.e. on the probability with which the intended functions of the system are performed. Technical safety is, therefore, objective. The same pays in the case when integral safety is understood as human safety in the human system. However, in the objectivization of safety, it is necessary to have the courage to exceed the traditional concept of safety.

Most humans perceive safety and security very subjectively, because they are more concerned about the difficulties of everyday life than the concerns associated with occurrence of extreme disasters and issues related to global aspects. Humans have always been threatened in some way by natural disasters, plague, slavery, etc., and therefore, issues related to the human security, i.e. human safety, are not just some today's specificity. Throughout historical developments, they have been changed knowledge and requirements and institutional options to mitigate anything that may threaten safety.

The concept of human security has intellectual roots in the psychological theories of [1]. Theoretical foundations of human safety are derived from the Globalist School of Thought, the paradigm of which argues that the complex processes of globalization entail a number of completely new safety-related problems. Human safety thus understanding differs from the traditional concept of safety in the following points:

1. Safety focuses on individuals and human society rather than the State as a basic unit of humans' organisation.
2. The purpose of human safety is to ensure the security of human society as a community of individuals and not just to ensure the mere protection of territorial sovereignty, with it pays that both objectives are not mutually exclusive.
3. The traditional concept of safety, used in building the executive components of the State to ensure internal and external safeties, emphasizes structured violence. On the other hand, human safety considers a dynamic system that changes and develops over time, i.e. it is accepted that there are previously unknown processes that still need to be explored and to which management of society needs to be adapted.
4. The traditional concept of safety sees the States as competitors, the inter-reactions of which lead to a zero-sum game (victory at the expense of someone else). In promoting the concept of safety in the human system, emphasis is placed on cooperative processes and cooperation.

Although many sources of traditional conflicts remain (State prestige, economic rivalry, nationalism, legitimacy of power, religious intolerance), in broader human community the understanding the human safety, which relies on the human system safety, is expanding, because the sources of danger they are not only phenomena that directly affect humans, but also phenomena that affect them vicariously through a complex network of links and flows in the human system and they are sometimes shifted differently over time.

The concept of "human safety" was competently elaborated in a UN report [30]. The concept of human security explains the security as a feeling of "fearless", "scarcity". In the followed document, it promotes that human security cannot be only reduced to the loss of fear of weapons and war, because it also includes human dignity and the quality of human life. The report formulated the following basic features of human safety:

1. Human safety is universal/integral.
2. Components of human safety are linked to individual protected assets and are interdependent, i.e. they are interconnected by different types of links and flows.
3. Early prevention is more important for the development of human safety than later self-qualified intervention (response).
4. Human safety is focused on humans, not on coping with disasters of all kinds and on the risks and threats arising from them.

5. Human safety is a summary of the conditions and circumstances offered by the institutional environment, but it is not and cannot be the result of mere administrative measures.

These principles show that human safety focuses on protecting the fundamental rights of human life (survival, dignity, existence) from critical and pervasive risks and threats (direct or indirect) in a way that is consistent with long-term human development. I.e., its aim is also protection against harmful phenomena of everyday life, i.e. against disasters of all kinds.

The current concept of human safety creates an extension of the ideas set out so far and is based on the theory of systems [22]. Human safety is understood as the safety of the human system, which is defined and for which sustainable development is ensured through the tool "safety management". The process model for the sustainable development of the human system—the basis is the system safety; the fundamental pillars are the protected assets of the human system, i.e.: the human lives, health and security, the environment, property and the public welfare; and technology and infrastructure; the outputs are the security and development of human society.

The human system is a system of systems, i.e. a set of systems that intersect with each other [22]. The main systems are shown in Fig. 1. It is important for humans that systems have interoperability (the ability to cooperate with each other) that ensures humans' goals. Since humans are not the ruler of the human system, they can for ensuring their goals only manage their behaviour by the way so that their actions and activities might not lead to a violation of interoperability and subsequent degradation of the living system.

Therefore, the safety management [22, 23] needs to consider:

– a systemic concept of community,
– links and flows among many open systems,
– dynamics of development versus human ways of solving problems that are solved on several levels: technical; functional; tactical; strategic; and political; Fig. 2.

The basic levels of governance: political, strategic, tactical, operational/functional and technical, need to be harmonised. The political level is often influenced by the

Fig. 1 Main systems of human system [22]

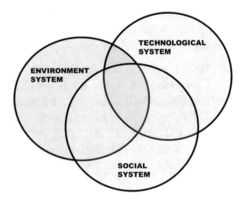

Fig. 2 Main levels on which the problems are solved in human society

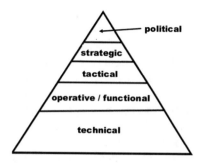

ideas and power objectives of the ruling political representations, and thus is sometimes far removed from the objectives of knowledge-based process management. However, it is important because it implements other levels. It is strongly influenced by phenomena such as: corruption, power relations, abuse of power and lobbyism.

In practice, safety management is represented by a risk management that includes the precautionary principle and seeks the optimal solution for all disasters in a given community [23].

2 Targeting to the Safety

The presented work is a contribution to the examination of safety in its comprehensive form, i.e. it integrates all the attributes of partial safeties, which have been examined as the starting points of safety sciences, as practical sciences [21] in order to develop a new integral theoretical model that will allow application to specific forms of risks and threats to human safety. It draws knowledge base and methodology from a number of scientific disciplines, and therefore, it has a multidisciplinary character. The new findings to which the authors come, are verified using the technique of secondary analysis and synthesis of research data and outputs of safety projects, empirical-analytical methods and generalized examples of good practice in the field of safety management.

Although the European Union FOCUS project [13] has shown that corruption and abuse of power have proved to be the biggest drag on progress, based on the knowledge of the academic environment, we would like to draw attention to the too narrow specialisation of a number of experts, which is still supported by the idea of research agencies, which is expressed by the fact that the software tool is the top result of the project, and therefore, they do not study the quality of professional background of the software and the way of use of the experience of good engineering practice for project proposals, which predetermines the ability of the software to solve real problems.

The above-mentioned facts are surprising at a time when world experts in all fields are focused on risk settlement and point to the existence of random and knowledge (epistemic) uncertainties in solving the problems of risk management and trade-off with risks that need to be well addressed in order to achieve the required expected objectives in practice—see list of reference given in Annex.

Based on recent professional work, standards and standards ([3, 5–8, 19, 21, 25, 28], references in Annex [32]), EU documents [9–12] and the UN [30], UNEP [31], the safety is rather focused on human and human society, i.e. not on the State as a basic unit of human organisation. As mentioned above, in a systemic sense, human safety is understood as the integral safety of a human system that has the dimensions of political, environmental, economic, technical, food, health, social, personality and community, which means that it is multidimensional. This is the system attribute on which the system existence depends. Due to its nature, integral safety is not limited to unilateral solutions such as repression, but deals with situations affecting a certain level of safety through the so-called safety chain, which consists of the following parts: proactivity (elimination of structural causes of random and knowledge uncertainties which disrupting the safety), prevention (elimination of direct causes of precarious situation violating the current state of security); preparedness (to deal with a situation in which the security is disturbed); response; and renewal [22].

Processing the further paragraphs uses a knowledge base, which is made up of data from the above-mentioned works and other works collected in [22, 23]. The data are analysed, logically compared on the basis of a systemic understanding the reality, and evaluated in order to find mutual connections and conflicts. Then, by synthesis, a model is created that meets the requirements for the human security and development and respects the existence of several systems that have different goals and mutually interconnect, i.e. the conditions for their coexistence are specified in accordance with the work [2].

3   Analysis and Evaluation of Basic Knowledge

Humans' long-term goal is a safe world with sustainable development. Because human system is open a system of systems, it is interconnected and dependent on the system of the Earth planet and the higher systems with which the planet's system is interconnected. Because main systems of human system, namely social, environmental and technological (Fig. 1) have sometimes different natures and different objectives, which leads to conflicts, and because humans have a certain potential to influence and direct the behaviour of a complex system, they need to be aware of the fact in question and their aim needs to be coexistence of systems, i.e. the o implementation of such measures and activities that will limit the emergence of conflicts and, where appropriate, to resolve conflicts among those systems and those protected assets in favour of a safe human system with potential for development.

On the basis of contemporary knowledge, two basic requirements for discipline aimed at ensuring a safe human system arise from the systemic conception of the issue:

- to use certain basic terms: security and danger; disaster, hazard and risk; safety and recklessness (criticality),
- to apply a comprehensive approach, which means not only addressing the problem of disaster management, i.e. emergencies, accidents, extra ordinary events, etc., but a whole chain of management sections, i.e. prevention, preparedness, response and recovery, with great emphasis on lessons learned from emergency response and recovery problems, and in particular after critical situations that cause humanitarian crises; i.e. to ensure the application of a strategic, systemic and pro-active approach based on global expertise and experience.

Security or danger depends on the processes, events and phenomena that take place in human society, the environment, the planetary system, the galaxy, and other higher systems. As it was said above, security is a condition in which harm to humans and other protected assets is unlikely, and danger is the condition for which the contrary claim applies. A disaster refers to all phenomena that, from a certain size, cause harm, loss and damage to humans and/or other protected assets. A hazard is the normative size of a disaster expressed in the relevant physical or other units given by the nature of a disaster, which determines the limit to which humans take measures and actions to ensure that they and publicly protected assets are protected from the impacts of disasters with a size lower or equal to this value. The risk is the likely size of unrequired impacts (losses, damages and harms) caused by a disaster size equal to the hazard to protected assets over a specified time interval, standardized per unit of territory and/or to a certain number of humans. The size of the risk posed by a disaster depends, on the one hand, on the size of the hazard at the site and on the amount and vulnerability of the assets protected at that location. Safety is a set of measures and activities aimed at ensuring the security and sustainable development of humans and other protected assets, i.e. an instrument that negotiates with the relevant risks. Recklessness is then a set of all the characteristics, activities and processes in the human system that pose a danger to humans and other protected assets. The most accurate term is "criticality" that has been used in technological domains.

A comprehensive approach means to apply strategic management that focuses on long-term sustainability. Its goal is system integrity because system services support life-supporting functions. It considers human to be part of the system, integrates human activity with the protection of the natural environment and responds sensitively to humans' needs in the context of ecosystems. Quality and qualified decision-making is an important part of any management process, and therefore, the decision support systems need to be created for the governance system, because decision-making against systems is complex and multidimensional. It is always necessary to work in the knowledge that sustainable development is not only about increasing and maintaining material well-being, but also about environmental vigilance, because most natural resources are not infinite (quantity), and some natural resources are also constantly contaminated (quality), which mainly concerns water and soil. And other impacts can be expected from potential climate change or from anthropogenic activities aimed, for example, at saturating nearly 7 billion humans. Water, soil scarcity, chemical and biological contamination show that the problems are complex in the biophysical area and controversial in the socioeconomic field.

Monitored systems are complex and many processes cannot be directly observed. In the socio-economic field, all environmental decisions can be characterised by a number of conflicting objectives. In order for the relationship between human settlements and the biophysical environment (landscape) to be balanced in the future, a new approach based on the integral safety management needs to be applied to address the problems of so-called 'grey' (i.e. man-made) and 'green' (natural) infrastructure.

Since there is still no general consensus on the formulation of the problems of sustainability of the public good (well-being) of human society in the context of systemic services, any solution to date is temporary, as it is constantly balanced between competing interests and social objectives (if established). It is difficult to solve decision-making problems clearly due to the changing nature of the decision-making process. The following dilemmas are addressed in the decision-making process:

- the relationship between risks and benefits (often greater benefits for humans means an increased risk to ecosystems; benefits for ecosystems mean for humans' shortages of food, energy, etc.),
- time conflict between current and future needs,
- social conflict (the relationship between the need of the individual and the whole).

It is difficult to solve inverse problems for the complexity of systems. If any risk-related symptoms are determined and sorted, new symptoms will emerge. Therefore, the practical approach to sustainability management need to be iterative, interactive and adaptive. The analysis of environmental and political, social and economic developments in the world shows the need to prepare to deal with cases and actions that, by their intensity of impacts, create critical situations, which requires that, in terms of human security, the development of the human system, the existence, stability and development of the State and each territory, the human safety management system might be proactive, strategic and includes sustainable development. Within this modernised safety management system, there need to be emergency management and, within it, crisis management (Fig. 3), i.e. reactive types of management

Fig. 3 Main management types [22]

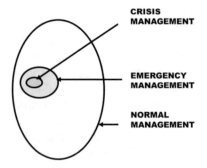

that ensure an immediate response to situations that endanger humans and the assets they need to live.

The aim of comprehensive management is to ensure the protection of lives, the health and security of humans, property, welfare, the environment, infrastructure and technologies necessary for human survival, i.e. always to ensure the mobilisation and coordination of the use of national resources (energy, labour, productive capacity, food and agriculture, raw materials, telecommunications, etc.), coordination of activities such as the system of communication, the rescue system and health services, which reduce the impacts of disasters, as well as the continuity of government activities and compliance with the law. The types of planning constituted by the basic methodological tools of each interrelated type of management need to form the basis in which the above objectives are enshrined.

For the objectives of human society, i.e. in particular for its sustainable development, they need to be combined measures and activities to reduce vulnerability and to increase resilience and adaptation capabilities that respect all fundamental protected interests, both individually and as a whole. In the light of current knowledge and experience, a proactive safety management system should be implemented at all levels of management, and migh be included risk assessment to a form that respects all protected assets and takes into account existing and proven internal dependencies. With regard to current knowledge, research into internal dependencies should be carried out and monitored, because they facilitating the secondary and other impacts of disasters on humans' lives, health and security. Those facts show that the present matters belong to all the basic fields of science, i.e. social, sociological, economic, environmental and technical.

4 How to Ensure Security Objectives

Recent knowledge shows that the safety of the entity (territory, organization, object, state) depends on the risk in a given place (i.e. it depends on both, the possible disasters affecting the site and on local vulnerabilities of assets to individual potential disasters that affect the site) and on the methods of negotiation with risks that are the source of loss, damage and harm to humans and other protected assets. *The relationship between safety and risk is not complementary, e.g. by installing warning systems we will increase safety, but the risk will remain the same.*

For the need to manage risks in favour of safety, the steps of the process shown in Fig. 4 should be applied. As the world evolves dynamically, monitoring should be installed and corrective measures applied if necessary [27].

Risks continue to increase and human society does not have the resources, forces and means to prevent this and it needs, therefore, to manage risks in a targeted way. For the management procedure to be successful, it needs to focus on the priority risks and their aspects. Trade-off with risks is based on the current possibilities of human society and consists in dividing risk settlement into categories in which the relevant part of the risk is ensured by: reducing, i.e. preventive measures averting

Fig. 4 Risk management process for safety

the realisation of the risk; mitigating, i.e. by purpose-based preventive measures and preparedness (warning systems and other emergency and crisis management measures) shall reduce or avert unacceptable impacts in the implementation of the risk; insurance; preparation of response and recovery reserves and reserves to ensure human survival and continuity of State/territory/organisation operations; and preparation of a contingency plan in the event of risks that are incontrollable or too costly to eliminate or are have low occurrence frequency.

The basic objective of the State is to ensure the security of the human system and its protected assets, and therefore, the main objective of public administration programmes aimed at safe territory is to prevent disasters and, in the event of natural disasters that cannot be averted, to mitigate the unacceptable impacts of the disasters in question. In order to be effective in preventing and mitigating disasters, all those involved at all governmental levels need to work together. It is important in every community that technology and infrastructure owners, local government and the public work together to reduce the risks of all possible disasters. The cooperation needs to be based on an open and correct policy, which also helps to increase humans' confidence in public administration and owners of infrastructures and technologies, in the sense that the measures taken for limitation of risks from disasters that have extreme impacts.

Safety needs to be, therefore, an integral part of the business activities of infrastructure and technology owners. All undertakings need to be managed in such a way that the occurrence of accidents affecting the safety is minimal. All activities and efforts of managers and employees need to be directed towards this goal. The key elements for the objective are mutual cooperation, open communication and regular monitoring the achievement of safety objectives. Based on the current requirements enshrined in developed countries' legislation, owners of technologies and infrastructures need to:

- promote safety as a holistic part of their business activities and promote safe activities,
- actively search for information on safety,
- cooperate with administration offices and other entrepreneurs in order to improve safety,
- create, together with other undertakings, the conditions for joint response and mutual assistance,
- create professional organisations.

Public administration needs to set safety objectives, establish a clear and holistic (integral) safety management framework and ensure that all relevant safety requirements are met through appropriate inspections and enforcement measures. It needs to be pro-pro-active in stimulating those involved in promoting new approaches to prevention, in addition to its traditional efforts to ensure that the impacts of disasters are managed. It has a leading role in motivating all sectors of society to support disaster prevention and to identify tools for the development of a national culture that promotes disaster prevention. It needs also to ensure that the public receives and understands all relevant information on the extreme impacts of disasters in a timely manner. By this effort, it gains public confidence that its surveillance is the right thing to do. These facts show once again that the issues under discussion belong to all the basic fields of science, i.e. social, sociological, economic, environmental and technical.

5 Safety Management

Each safety management has four basic phases: prevention; preparedness; response; and recovery, which are interrelated, as shown in Fig. 5, and which logically are followed by lessons learned from experience [22]. Public administration determines the level of safety even in the case of private entities, because it is responsible for legal and administrative regulations, policies and practical measures and for enforcing them from citizens, owners, etc. For this aim, it needs:

- regularly to review and update the rules,
- to monitor legal and natural persons as well as citizens in the community to ensure that their approach to risks is correct,
- to ensure effective cooperation and coordination of all stakeholders in the community and, for this purpose, their open and effective communication,
- to know the risks in the community and ensure appropriate emergency planning,
- to prepare and be able to implement an effective response and recovery.

It is clear from the above that the public administration:

SAFETY MANAGEMENT SYSTEM HAS 3 LEVELS.

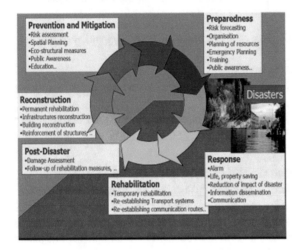

1.   Each level has its
     particularities.

2.   The levels are
     interconnected and
     directed to one aim.

3.   For safety
     management
     the appropriate bodies
     of public administration
     are responsible.

4.   All stakeholders have
     tasks.

**SAFETY MANAGEMENT
IS CONTINUOUS
PROCESS.**

Fig. 5 Time sequence of safety management phases

- sets out general community safety objectives,
- provides for: a clear and well-thought-out surveillance system; an appropriate inspection system; as well as a system for enforcing requests.

In order to ensure a safe community, a safe region, a State, the EU and a safe world, the risks of selected processes that are part of integral risk are systematically managed on the basis of the current knowledge in the EU [23]. The aim of their management is to apply only the right solutions to problems in practice. It deals with: security risks, construction-technological and design risks; credit risks; market risks; external risks; operational risks; management and decision-making risks. The list clearly emphasises the risks associated with management and decision-making, as their implementation creates organisational accidents.

Security risk:

- represents a level of disruption to the security of the monitored system, i.e. generally the human system, landscape, human settlements, civil or technological objects, infrastructures, human society, which are the subject of the monitoring,
- is expressed in terms of the likelihood of the occurrence of the security disruption size, which is represented in the case in question by the sum of losses, damage and harms to the assets monitored (expressed, for example, by money or losses of human lives).

The partial risks that belong to the integral risk are: health risk; risks of damage or destruction of the property; risks of damage or destruction of a particular part of the environment; risks of damage or destruction of infrastructures or technologies; and risks associated with disruption of the public good caused by conflicts in society or lack of life-giving needs and services (e.g. water and energy), and as well as all of the above-mentioned risks. A number of partial risks are defined in the legislation and in the methodologies or directives related to them. Their disadvantage is that they are based on different concepts that are not interrelated.

5.1 Systemic Approach to Safety Management

The development of a system approach within the concept of management was fundamentally influenced by the origin and development of cybernetics, in particular by the definition of system terms, which became the cornerstones of modern system thinking (system, subsystem, its elements, target behaviour, links, feedback, information, transformation, etc.). The task of system theory is theoretical solution of system description, classification and identification of systems, analysis and synthesis of systems, control and regulation, optimization of systems, learning, adaptation and reliability of systems, stability and controllability, etc. System theory, as separate scientific discipline processes general theoretical foundations and a unified conceptual apparatus for examining the behaviour of different types of systems. The cyber concept of systems has contributed to the analysis of different systems, in particular in terms of management and communication principles, which are important characteristics of systems. System control as a management concept emphasizes system characteristics—both self-control and the subject of control are studied, analysed and constructed as a system.

Safety management is methodologically based on a systemic approach to control, emphasising the theoretical and practical provision of the system of control. System interpretation of control creates prerequisites for the system application of IT technologies and integration of system security components control. This creates the possibility to combine scientific and theoretical, but also intuitive, experience-induced procedures [14]. On a systemic basis, design, identification, discomposing and composition, analysis and synthesis, design and implementation of the management system are created by algorithmic procedures, closer to the engineering approach in the environ of industry 4.0, artificial intelligence and machine learning. This concept will allow typing (standardization) of system problems, creation of methods of their solution and subsequent algorithmizing and implementation.

A constructively methodical approach to system management does not mean a completely new born-again experience, but it does create a new constructive approach, terminology for a systematically organized safety management system based on a combination of known principles, proven approaches and recommendations of management theory and practice that can be applied in the management processes of entities ensuring the safety. Information and knowledge are central to the concept of system management. However, their purposeful creation and proactive use in safety management processes is key. Systemic safety management is based on the principle of proactive management, i.e. processes in which ex-ante measures are organised to avert or at least mitigate certain unrequired (adverse) events and ensure preparedness to cope with unrequired events. Reactive management is based on processes in which ex-post measures are organised to address adverse events that have occurred.

Terms' categories, knowledge, postulates, methods and techniques of system analysis and synthesis, or system science as a whole, become the theoretical and methodological basis of holistic management. Highlighting the system nature is a response to the increasing scale, complexity and dynamics of operating and controlled systems (subsystems, elements) and their qualitative influence on IT technologies. The determining relevant framework for holistic (integral) safety management is the legislative and institutional framework, which sets out the tasks of the bodies responsible for managing activities in this area—in particular the top constitutional bodies (bodies) as well as the duties of organisations directly carrying out the national security activities.

The safety of the State (especially its inhabitants) is one of the fundamental functions of any sovereign State. As it was mentioned above, the basic function of the State is to ensure the protection and development of the protected interests of the State, i.e. the items that are protected as a matter of priority; the human lives, health and security, property, the environment, critical infrastructure, critical technologies and the existence of the State. Ensuring the national security and emergency management are expressed in the constitutional regulations of most countries of the world. They aim at the universal care of humans, their life, respect for human rights and freedoms, the protection of property and life sureness, and the preservation of the functions of the State as an institution that ensures safety.

A pandemic of a magnitude such as the Covid-19 of the early 2020s has plunged the Europe and the world into a very complex security situation and could become a very serious security risk in the future. In any case, it is necessary to apply quickly and professionally theoretically, methodologically and research-proven model of safety management.

5.2 Procedural and Knowledge-Based Approach to Safety Management

Safety management is based on process management, which is based on the consistent use of knowledge about the problem in and around the system and is, therefore, also called "knowledge management". Knowledge carriers are humans, knowledge cannot be taken away from anyone, but it can be expanded and multiplied indefinitely. In a knowledge society, it is intellectual capital that dominates and has a completely different position than before. All this requires a different view of the management of departments and units. *Process control based on control of managemental and implementation processes differs* from the operational approach commonly used in the decision-making process of classical management. Classical management is based on a functional approach that focuses mainly on outputs (results), which is actually an orientation on the consequences and not on the causes. It is clear that evaluating the results may not reveal the causes of the failure to meet the target. At the moment when we focus on the outputs, we neglect the principles of prevention.

Process management based on knowledge management is focused not on results, but on causes. It is based on the elaboration of concept and methodology. The application of knowledge management elements in the decision-making process of manager leads to a transition from individual decision-making to a group approach. What is important is the role of the manager, who needs to direct the process to make a quality decision. However, it should be considered that the described procedure is not only more time-consuming, but also more demanding for the preparation of individual members of the process team, including the managing director. Experience in the application of procedural management elements in the corporate sphere has shown that in routine decision-making, individual decisions are more advantageous, however, for the preparation of non-program decisions (i.e. complex and non-standard) it is necessary to choose the method of group decision-making (creation of a process team).

In both cases, however, the manager is always responsible for the decision. When making group decisions, an appropriate environment needs also to be created to support the group's creative abilities. It is important that the manager could be able to suppress the influence of incompetent, ignorant and lazy but ambitious individuals who attack the knowledgeable and hardworking to assert their ambitions. The manager needs to ensure in team decision-making: promoting the originality and unusualness of a solution that is timeless; group management in such a way as to separate the sources from the content of the information; ensuring the exercise of independent personal judgment and experience; maintaining the open communication; strengthening the self-confidence; avoiding the ridicule; preventing the quick solutions and short-term results; and consensus. If this is not possible, to take and implement the decision after a rigorous evaluation of any circumstances that may affect the achievement of the goal.

In knowledge-based process management (having the form shown in Fig. 2), the strategic level determines the basic development directions that indicate which processes need to be adjusted or created, what organisational changes will need to be made, where to acquire know-how, financial resources, etc. The tactical level of process management helps to sort out the activities necessary for the implementation of long-term projects. It searches of questions: how to set up processes; what state (condition) to maintain them in; and how processes need to work together. Operational (functional) management decides on the specific distribution of resources in the process (human, technological, financial) as well as on the performance of individual activities within the set processes (how to perform a specific operation). The aim is to ensure the transfer of knowledge and skills among workers. At the technical level, specific problems are addressed. It should be remembered that the most challenging is the trade-off with risks that is performed at this level; by this it is increased the resistance of elements, equipment, components and whole systems and according to data from practice data the success rate of technical measures varies between 40 and 80%. The significant effect and competitive advantage of the ***entity*** (territory, organisation) is achieved only by ***aligning all levels of management.***

The aim is to achieve a situation where processes are defined and managed on the basis of a strategy, operational management is not only a firefighting the emergencies, but is aimed at successfully achieving the development goals. The processes are having been improving on the basis of the knowledge transmitted from the operative. New knowledge resulting from the process management will then be quickly reflected in the strategy and will bring about further fundamental changes in the development of the entity.

Process management is based on the principle of integration of activities into complex processes, i.e. partial operations are consolidated into processes. Processes are controlled by process teams. Each process team manages processes at its level and gives subordinate groups tasks that lead to the achievement of the goal. At the same time, all process teams are motivated to achieve optimal results and all stages follow the achievement of partial results of the final goal. In process management, there are two management systems side by side, namely functional and procedural, which makes the management more complex. The process management uses the general process "Problem Solving Process", which is part of the best-practice and is widely used worldwide. This is a general process, which consists of ten points: identification of the problem; definition of the problem; analysis of the state of art; finding causes; definition of the target state; proposal for a solution; choosing a solution; validation of the solution; implementation; evaluation.

Safety support processes are in the areas of technical, economic, educational, human resources, communication, management, administration, documentation, surveillance, research, etc. In order to achieve the highest efficiency, so:

– processes in individual areas need to be coordinated, and therefore, a process safety management process (PSM) is set up in each area to ensure coordination and maximum efficiency,
– all areas need to be coordinated, and therefore, each entity has a safety management system (SMS) that ensures this requirement.

Safety management processes determine ways of trade-off with risks for the benefit of protected interests, i.e. identifying the prevention, preparedness, response, recovery and response measures and activities. As technical prevention measures are the most effective, it is clear that technical sciences need to prepare the relevant background. This claim supports the best available technology (BAT) requirement, which means that the equipment is designed and constructed, maintained, operated and decommissioned in such a way that the risk or potential to cause damage is settled for the benefit of the protected assets, regardless of the costs involved [27].

In order to ensure the second requirement, engineers from all technical fields, system engineers, IT specialists, economists, HR professionals and other specialists need to understand each other, because this is the only way how they can ensure the overall goal. That fact shows that, safety management in this sense belongs to all the basic fields of science, i.e. social, environmental and technical.

A separate task of safety management is to ensure a safe critical infrastructure [26], to which the following infrastructures usually belong:

- energy system (electricity; gas; thermal energy; oil and petroleum products),
- water system (supply of drinking and utility water; security and management of surface waters and groundwater sources; wastewater system);
- food and agriculture system (food production; food care; agricultural production),
- healthcare system (pre-hospital emergency care; hospital care; public health protection; manufacture, storage and distribution of medicines and medical devices),
- transport system (road; rail; air; inland waterway),
- cyber system providing the operation of communication and information systems (fixed telecommunications network services; mobile telecommunications network services; radio communications and navigation; satellite communications; television and radio broadcasting; postal and courier services; internet and data services),
- banking and financial sectors (public finance management; banking; insurance; capital market),
- emergency services (National Fire Rescue Service; fire protection units for area coverage; Police of the State; Army of the State; radiation monitoring; forecasting, warning and preaching service),
- public administration (state administration and self-government; social protection and employment; state social support and social assistance; justice and prison performance).

It needs to be remembered that the life-giving infrastructures of the human system are not only the infrastructures that legislation refers to as critical infrastructures, but they also include: the infrastructure of the food chain; the infrastructure for educating the population; the infrastructure for education; the infrastructure for research; the infrastructure for learning from experience. It should be remembered that only high-quality interoperability of all these infrastructures will ensure the existence, security and development of humans, i.e. the protection of the population under critical conditions.

To protect humans, it is necessary:

- to consider all disasters, including terrorism, corruption and abuse of power,
- to work properly with risks in a dynamically variable world,
- to have high-quality crisis management to ensure a correct and rapid response to critical situations.

However, there is one problem in managing the safety with goal security and development of humans, which is related to the need to take measures concerning the persons themselves, which can also interfere with their privacy. Therefore, when choosing the measures to ensure safety, it is necessary to protect the privacy of a person.

6 Disaster Management in Europe and Safety Culture

The FOCUS project carried out detailed research on disasters in Europe [13, 24], the results of which are the following:

1. The security situation in the Europe, the world and in each territory is changing continuously over time, and it is, therefore, necessary to formulate a new safety culture, which particularly considers the current knowledge and experience of internal dependencies among public assets that lead to extreme social crises. On the basis of historical development, there are: a number of preventive and mitigating measures that are put into practice by legislation, technical standards and norms and public rules; response systems; and recovery methods. However, their effectiveness decreases with time, as new risks arise and the vulnerability of humans and other assets grows with time.
2. In managing the disasters designed to protect the population, it needs to be respected that problems need to be managed at levels: technical, organisational, tactical, strategic and political. The levels of management and resolution and their interdependences need to be respected. Non-interconnection and omission are possible sources of organizational accidents.

On the basis of a critical evaluation of the lessons learned [13, 24]:

1. The list of monitored disasters should be supplemented by:

    − further natural phenomena: geomagnetic storms; soil salting; space body falling on the Earth's surface; sandstorms; and sudden weather changes (cold wave or heat wave),
    − technological phenomena: organisational accidents in technological facilities; biotechnology-related accidents; misuse of technologies—nuclear, nano and IT; misuse of genetic engineering; misuse of CBRNE substances,
    − imperfections in the management of human activities: failure of educational infrastructure; failure of research infrastructure; failure of public administration; supply chain failures,
    − phenomena in the environment (including humans): movements of the Earth's plates; the rapid Earth's surface drops; water circulation disorders in the environment; disturbances in the circulation of substances in the environment; disorders of the food chain of a person; disturbances of the Earth's body due to planetary processes; disturbances of the Earth's body due to interactions between solar and galactic processes; incurable diseases in humans, plants and animals,
    − phenomena associated with environmental responses to human activities: surface decreases due to underliving; interactions caused by the militarization of space,
    − phenomena in human society: illegal production and distribution of narcotics and psychotropic substances; illegal migration; proliferation of weapons of mass destruction.

2. The order of disasters according to the criticality of their impacts in the Europe is:

    - natural: impact of a large cosmic body; extreme earthquakes; extreme flooding; extreme forest fire; extreme drought,
    - technological: a beyond design (severe) accident with the presence of radioactive substances; a beyond design (severe accident with the presence of mutagenic, carcinogenic and threatening reproduction substances,
    - imperfections in the management of human activities: corruption; abuse of power; abuse of power; lack of respect for the public interest; failure of educational infrastructure; failure of research infrastructure; public administration failure (organisational accident); supply chain failures; the small robustness of technical and financial infrastructure resulting in long-term shortfalls in energy supply, drinking water, food and financial market disorder,
    - in the environment (including humans): water circulation disorders in the environment; disturbances in the circulation of substances in the environment; major pandemics and epidemics, and incurable diseases in humans, animals or plants,
    - associated with environment responses to human activities: contamination of air, water, soil and rock masses, uncontrollable population explosions, migration of large groups of people, militarisation of space and climate variation,
    - in human society: abuse of power; disintegration of society into intolerant groups; misuse of technology; abuse of power; illegal entry into information systems; cybercrime; terrorist attacks; corruption in government and public administration, including in the political sphere; serious economic crime involving money laundering and tax evasion; trafficking in human beings and illegal migration; illegal production and distribution of psychotropic substances; extremism; all forms of discrimination and intolerance.

The works [13, 24] contain the main conclusions:

1. There are total of 15 challenges for security research in the field of prevention, preparedness, response and recovery—e.g. to create standards for robust and flexible critical infrastructure.
2. It is identified 16 serious vulnerabilities that should be monitored in research—e.g. insufficient targeting the crisis management to extreme situations, long-term power outages in the cold weather, extreme pandemics.
3. It is revealed 7 main knowledge gaps that should be targeted by security research—e.g. poor data collection, poor data processing on the basis of which the decision-making is done; mainly from the social domain.
4. It is proposed 18 types of future security research—e.g. how to implement strategic integral safety management in a dynamically variable world.
5. The 5 most necessary topics for future security research—e.g. how to systematically implement professional knowledge for the benefit of the public interest—have been identified.

6. The originators of disaster management failures, i.e. the originators of organisational accidents, have been found in many areas—examples:

    – senior management—e.g. governance is predetermined by political and military aspects,
    – technical area—e.g. decision-making scenarios are prepared only on the basis of simulations, without verification on real data,
    – area of the organisation—e.g. it is not an effective tool against corruption,
    – area of knowledge—e.g. creeping disasters are neglected.

It follows from the above facts that it is necessary to improve human' management because the level of safety is determined by the competence of human measures and activities and the competence of their implementation into practice; the tool is the safety culture.

The safety culture is an expression of the sharing the values and measures of the safety management system and is an essential element for safety management. It reflects the concept of safety and is based on the values, opinions and actions of senior executives and their communication with all stakeholders. It is a clear commitment to actively participate in addressing the safety issues and advocates that all those involved act safely and comply with relevant legislation, standards and norms. Safety culture rules are integrated into all activities in a territory or other entity. Their basis is not the concentration on punishing the culprits/the originators of errors, but the lessons learned from mistakes and the introduction of such corrective measures, so that errors cannot be repeated or at least significantly reduce the frequency of their occurrence.

In the context of a safety culture, the concepts of loss prevention and process safety are often used in current technology-related literature. These are tools that are used in connection with technologies to protect people and property. Loss Prevention is a systematic approach to preventing or minimising the accidents [27]. It includes means to eliminate sources of risk or reduce the likelihood of their implementation and to mitigate the impacts associated with such implementation (preventive and follow-up measures). It shall also include the identification of appropriate control measures, the identification and application of appropriate remedial measures to ensure a safe entity with an appropriate level of security and sustainable development and not presenting an unacceptable danger to its surroundings.

Process safety is a safety related to safety in industry in which there are a number of manufacturing and additional processes that are required to create the final product of that industry. This is about preventing the accidents which have specific and characteristic features for the specific industry concerned. It deals, for example, with the prevention of imminent releases of chemicals or energies in harmful quantities and, in the event of such leaks, by limiting their size, impacts and consequences. It does not cover issues of classic occupational safety and health, i.e. it deals with purely technical problems, by which it differs from the integral safety of the system. Once again, the facts show that the issues under discussion belong to all basic fields of science, i.e. social, environmental and technical.

7　Conclusion

It follows from the above that the main thing to ensure safety is: monitoring the situation in the territory; correct prevention; and timely, correct and rapid response to emergencies of all kinds. In order to manage safety properly, it is necessary: permanently to collect knowledge; to apply knowledge considering the local specificities; to be aware of the existence of conflicts when deciding on disaster prevention measures; to learn from experience; proactively to manage the territory; and to have sufficient resources, forces and means.

Professional surveys show that security activities contain strong elements of social, especially legal and economic, sociological and psychological, science and technical, as well as managerial and informatics. They cannot be overlooked research trends and the work of Czech and Slovak theorists e.g. [15, 21, 29] focused on the constitution of independent security science, in which a number of opinions on the constitution and development of security science have been formed. Some authors consider security science to be pre-existing, constitutional, characterised by relevant methodological features, other authors talk about the process of forming the separate security science without further describing the state in which the constitution of this science is located, others authors do not question the importance of constituting security science and wonder which of the existing sciences or sciences that still arise in the process of differentiation could fulfil the integrative function in its creation, approaches and doubts about the possibility of building a "unified" security science or an integrated set of security sciences cannot be ruled out.

From the pending discussions and questions that await an answer, from the current knowledge, interpretation of research data and the above-mentioned facts, it follows that safety and safety management are multidisciplinary and interdisciplinary disciplines, all their affairs belong to all basic scientific disciplines, i.e. social, sociological, economic, environmental and technical. The basic reason is that in order to ensure safety and its qualified management, the cooperation of engineers from technical fields, system engineers, IT specialists, sociologists, economists, HR specialists, public officials and politicians is necessary, because this is the only way to ensure the overall objective pursued by the disciplines in question in the interests of the people, in cooperation with the inhabitants as citizens, employees and activists.

The ultimate goal is to create a systemic environment without chaotic changes in risk management and trade-off with risks that can help organizations to have risks "under control" and increase the likelihood of survival and success in installation of sustainable development. An integral part of the system approach in the field of safety is the area of improvement of the information safety management system based on the Deming cycle [4], i.e. the Plan-Do-Check-Act concept.

The concept in question can be applied to all processes of the information safety management system, but the variant is also usability for the process implementation of the selected quality model. The standard in detail characterized in [16] is linked and harmonised with standards [17, 18] in order to support their consistent and efficient implementation, operation and development. At present, these methodologies are used for processes of gradual, continuous improvement of quality management in the field of safety systems and are the basis of the integrated safety management system and its individual subsystems.

The transformations of society and human civilization towards greater complexity, requiring systemic solutions, place significantly different requirements on the reflection of security in the form of new scientific knowledge and theory and scientific organization of security activities that meet the needs of human civilization in a new stage of its development. This means not only the transformation of society, but also the modification of scientific disciplines and the emergence of new ones, derived from societal needs in connection with technological development and current problems of manifestations of diversified security threats. The need and emergence of security science is a necessary response to the development of human civilization and modern society and will evolve towards sociological, economic, ecological and information contexts based on contemporary knowledge and interpretation of research. On the contrary, the dominance of the mechanical-military view of security issues will recede.

Finally, a thorny issue needs to be addressed, namely the definition of science on safety and security. Until now, it has been understood as a summary of traditional scientific disciplines such as legal sciences, natural sciences, criminology, criminology, informatics, systems theory, etc. The permanent risk to humanity and the planet is not only weapons systems, but the entire technological development of human civilization, the impact of industrial production and other interventions in nature on the planetary environment with potentially disastrous consequences. From the above-documented facts, it should be noted that this is an interdisciplinary and multidisciplinary discipline aimed at ensuring a safe human system. Therefore, we give a short definition:

The science on safety and security of the human system is the science on the way how by human measures and activities the human security can be ensured in a variable world. It is a continuous, critical and methodical pursuit of a true and general understanding the safety and security of the human system. It aims at a continuous social process of continuous rational cognition of the human system in order to ensure its requirable state and sustainable development in the requirable direction. It includes the general theory of safety (in an integral/complex sense), the theory of the art of safety management (i.e. methods and forms of application of measures at individual stages), the theory of construction and operation of the safety management system, i.e. the construction, operation and behaviour of the system of organs, components and all other stakeholders to ensure the safety of the human system in the territory, the preparation, functionality and capability of the safety management system (i.e. its elements, links and flows) to ensure safety and development under normal, abnormal and critical conditions.

References

1. Blatz (1966) Human security-some reflection. University of Toronto, Toronto
2. Bossel H (2004) Systeme, Dynam, Simulation – Modellbildung, Analyse und Simulation komplexer Systeme. Books on Demand, Norderstedt
3. Bundesministerium des Innern (2006) Protection of critical infrastructures—baseline protection concept. Recommendation for Companies. Bundesamt für Bevölkerungsschutz und Katastrophenhilfe, Zentrum Schutz Kritischer Infrastrukturen, Bonn. www.bmi.bund.de
4. Deming WE (1986) Out of the crisis. Massachusetts Institute of Technology, Cambridge, MA, 88 pp. ISBN 978-0911379013
5. DoD US (2006) DoD security engineering facilities planning manual. Department of Defense, Washington. http://www.wbdg.org/ndbm/DesignGuid/pdf/FINAL%20DRAFT_UFC_4-020-01.pdf
6. Dunn M, Wieger I (2004) Critical information infrastructure protection. International CIIP handbook. ETH, Zuerich, 405 pp
7. EEA (2001) Late lessons from early warnings: the precautionary principle 1896–2000. European Environmental Agency. Environmental issue report No 22, Copenhagen. http://reports.eea.eu.int/environmental_issue_report_2001_22/en/tab_content_RLR
8. EMA (2003) Critical infrastructure emergency risk management and assurance. Handbook emergency management Australia. www.ema.gov.au
9. EU (2005) Green paper on european programme for critical intrastructure protection. Brussels 17.11.2005, COM(2005) 576
10. EU (2006) ESRAB report: a report from the European Security Research Advisory Board. EU, Brussels, 95 pp
11. EU (2006) The seventh frame research programme 2007–2013. EU, Brussels
12. EU (2009) ESRIF final report. EU, Brussels, 319 pp
13. EU (2013) FOCUS project study. http://www.focusproject.eu/documents/14976/-5d763378-1198-4dc9-86ff-c46959712f8a
14. Gupta M, Sharman R (2009) Handbook of research on social and organizational liabilities in information security. Information Science Reference, New York. ISBN 978-1-60566-132-2
15. Holcr K, Porada V et al (2011) Policajné vedy. Úvod do teorie a metodologie. Aleš Čeněk, Plzeň, 248 pp. ISBN 978-80-7380-329-2
16. Hoyle D (2009) ISO 9000. Quality systems. Handbook: using the standards as a framework for business improvement. Elsevier, Ltd., London. ISBN 978-1-85617-684-2
17. ISO (2018) ISO/IEC 27000:2018. Information technology—security techniques—information security management systems—overview and vocabulary. ISO, Genève
18. ISO (2015) ISO 9001:2015 (Quality management systems—QMS). ISO, Genève
19. Konersmann LJ, Peinelti R (2002) Safety considerations for the transport and storage of dangerous goods, based on the example of pipelines. Federal Institute for Materials Research and Testing, Berlin. Publ. NATO-Committee on the Challenges of Modern Society
20. Maslow (1954) Motivation and personality. Haper, New York, p 236
21. Porada V et al (2019) Security science. Aleš Čeněk, Pilsen, 780 pp. ISBN 978-80-7380-758-0
22. Procházková (2011) Strategic management of the safety of the territory and organization. CTU, Praha, 483 pp. ISBN 978-80-01-04844-3
23. Procházková (2011) Risk analysis and management. CTU, Praha, 405 pp. ISBN 978-80-01-04841-2
24. Prochazková (2013) Study of disasters and disaster management. CTU, Praha, 202 pp. ISBN 978-80-01-05246-4
25. Procházková D (2014) Fight against terrorism. Project EU: improving security by democratic participation—ISDEP. ČVUT, Praha, 200 pp. ISBN 978-80-0105568-7
26. Prochazkova (2014) Challenges connected with critical infrastructure safety. Lambert Academic Publishing, Saarbrucken, 218 pp. ISBN 978-3-659-54930-4

27. Procházková (2017) Principles of risk management of complex technological facilities. ČVUT, Praha, 364 pp. ISBN 978-80-01-06180-0, e-ISBN 978-80-01-06182-4. http://hdl.handle.net/10467/72582
28. PSEPC (2004) Assets criteria. Public Safety and Emergency Preparedness Canada, Ottawa. www.psepc.gc.ca/prg/em/nciap/assets_criteria-en.asp
29. Sak P (2018) Introduction to theory on security. Unconventional views on past, present time and future. Petrklíč, Praha, 270 pp. ISBN 978-80-7229652-1
30. UN (1994) Human development report. UN, New York. www.un.org
31. UNEP (2006) Caring for the earth. A strategy for sustainable living. IUCN/UNEP/WWF, Gland
32. US (2001) US critical infrastructure conception. Washington

Annex—Further References Used for Paper Processing

33. Adger NV (2000) Social and ecological resilience. Prog Human Geogr 24:3
34. Althoff J (2001) Preface. In: Safety of modern systems. Congress Documentaion Saarbruecken 2001. TÜV-Verlag GmbH, Cologne, pp 5–6. ISBN 3-8249-0659-7
35. Anderson R (2008) Security engineering—a guide to building dependable distributed systems. Willey, 1001 pp. ISBN 978-0-470-068552-6
36. ASCE (2001) Global blueprints for change—summaries of the recommendations for Theme A "Living with the potential for natural and environmental disasters", Summaries of the recommendations for Theme C "Learning from and sharing the knowledge gained from natural and environmental disasters." ASCE, Washington, p 1089
37. AS/NZS (2004) Australia and New Zealand Standard: risk management, Issued by Standards. Australia, Guideline 4360. http://www.riskmanagement.com.au/Default.aspx?tabid=148-116pp
38. Bérenguer C, Grall A, Soares CG (eds) (2011) Advances in safety, reliability and risk management. Taylor & Francis Group, London, 3068 pp. ISBN 978-0-415-68379-1
39. Bris R, Soares CG, Martorell S (eds) (2009) Reliability, risk and safety: theory and application. CRC Press/Balkema, Leiden, 2367 pp. ISBN: 978-0-415-55509-8
40. Centre for International Security Policy (2005) Workshop on critical infrastructure protection and civil emergency planning-dependable structures, cybersecurity, commnon standard. Zurich. www.eda.admin.ch
41. COMAH (2002) Safety report assessment manual: COMAH. UK-HID CD2 London, 570 pp
42. Dow K (1991) Exploring differences in our commnon future. Geoforum 23(3)
43. EMA (2006) Process of process safety management. www.ema.gov
44. EU (2010) Risk assessment and mapping. Guidelines for Disaster Management. Working paper SEC(2010) 1626. Brussels 2010
45. FEMA (2005) Promoting critical infrastructure protection by emergency managers and first responders. Nationwide. www.usfa.fema.gov
46. Filippini R, Silva A (2011) A modelling language for the resilience assessment of networked systems of systems. In: Advances in safety, reliability and risk management. CRC Press, Leiden, pp 2443–2450. ISBN 978-0-415-68379-1
47. Fullwood RR (2000) Probabilistic safety assessment in the chemical and nuclear industries. Butterworth Heinemann, Boston, p 514. ISBN 0-7506-7208-0
48. Geysen W (2001) The acceptance of systemic thinking in various fields of technology and consequences on respective safety phylosophies. In: Safety of modern systems. Congress Documentaion Saarbruecken 2001. TÜV-Verlag GmbH, Cologne, pp 19–27. ISBN 3-8249-0659-7
49. Glantz M (1992) Global warming and environmental change. In: Global environmental change, vol 2
50. Gunderson L, Holding CS (2002) Panarchy: understanding transformation in human and natural systems. Iceland Press, Washington

51. Gustin JF (2002) Disaster recovery planning: a guide for facility manager. The FairMont Press, Inc., Lilburn, 304 pp. ISBN 0-88173-323-7

52. Hale AR (2001) Safety management in production. In: Safety of modern systems. Congress Documentaion Saarbruecken 2001. TŰV-Verlag GmbH, Cologne, pp 383–392. ISBN 3-8249-0659-7

53. Holling CS (1973) Resilience and stability of ecosystem. Annu Rev Ecol Syst 4(1)

54. IAEA (2020) Safety guides. IAEA, Vienna, pp 1954–2020

55. ISM (2020) International Safety Management (ISM) Code 2002. IMA, London

56. ISO (2019) ISO 31000:2018. Risk management. Guidelines. BS EN ISO 22301:2019. Business Continuity Management System. Requirements

57. Kossiakoff A, Sweet WN (2003) Systems engineering. Principles and practices. Wiley, New Jersy, 459 pp. ISBN 0-471-23443-5

58. Kuhlmann A (2001) Does safety science fulfill the requirements of modern technical systems? In: Safety of modern systems. Congress Documentaion Saarbruecken 2001. TŰV-Verlag GmbH, Cologne, pp 9–17. ISBN 3-8249-0659-7

59. Langeweg F, Espeleta EE (2001) Human security and vulnerability in a scenario context. HDP Update 2

60. Lees FP (1980) Loss prevention in the process industries. Butterworths, London

61. Lucas C (2006) Quantifying complexity theory. www.calresco.org/lucas/quantity.htm

62. Mayers RA (2009) Encyclopedia of complexity and systems science. Springer, Berlin. ISBN 978-0-387-75888-6

63. McGuinwess E, Utne IB, Kelly M (2011) Development of a safety management system for small and medium enterprises (SME's). In: Advances in safety, reliability and risk management. CRC Press, Leiden, pp 1791–1799. ISBN 978-0-415-68379-1

64. Moteff J, Copeland C, Fischer J (2003) critical infrastructures: what makes an infrastrucuture critical? Report for Congress, CRS Web, Order Code RL31556

65. OCHA (2000) OCHA orientation handbook on complex emergencies. OCHA, Geneve

66. OECD (2002) Guidance on safety performance indicators. Guidance for industry, public authorities and communities for developing SPI programmes related to chemical accident prevention, preparedness and response. OECD, Paris, 191 pp

67. OECD (2003) Guiding principles on chemical accident prevention, preparedness and response. OECD, Paris, 192 pp

68. OECD (2006) Assessing societal risks and vulnerabilities. In: OECD studies in risk management. OECD, Paris, 276 pp

69. Pasman HJ, Vrijling JK (2001) Social risk assessment of large technical systems. In: Safety of modern systems. Congress Documentaion Saarbruecken 2001. TŰV-Verlag GmbH, Cologne, pp 151–162. ISBN 3-8249-0659-7

70. PetroChem (2004) Loss prevention. PCHE—PetroChemEng, Praha. ISBN 80-02-01574-6

71. Porada V et al (2017) Bezpečnostní vědy. Potřeba a transfér vědeckých poznatků. Aleš Čeněk, Plzeň, p 132. ISBN 978-80-7380-663

72. Porada V et al (2017) Bezpečnostní vědy. Úvod do teorie a metodologie. Aleš Čeněk, Plzeň, p 136. ISBN 978-80-7380-658-3

73. Rinaldi SM (2004) Modeling and simulating critical infrastructures and their interdependencies. In: Proceedings of the 37th Hawaii international conference on system sciences—2004. Sandia National Laboratories, Sandia. http://ieeexplore.ieee.org/xpl/freeabs_all.jsp?arnumber=1265180

74. Rinaldi SM, Peerenboom JP, Kelly TK (2001) Critical infrastructure interdependencies. (Identifying, understanding, and analyzing). IEEE Control Syst Mag 21:12–25. www.ce.cmu.edu/~hsm/im2004/readings/CII-Rinaldi.pdf

75. Ropohl G (1999) Philosophy of socio-technical systems. Soc Philos Technol 4(3)

76. Roland HE, Moriarity B (1990) System safety engineering and management. Willey, 321 pp. ISBN 0-471-6186-0

77. SAIC (2002) A guide to highway vulnerability assessment for critical asset identification and protection. National Cooperative Highway Research Program Project 20-07/Task 151B, Science Applications International Corporation-Transportation Policy and Analysis Center, Vienna
78. Smithers J, Smit B (1997) Human adaptation to climatic variability and change. Glob Environ Chang 7:2
79. Stein W, Hammerli B, Pohl H, Posch R (eds) (2003) Critical infrastructure protection—status and perspectives. Workshop on CIP, Frankfurt am Main. www.informatik2003.de
80. US (2005) The national strategy for the physical protection of critical infrastructures and key assets. http://www.whitehouse.gov/pcipb/physical_strategy.pdf
81. US (2005) Federal response plan 9230.1-PL
82. US NAS (2010) Framework for vulnerability analysis in sustainability science. Proc Natl Acad Sci 100(2010):14
83. WHO/Europe (2003) REHRA methodology (Rapid environment and health risk assessment). http://www.euro.who.int/watsan/CountryActivities/20030729_11

Concept of Science on Safety and Security

Dana Prochazkova and Hana Bartosova

Abstract In the present period, the world has been damaged by many disasters that have highly destructive potential. Due to the growing connectivity of the world, the disasters affect large territories and many domains needed for the human security and development. In order to protect human society, the need for specific science is greatly increased, which would formulated theoretical foundations and show the starting point. Because of the complexity of the system in which human society lives, the science in question needs to include elements of sciences: social, legal, sociological, psychological, natural sciences, technical, economic, managerial and informatics [9]. Submitted article shows object and subject, goal and tools of science on safety, which are based on a systemic concept and has a multidimensional character.

Keywords Security · Safety · Integral safety · Concept of science on safety and security · Object and subject of research · Methods and tools

1 Introduction

The Third Millennium opens up to humans unsuspecting possibilities in the field of science, techniques and new ground breaking discoveries, but on the other hand, it puts the humankind at the challenge of the complex problems dealing with the human security and protection with regards the impacts of natural disasters, major accidents, rising global terrorism, terrorist and cyber-attacks, socio-economic consequences of financial crises and, last but not least, the weakening the natural self-regulation of the environment in the form of heat waves and subsequent droughts, solar and volcanic eruptions.

D. Prochazkova (✉)
Czech Technical University in Prague, Technická 4, 166 00 Praha, Czech Republic
e-mail: prochdana7@seznam.cz

H. Bartosova
Ambis College, Lindnerova 575/1, 180 00 Praha 8, Czech Republic
e-mail: hana.bartosova@ambis.cz

The history of the world is a history of disasters of varying scale, from global wars to earthquakes, floods, terrorist attacks and accidents of local importance. A pandemic of such magnitude as the Covid-19 put the Europe and the world in a very difficult situation in early 2020. The high and rapidly spreading danger caused panic, mainly because there was not prepared response, although the entire scientific teams had been expecting a global pandemic since 2007 [6].

Based on historical experience (e.g., pandemics of the plague in Europe in the Middle Ages) were and still are a very serious risk to humans, which was underestimated in the euphoria of well-being in the second half of the twentieth century. For the further development of human society, it is necessary to find a way of managing the risk in question, which will ensure its successful management and thus the humans' safety on acceptable level [1–5, 8, 10, 13–15, 19, 23]. This fact significantly increases the need for specific science, which to form theoretical foundations and show possible actions. Because of the complexity of the system in which human society lives (further called the human system), the science in question needs to include elements of sciences: social, legal, sociological, psychological, natural, technical, economic, managerial and informatics.

Since humans are an inherent part of the world, they can only in the context getting over the risks influence their behaviour and their activities to ensure their security and development. In drawing up a risk management strategy for the benefit of safety, it is important to be aware that everything that exists cannot be developed and protected, but that key priorities need to be identified, to focus attention on them in detail, while monitoring the wider context in order to prevent the invocation of phenomena that would disrupt the degradation of the human system.

The state's core function is to ensure the protection and development of the state's protected interests (assets), i.e. items that are primarily protected; it is about the lives, health and security of humans, the property, the environment, the public welfare, the critical infrastructure, the critical technologies and the existence of the State. Such understood safety can only be achieved on the basis of knowledge, qualified monitoring the human system conditions and application of qualified measures within the management process [16, 17].

The transformation of society and human civilization towards greater complexity (i.e. the interconnectedness of the world) requires systemic solutions, thus placing the significantly different requirements on the reflection of safety in the form of new scientific knowledge and theory and scientific organization of security activities that meet the needs of human civilization in a new stage of its development. This means not only the transformation of society, but also the modification of scientific disciplines and the emergence of new ones, derived from societal needs in connection with technological development and the current problems of manifestations of diversification and security threats. From the point of view of current knowledge [1–5, 8, 10, 13–15, 19, 23], it is mainly about connecting the knowledge of scientific disciplines, which is possible only by using the same terms and mutual efforts of all after achieving the common goal, which on the day includes the concept of a safety culture.

2 Object and Subject of Safety Research

The object of safety research is the world concentrated on planet Earth, i.e. the space in which humans live. From the point of view of knowledge, it goes on an open system of systems (interconnection of many open systems) that is evolving. Its safety was formulated by the UN document [21] as the safety of the human system. In the same document, the terms of safety and security were distinguished and their link was indicated by the way that the safety means a set of anthropogenic measures and activities and it is a tool for achieving a certain human system conditions, i.e. human security. It means that the security is the level of human system conditions.

In the human system, in conjunction with developments within space, planet Earth, the environment, human societies, they are under way processes, events and phenomena, the causes of which lie outside or within the human system. The time scales in which processes, events, phenomena and events take place range from milliseconds for the microworld, through periods comparable to the duration of human life to geological scales based on millions of years [15]. Their common manifestations, which are harmful always or from certain sizes, can harm humans or their protected interests, or only in certain circumstances can harm, i.e. they may cause losses, damages or harms to human or human society, we refer to as disasters.

As mentioned above, the humans cannot be controlled the human system because they are an inherent part of it. However, they have the capability to manage human safety, i.e. their set of measures and activities aimed at protecting themselves and public protected interests so that their security and potential for development are on the acceptable level. The aim of anthropogenic management is a safe community, a safe territory, a safe Europe and a safe world, i.e. a safe space in which humans live.

The role of reasonable humans is understood in such a way that, thanks to their intellect, humans seek to understand and guide processes, events and phenomena in the world in order to provide a safe space for humankind with sustainable development. To do this, they use the "management" tool and the approach that we refer to as pro-active [22]. The concept pursued differs significantly from the concept in which the State or the public administration of the territory or individual relies on the fact that disasters occur only exceptionally, and that to protect humans and other protected interests, it is enough to build executive forces to save humans and eliminate impacts, when a reactive approach is preferred.

2.1 The Human System and Its Basic Public Assets

The human system is the minimum space for human life and human society, i.e. it includes the elements that make up humans, the parts of the environment necessary for the life of humans, parts of planet Earth necessary for human life, property, technology, infrastructure and flows and flows among these elements [15]. The human

Fig. 1 Model showing the process from human system safety to humans' security and development [15]

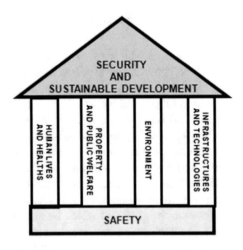

system as an object, is complex by its nature and by its spatial and temporal dimensions. From a modern point of view, it is a system of systems [15] in which the links and flows are both, in the components that represent the subsystems and across the components (e.g. water cycles in nature, geological rock cycles, photosynthesis, etc.) as well as links and flows connected to the planetary system (e.g. changes in the ozone layer, anomalies of hot plasma in the ionosphere, etc.).

The safe space established by the European Commission [7] is an analogy of a safe human system [21] and its process model is shown in Fig. 1.

Humankind as a whole and its individual communities in trying to solve the problems of the object need to divide the object into substructures and gradually solve the problems according to their possibilities at different levels. Although local problems tend to affect individuals of human society, it is not possible to be confined only to them, because the global problems will manifest themselves later, but they will end up with much greater power [15].

A safe human system is a system in which safety is at an acceptable level and that ensures the human security and the public welfare. Safety in this sense is an integral term that combines all the attributes of individual safeties that have already been defined (e.g. external safety, internal safety, nuclear safety, health safety, chemical safety, etc.). It is a set of measures and activities to ensure the security and sustainable development of the human system, i.e. to the security and development of protected interests.

2.2 Terms

In many countries, the situation in the terminology used has not been pleasant yet, because there are frictions between specialists from different areas of competence, which is documented by the diversity of definitions of terms used at work [14].

Despite all the efforts since the turn of the millennium to agree on certain professional terminology for all areas of social life based on a particular professional concept, this has not yet been achieved. A set of unified concepts based on the theory of process models [18] is used in the world based on the UN concept [21] and scientific activities of the EU, OECD, IAEA, WB and others [1–5, 8, 10, 13, 18, 19, 23]. In the sample we only give the definition of the most important ones:

1. The basic function of the State is to ensure the security and sustainable development of the protected assets (interests) and the State.
2. Protected assets (interests) of the State mean state assets that are primarily protected (humans' lives, health and safety, property, environment, public welfare, technology, infrastructure). They are defined in basic legislation/constitution and are subject to emergency planning.
3. Security means the state (condition) of the system in which damage to protected assets has an acceptable probability (i.e. it is almost certain that harm will not occur).
4. The danger means the state (condition) of the system in which the occurrence of harm to protected assets is highly likely (i.e. it is almost certain that the harm will occur).
5. Safety means a set of anthropogenic measures and activities to ensure the security and sustainable development of the system, i.e. to ensure the security and sustainable development of protected assets.
6. System safety integrity means a feature of the system that expresses the degree of ability of the system to ensure its security.
7. Recklessness (contrast to safety) means a set of properties and characteristics of elements, substances, disasters, processes and activities that act on protected assets or, under certain conditions, may cause harm (source of injury, damage, loss).
8. Criticality means a feature of the system, which is measured by the quality of a set of measures and activities with regard to the security of the system; at smaller values, the system condition is without a problem and at higher values there is a high probability of accidents or system failures. It is a complementary quantity to safety and includes the dysfunction and unreliability of infrastructures and technologies.
9. A secure system means a system that is protected against disasters of all kinds (internal, external, human factor).
10. A safe system means a system that is protected against disasters of all kinds (internal, external, human factor) and does not endanger itself or its surroundings even in critical conditions.
11. The safe system of systems means a system that is protected against all disasters, unrequired effects of the human factor and harmful interconnections, and does not endanger itself or its surroundings, even in its critical conditions.
12. The complexity of a system is a system feature that causes under certain conditions the unexpected interconnections among certain elements or subsystems that lead to failure of a system or its part.

13.   Disaster is a phenomenon that leads or can lead to harm and significant damage to protected interests. It is a phenomenon that leads or may lead to an unacceptable impact on protected assets.

14.   Vulnerability indicates the sensitivity of a protected asset or system to damage in the event of a disaster.

15.   Resistance indicates the capability of an asset or system to cope with the impacts of a given disaster that does not exceed a certain limit; resilience (elastic resistance) is the ability of an asset or system to cope with the impacts of a given disaster, the size of which is caused by impacts on sizes that are around the limits of system resistance.

16.   Adaptability means the capability of a system to cope with the impacts of a given disaster by certain regulation that do not change the properties of the system.

17.   The impact means the adverse effect of a disaster at a given site and time on protected assets or systems.

18.   The human factor includes the innate ways in which humans respond to stimuli, i.e. the conditional response and the purposeful will of the controlled action. The failure of the human factor indicates either the wrong execution of the act or the execution of the wrong decision.

19.   An organizational accident means an entity accident or failure caused by a bad decision by the human making the decision.

20.   A safety culture means a set of rules aimed at creating a safe entity system.

21.   A hazard associated with a disaster means a set of maximum disaster impacts that can be expected at a given site over a specified time interval with a probability equal to the specified value. According to standards and norms, it is usually determined by the size of the disaster, which occurs with a probability of equal to or greater than 0.05 with regard to the frequency distribution for the time interval of a hundred years.

22.   The risk means the probable size of damage, loss and injury to protected assets, which corresponds to the hazard associated with the disaster, which is prescribed by normatively. It depends on both, the size of the disaster and on the characteristics of the territory or other monitored entity, which predetermines the vulnerability of the entity to the disaster, as well as on the characteristics of protected assets in the territory, which also predetermine vulnerability to disaster.

23.   A threat means the rate of attack (terrorist or military) at a given location. This is the probability that an event or set of events, completely different from the required state (condition) or development of protected assets, will arise or can arise in terms of their integrity and function.

24.   An emergency situation means a situation that is caused by a disaster in the territory or object. The emergency category is a measure of the severity of the emergency in terms of their impact on protected assets. The following categories are distinguished: 0: impacts negligible in terms of citizen life; 1: impacts unimportant from the point of view of the citizen; 2: citizen-important impacts; 3: impacts important from the point of view of the society; 4: very

serious impacts from the point of view of the society; and 5: impacts threatening the existence or substance of society. Categories are marked with colours for simplicity (the highest in colour sequences—yellow, orange, red). The State education system needs to ensure that every citizen is able to cope with emergency categories 1 and 2 through their education and training.

25. Entity's tools to ensure the sufficient level of safety, i.e. the security and sustainable development of protected interests are: managing all levels of the entity based on qualified data and the right decision-making methods; education of citizens or employees; specific training of technical and management staff; standards, norms and regulations, i.e. regulation of processes that may or could lead to the occurrence (occurrence) of a disaster; inspections; executive forces to manage emergencies (firefighters, police, paramedics and support services and resources to ensure their operation); planning (security—strategic, emergency and crisis).

26. Risk management means the management of a set of anthropogenic measures and activities so that damage and loss on assets are below the specified level (usually set levels—ALARP and ALARA—Fig. 2). It is a principle that sets out the rule that, from the possible impacts of disaster, the small value that can be achieved by applying reasonable mitigating technical measures is acceptable to society.

27. Safety management means the management of a set of anthropogenic measures and activities so that damage and loss on assets or systems may be acceptable.

28. Crisis management means a procedure aimed at ensuring the management of possible critical situations within the scope of the crisis management authority and the implementation of the measures and tasks imposed by higher crisis management bodies (legal measures "declaring a crisis or crisis situation" are generally used to cope, which allows for the temporary restriction of human's rights and freedoms, the use of above-standard resources, etc.), including ensuring preparation to cope with possible critical situations.

29. Pro-active management means a type of management in which we perform advance anthropogenic measures and activities to avert or at least mitigate certain unrequired phenomena and ensure readiness to cope with expected

Fig. 2 ALARP region and upper limit [18]

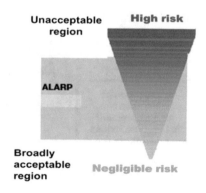

Unacceptable region | High risk

ALARP

Broadly acceptable region | Negligible risk

adverse phenomena. Reactive control is the type of control in which we solve problems only when they occur.

30.　　Limits and conditions mean tools for managing the entity safety. Their compliance guarantees the safe operation of entities (widely used in technological facilities and infrastructures).

2.3　Disasters and Risks

A disaster is a sudden, usually rapid and unexpectedly coming phenomenon, often rapidly disappearing, which in many cases leaves permanent or long-lasting, severe impacts, the size of which depends on the size of the phenomenon and the vulnerability of the assets. Today's knowledge shows that any significant change in the conditions set in the human system is, in its own way, a disaster. Therefore, in the professional field, the term expanded from the concept in which it referred only to natural disasters to other phenomena with the same impacts on humans and their protected interests. This means that in the observed concept of safety and security, every phenomenon that has unacceptable impacts on humans and their protected interests is a disaster. From a global perspective, the disaster is an event that changes the way existing up to date and sets the way for a new one. Therefore, there is for example a disproportion between the goals of humankind and the goals of the evolution of planet Earth, from the point of view of which the natural disasters are not destructive phenomena, but products of its development.

Disasters are divided into several groups, e.g. according to the types of processes taking place inside and outside the Earth as planets that cause them; Fig. 3 [15].

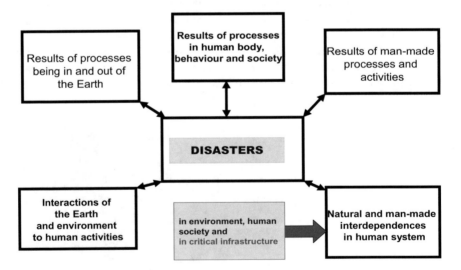

Fig. 3 Sources of disasters [15]

They have different places of occurrence and different characteristics. Based on the current knowledge of their possible sizes depend on regional processes and the sizes of their impacts depend on both, the regional processes and the local conditions. The causes and characteristics of disasters of different types are immeasurable. From the point of view of protected interests, they have one thing in common, and thus their capability to destroy, i.e. to cause them losses, damages and harms.

Disasters have certain characteristic properties that are sources of impacts that causes losses, damages and harms to important elements, links or flows of the human system, namely from a point of view of humans, because only that is what the de facto human is interested in (human is interested in surviving). Impacts include, for example, the vibrations; directed rapid flow of air, water or soil; violation of the stability and cohesion of rocks or soil; mass movements; sprats of fluids; temperature anomalies, etc. They act directly or vicariously through links and flows in the human system, Fig. 4.

Since historical times, humans consciously build the resilience of territory, objects, infrastructures and technologies to disasters by selecting elements, links and flows, their interconnections and specific preventive measures and activities up to a certain size of disaster, which is due to their knowledge, capabilities and possibilities financial, technical, etc. [17], the interdependences indicated in Fig. 3 will only be reflected in beyond design disasters, which are in size above the disaster limit to which resilience is systematically ensured.

2.4 The Possibilities and Capabilities of Humans

The evaluation of credible data, knowledge and experience [1–5, 8, 10, 13, 15, 18, 19, 23] shows that the knowledge and capabilities of humankind are:

– small to prevent the occurrence of disasters that are a manifestation of the evolution of the Earth's planetary system,
– adequate to mitigate some of the impacts of disasters that are a manifestation of the evolution of the Earth's planetary system;
– sufficient to prevent the occurrence of disasters associated with human activities and the development of human society.

Since humans are part of the world, i.e. the complex system of systems, their possibilities to ensure their security are limited because they have no competence to interfere with the management processes taking place in the planet and higher systems. That's why they need to be clear about what they need and what they can do for it. Figure 5 summarises items relevant to the safety of the human system.

To achieve the set goal, humans need to determined goals and rely on the expansion of knowledge of the world and the rational management of their activities so as not to cause phenomena that harm them. On the basis of the collection of information, they need to develop knowledge of disasters and their risks and, according to them, to create procedural models for qualified negotiations with risks to sufficient human

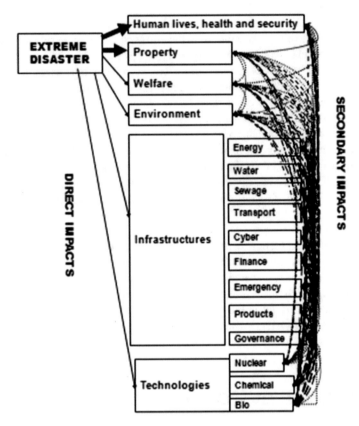

Fig. 4 Impact of beyond design (extreme) disaster on the human system. They are shown direct impacts on protected interests and the significant secondary impacts mediated by links and flows in the human system, as well as cases for which protective measures are systematically developed [17]

system safety, i.e. to humans' security and sustainable development [18]. The process model for negotiation with risks is shown in Fig. 6.

The risk negotiation is based on the current possibilities of the human society and consists in dividing the settlement of risks into categories in which the relevant part of the risk is:

– decreased i.e. by preventive measures is averted the implementation of the risk,
– mitigated, i.e. by the purposeful preventive measures of response and preparedness (warning systems and other emergency and crisis management measures) are reduced or averted unacceptable impacts in the implementation of the risk,
– insured,
– provided for reserves for response and recovery and stocks for ensuring the survival of humans and continuity of State/territory/organisation operations,

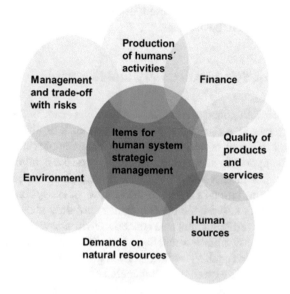

Fig. 5 Items important for a safe human system [18]

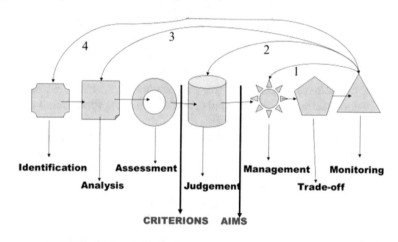

Fig. 6 Process model of working with risks. Criteria = conditions that determine when a risk is acceptable, conditionally acceptable or unacceptable. The targets indicate required states (conditions). 1, 2, 3, 4 denote feedbacks which are used when monitoring shows that set requirements on safety are not fulfilled [17, 18]

– prepared a contingency plan for the event of risks that are non-controllable or too costly to eliminate, or a few frequent.

3   The Target of Science on Safety and Security

Safety in the monitored concept is a tool to ensure the security and development of the human system in a dynamically variable world. It is about the long-term sustainability and integrity of systems that make up a complex human system, as system services support life-supporting functions. It considers human to be part of the system, integrates human activity with the protection of the environment and responds sensitively to human's needs in the context of ecosystems. In doing so, it needs to respect that sustainable development is not only about increasing and maintaining material well-being, but also concerns environmental vigilance, as most natural resources are not infinite (quantity), and some natural resources are also constantly contaminated (quality)—this applies in particular to water and soil.

Further impacts can be expected from potential climate change, or from anthropogenic activities such as the saturation of more than 7 billion humans. The lack of water, soil, chemical and biological contamination show that the problems are complex in the biophysical field and controversial in the socioeconomic field. The monitored systems are complex and many processes cannot be directly observed. In the socio-economic field, all environmental decisions can be characterised by a number of conflict objectives. In order to balance the relationship between human settlements and the biophysical environment (landscape) in the future, it is necessary to solve the problems of the so-called "grey" infrastructure (i.e. humans-created) and "green" (natural) infrastructures by application of a new approach based on safety management in an integral concept [18].

Since there is still no general consensus on the formulation of the problems of the sustainability of the public welfare (well-being) of human society in the context of system services, any solution to date is temporary, as it is constantly balancing between competing interests and societal objectives (if established). It is difficult to solve decision-making problems clearly due to the changing nature of the decision-making process. In decision-making, dilemmas are solved:

– the relationship among risks and benefits (often greater benefits to humans mean an increased risk to ecosystems),
– the time conflict among current and future needs,
– the social conflict (relationship of the need of the individual and the whole).

It is difficult to solve inverse problems for the complexity of systems. If any symptoms associated with the risks are determined and sorted, new symptoms will emerge. Therefore, a practical approach to sustainability management needs to be iterative, interactive and adaptive.

The analysis of environmental developments and the development of the political, social and economic situation in the world shows that it is necessary to prepare to deal with cases and actions that, by their intensity of impacts, will trigger critical situations. Therefore, in terms of human security, the development of the human system, the existence, stability and development of the State and each territory, the human safety management system needs to be proactive, strategic and involve sustainable development. Within this modern designed safety management system, there need to be emergency management and crisis management within it [20].

Comprehensive management aim is, in all situations, to ensure the protection of the human lives, health and security, property, welfare, environment, infrastructure and technologies, which are necessary for the survival of humans, i.e. always to ensure the mobilisation and coordination of the use of national resources (energy, labour, production capacity, food and agriculture, raw materials, telecommunications, etc.), coordination of activities such as the notification system, the rescue system and health services that reduce the impact of disasters, as well as the continuity of government and compliance with laws in force. The types of planning that make up the basic methodological tools of each interconnected management type need to create a basis in which the above objectives are anchored.

For the objectives of human society, i.e. above all for its sustainable development, they need to be combined measures and activities for reduction of vulnerability and for increase of resilience and adaptation capabilities that respect all fundamental protected interests in individual and whole. The current tool based on knowledge and experience is to implement a proactive safety management system at all levels of management, in which risk assessment is adapted to a form that respects all protected interests and considers existing and proven internal dependencies. In view of current knowledge, research on internal dependencies that mediate secondary and other impacts of disasters on human's lives, health and security [17] needs to be carried out and monitored.

Quality and qualified decision-making is an important part of each management process, and therefore, the decision-support systems need to be set up within the management system in place, as decision-making against systems is complex and needs to be multidimensional in nature [18].

3.1 Partial, Integrated and Integral Risk and Safety

The history of risk estimation is very long and comparable to the history of banking and insurance. E.g. without knowledge of risk, loans, bank guarantees and other financial services cannot be provided. Countless auxiliary work tools, methodological instructions, user manuals and software have been developed to assess risks. Their structure is extensively vertically and horizontally differentiated and the exhaustive classification is difficult [18].

The risks vary according to:

- what protected interests are chosen and whether one protected interest (partial risk) or set of protected interests (integrated or integral comprehensive risk) is pursued,
- what disasters/sources of risk are considered. For some tasks, a limited number of disasters are sufficient, e.g. in the area. only those that may have unacceptable impacts in the monitored area, for example, twice in 100 years, etc.

The integrated risk and integral risk differ in the determination approach. An integrated risk is the aggregation of partial risks (sum, weighted sum, etc.; it is usually defined in standards or legislation) for the protected interests under consideration. An integral risk is the risk determined by the systemic approach, i.e. the effects of links and flows among protected interests are also considered.

The partial risks are varied, e.g. health risks, technological risks, fire risks, etc. There are already a number of legal rules, norms and standards and related support software to calculate partial risks. In order to determine the integral risk, it is necessary to use specific procedures, which are methods using the multiple criteria, i.e. methods of operational analysis and specific multi-criteria methods based on decision support systems [18].

Because of the dynamic variability of the world, the size and nature of the risks in time and space change, and therefore, the risks need to be managed; the process model of risk control [18] is shown in Fig. 7. Risk management ensures conditions for entity safety; Fig. 8.

Fig. 7 Entity risk management model [18]

Fig. 8 Logical procedure for ensuring the safety of an entity [17]

3.2 Integral Safety and Security

Globalisation, on the one hand, and regionalisation or decentralisation (e.g. the idea of 'Europe of the regions') on the other hand mean mutually complementary processes that are often expressed by the slogan "think globally, act locally". However, their implementation requires that the attitude to security and safety might be reconsidered, on the one hand in the context of the growing complexity and vulnerability of contemporary society (critical processes, critical elements, critical objects, critical infrastructure and its functions) and on the other hand in the context of the undeniable changes that we observe (and may expect) in the human system, e.g.: in the environment, it goes on climate changes, landscape changes, etc.; and in the human society, it goes on dehumanization, great dependence of individuals on property, loss of such values as friendship, etc.

Considering these contexts, it is clear that security and safety need to have a wider social dimension, i.e. they need to express social, economic, cultural and ethno-political factors, and all government offices need to deal with them. This pays not only for central public authorities, but also for local public authorities and, in fact, for all those involved [15]. The public administration's position on security and safety for the citizen legitimizes its activity. The public administration is responsible for security and safety in the entrusted territory, namely for all facilities inserted in it, i.e. the safety should be continually a public service that does not deregulate or privatise. Thus, the starting points for the present concept of safety have a much broader basis than previously formulated safety on the state level.

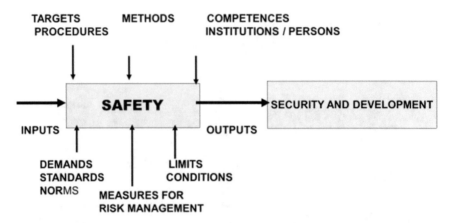

Fig. 9 Process model for ensuring the security and development of entity [17]

At present, the division of safety into external and internal is no longer sufficient, but safety needs to be understood from a systemic point of view [15]. From the system viewpoint, ensuring the safety is the basic requirements on system as a whole, not only demands on its components; system scheme of safety management at certain situation is shown in Fig. 1. From the process model of building the safety and security in Fig. 9, it is clear relation between safety and security; their often-discussed conflict is removed.

The requirement for a systemic concept of safety complies with the concept of integral safety introduced by the United Nations in 1994 [21]. Wever [24] supported the introduction of the term in 1995 for the following reasons:

1. A way of perceiving the safety by a citizen. Unlike the central public admin-istration institutions, the citizens see the safety primarily as a local problem, and therefore, they expect the local solutions that may vary from case to case. In other words, the citizens are particularly interested in their security, i.e. in security in the place where they live.
2. Security policy should cover a causal chain that solves the safety issues. The integral safety is not limited to unilateral solutions in the event of problems such as repression, but it deals with situations affecting a certain level of safety through so-called "the safety chain", which consists of the following parts:

    – proactivity (it eliminates the structural causes of uncertainty that undermines the safety, i.e. which they threaten security and sustainable development),
    – prevention (it eliminates direct causes of precarious situations infringing the present safety, if possible),
    – preparedness (it addresses to situations in which safety is impaired),

    repression—response (it manages faults of safety, stabilises the situation and ensures conditions for recovery and growth of safety).

3. The level of danger is territorially dispersed, and this dispersion is not even. Some safety problems are concentrated in certain areas, with types of safety problems (i.e. in terms of work [1] (disasters)) may not be and in practice are usually not the same.

4. Public administration often faces ineffective and inefficient solutions to safety problems. This fact is the result of the so-called "safety bureaucracy", which does not deal at all with the causal chain of safety. It is the result of a lack of understanding the concept of safety in reality (in a given case), i.e. it is the consequence of misunderstanding the links associated with the creation of safety and security as shown in Fig. 1, which shows that the level of safety predetermines the level of security of the system (i.e. the territory or technical facility which we monitor).

However, the concept of integral safety is slowly expanding in practice for the reasons set out in [15].

1. Integrity is understood more as an organizational aspect with horizontal and vertical connection among components/organs, i.e. not in the concept of a system with components, linkages and flows, and its understanding is mainly associated with police forces or the military.

2. There is still no satisfactory and generally accepted definition of integral safety in legislation.

3. Implementation of the concept of integral safety is in practice time-consuming (especially in domain of data collection and their analyses).

4. Local public authorities do not know "to deal with safety problems" because they focus too much on local problems.

However, the safety as a quantity/measure expressing the certain system behaviour, is not and cannot be isolated from its background. Each system and its surroundings are in interdependent relationship, which is due to the fact that each system is open system. The relationship in question can be characterized by some attribute of the system, such as adaptability, durability, flexibility and reliability [17].

To the concept of integral safety, they belong life-supporting functions, the risks of which with regard to human health, ecosystems and system safety are minimized. These are, in particular, possible non-demanded and unacceptable impacts, e.g.:

- industrial agriculture with regard to food safety,
- contamination of the environment,
- climate changes,
- lack of natural resources, energy and water,
- poverty and migration of humans,
- social discrimination,
- industrialisation and misuse of technologies,
- and gene manipulation.

It is, therefore, apparent that the security (in other words the system condition and its protected assets conditions) in relation to the environment needs to be specified in the context of sustainable development, i.e. to ensure its provision, the disasters should be monitored in the concept defined at Fig. 2.

The Johannesburg World Summit on Sustainable Development has pointed out that the development in question needs to be carried out primarily at local level and should be focused on the following objectives:

– environmental quality protection,
– quality of human life (health and human security, social justice),
– resilience to disasters,
– and economic vitality.

Sustainable development is not a static state (conditions) of harmony of society and the environment, but it is a process of changes in resources use, technologies' focuses and institutional transformations in order to avoid possible irreversible difficulties. It is just one of the possible dynamic models of the development of the human system. However, in practice, especially in public administration decisions, the concept of sustainable development is not more pronounced. Intuitively, however, it can be assumed that development requires a certain degree of sureness and stability, which are significant attributes of safety and security.

Integral safety is directly linked to the concept of sustainable development, as it can be characterised as a set of conditions under which humans are protected. By these conditions, it is strengthened the humans' ability to cope with serious and sudden threats to their survival (biological and social) and existence (health and housing), namely including the access to society's resources and the respect of human dignity [15]. Pillars of sustainable development are:

– environmental protection being related to environmental, technological and health safety,
– economic development being in relation to social, economic and technological safety,
– social development being linked to social, cultural, legislative and political safety.

Integral safety is measured using the indicators that already have a large number [15]. Indicators relevant to sustainable development and to technical facilities were introduced by the OECD [11, 12]. In practice, it is always necessary to select indicators that are relevant to the objective of the task addressed, choice is a critical activity and the success of the solution is dependent on it. It should be noted that in practice the following types of indicators are used:

– contextual (input and output relationship),
– causal,
– trending,
– and stative (measuring the conditions).

According to the works [12, 15] for the assessment of indicators, they are used the criteria for assessing:

– the validity, where there are evaluated aspects such as:

- relevance and importance,
- appropriate measuring scale,
- correctness (relation to the system examined),
- sensitivity (how system responds to changes),
- distinguishability (resolution of natural variability from mand-made changes),

– the clarity, when there are evaluated aspects such as:

- understanding (appropriateness of indicators for decision-making),
- simplicity,
- compliance with the interests of the public,
- the possibility of presentation and documentation,

– the interpretation, when there are evaluated aspects such as:

- robustness (the calculation is transparent and defensible),
- interpretability (to current status, changes and trends),
- credibility (the direction of change reflects certain experiences),
- trend evaluation,

– the information richness,
– the data availability, when there are evaluated aspects such as:

- sources for immediate use,
- time series,
- the possibility of updating,
- updating,
- topicality,
- anticipation and symptoms of warning,
- cost-check and feasibility,
- comparison of the costs and benefits of the indicator,
- ease of quantification
- the cost of collecting data,
- the ease of calculations

– the procedure of work with indicators.

This overview may be supplemented by a selection of appropriate measuring and evaluation scales and a description of the data type: time series, spatial data from GIS, relative or aggregated data, average, median, percentile, distribution function, etc. [18].

4 Methods and Tools of Science for Safety Management

The instrument for ensuring a safe entity [1–5, 8, 10, 12, 13, 17, 19, 23], i.e. an entity in which security is at the required level, i.e. it is an effective safety culture, it is an entity safety improvement program [12]. The procedure for creating an entity safety improvement program consists of the following steps:

1. To define tasks (sub-goals) and strategic goals of an entity with respect to safety.
2. For each section of the entity, to select the appropriate target and running indicators to assess the level of safety.
3. To create a dictionary for integral safety management needs.
4. To align standards, good practice methods and local practices.
5. To edit the list of target indicators according to the conditions in the entity.
6. To edit the list of running indicators according to the conditions in the entity.
7. To determine how the target indicators (i.e. value system) are evaluated according to the conditions in the entity.
8. To determine how running indicators (i.e. value system) are evaluated according to the conditions in the entity.
9. To determine the method/scale for measuring a set of indicators (i.e. system values) and marginal limits according to the conditions in the entity.

In practice, this means that for each section of an entity within the selected scope, target and running indicators are identified, which take the form of limits and checklists [12, 15]. In practice, they are assigned by evaluation criteria and scales to determine in which cases the target is achieved and in which it is not.

The Safety Improvement Program [17] includes a strategy for dealing with risks and ways of settling the risks and ensuring them in terms of technical, financial, organisational and personnel. In view of the world's changingness over time, its process model is in Fig. 10. The process model includes the organisational structure, responsibilities, practices, regulations, procedures and resources for identifying and applying disaster prevention or at least mitigating their unacceptable impacts. As a rule, it concerns a number of issues, including the structure of the human system, humans, the identification and assessment of hazards and the risks involved, the management of human society, the management of changes in human society, emergency and crisis planning, safety monitoring, audits and review.

Process coordination is aimed at ensuring a safe human system under normal, abnormal and critical conditions. Coordination is understood in the context as a controlled process aimed at ensuring the safety of the human system of the necessary quality; it monitors processes in space and time with respect to humans, resources, human's needs, finances and limits and conditions for human's activities.

Figure 10 shows the essential role of the concept of object safety, continuous evaluation of integral risk and serious partial risks. In the event that the assessment finds that the risk is unacceptable, changes should be made, as indicated by the feedbacks in Fig. 9. Since changes require resources, forces and means, feedback 1 is implemented first, and only when it does not bring the required state, it is implemented feedback 2; then feedback 3, and when the result is not required, then feedback 4. In the event of extreme events with catastrophic impacts, it is immediately taken on feedback 4.

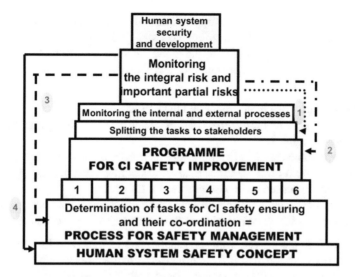

Fig. 10 Process model of human system safety management. Processes: 1—concept and control; 2—administrative procedures; 3—technical matters; 4—external cooperation; 5—emergency preparedness; 6—documentation and investigation of accidents; numbers in a yellow field indicate the feedbacks that are used in the control of entity [17]

5 Conclusion

Increasing density of the world population, increasing risks caused by increasing vulnerability of buildings and technologies and concentrations of inhabitants (transport systems, power plants, chemical plants and warehouses, etc.), large agglomerations (cultural events, sports events, politically oriented events, etc.), connection of growing populations to source units characterized by increasing unit power (power plants, water systems, transport systems), the rapid financial and economic interconnection of institutions in a globalised world and the associated transfers of huge funds, the extreme and growing potential and reach of weapons, the increasing hardness of extreme population groups and their internationalisation, and the extremely rapid spread of terror and violence between continents, the effects of technology and human factor failures, the potential of endogenous and exogenous natural forces operating on increasingly densely populated areas and other risk factors threaten more than before, locally and globally human society in contemporary civilization and the environment I.e. the whole human system. To this new situation, it is necessary to adapt the historically established systems for the protection of human lives, health and security and also for all other protected interests. Humans need to do their best for a good level of human system safety, understood as the integral safety of a complex system of systems.

The above knowledge and experience demonstrate that the issue of risk management for safety is essential and requires that we be able to move along the axis of "anticipation of the future" and "appreciation of the past". Since the risk in each entity depends on both, the immediate size and type of potential disaster and the vulnerability of the entity, which is due to the vulnerabilities of the protected interests (assets) of the entity, their quantity and the interconnections, it is necessary to:

- determine the assets in the entity and their values,
- specify the type of entity, the risks of which need to be controlled (simple, composite, complex),
- identify possible disasters that may have an impact in the monitored entity,
- identify for each possible disaster the impacts that cause losses, damages and harms and the subsequent consequences for the entity,
- determine how often emergencies occur with unacceptable impacts,
- determine how it will be negotiated with risks that are not acceptable.

No process in a complex world is usually not so simple that its outputs can be affected by one criterion. It is, therefore, necessary to use more criteria when deciding on the world and to focus on methodologies that allow the use of several criteria [18]. A multi-criteria approach to decision-making offers some advantages, namely:

- structured decision-making (definitions of criteria, impacts' assessment);
- providing a framework for examining the objectives, concerns and priorities of interest groups,
- allowing to work with stochastic uncertainty (the future situation cannot be accurately determined) and the so-called fuzzy (vague) uncertainty, i.e. vagueness for which information can only be expressed verbally.

The entity safety management process is a continuous process of solving the problems of, i.e. decision-making process. In the process in question, it is necessary to correctly diagnose or specify the problem, to make rational decisions, to accept and implement the decision in specific conditions. The decision-making process is a logically interfaced, effective sequence of decisions by the decision maker from finding a decision problem to formulating a decision. It consists of the following steps:

- the collection and processing of information, the processing of which needs to be adequate to the problem that we monitor (e.g. that means that data processing methods for safety management purposes need to respect that large disasters with destructive force rarely occur, and therefore, the procedures respecting the law of large numbers, i.e. algorithms based on extreme or limit estimates, need to be used,
- detection of variants of problem solving,
- finding the optimal solution to the problem,
- own decision.

The essence of decision-making is that the entity managing subject selects from a set of possible solutions that are a response to the stimuli, obstacles and problems they encounter. The design of several variants of the solution is actually a scientific matter, because the problem needs to be analysed, evaluated and predicted in possible variable conditions. The *design of* control options to enable the effective implementation of the proposed measures need, therefore, be included in all options. This step may be seen as administrative, but cannot be underestimated or omitted, as it is sufficiently known that the best-intentioned executive (implementing, executive) measures are completely passing through without effective control and, where appropriate, sanctions.

A decision-making process is a sequence of steps that lead to a particular decision. The subject/object of decision-making is a problem on which the decision-making process is aimed. Decision-making factors are real factors that the decision-making body recognizes and gives them importance during the decision-making process. It is divided into internal and external. Decision criteria are measures or aspects to measure the degree or rate of achievement of the objectives.

When selecting decision-making methods, it is necessary to respect the nature of the problem solved, the stated objectives of the solution, the criteria of the solution and the possibility of gathering the necessary input information. I.e. in the area of territorial safety management, it needs to be respected that most problems are associated with uncertainties and vagueness's, which are also caused by the fact that the human system is developing dynamically in an ever-changing external environment, and that, therefore, the choice of the right strategy to ensure the security and sustainable development of the human system is necessary to meet the safety management objectives.

When choosing the optimal option in a given particular case, it plays a role:

– the level of security and sustainable development achieved in applying the option,
– the technical feasibility of the measures, considering the appropriateness of the measures for the given system,
– material demandingness and energy demandingness,
– speed of implementation,
– demands to qualified personnel,
– claims on ensuring the information,
– claims on finances,
– claims for liability,
– demands to management/organisation in the territory, etc.

To ensure humanity's goals, the science on safety need to be super departmental, develop and put into practice qualified risk management tools for the benefit of the human system safety.

Acknowledgements The research results presented have been supported by projects: EU—FOCUS, grant No 261633, EU—ISDEP, MŠMT—RIRIZIBE-CZ.02.2.69/0.0/0.0/16-018/0002649.

References

1. Ale B, Papazoglou I, Zio E (2010) Reliability, risk and safety. Taylor & Francis Group, London, 2448 pp
2. Beer M, Zio E (2019) In: Proceedings of the 29th European safety and reliability conference. ESRA, Singapore. enquiries@ rpsonline.com.sg
3. Bérenguer C, Grall A, Guedes Soares C (2011) Advances in safety, reliability and risk management. Taylor & Francis Group, London, 3035 pp
4. Briš R, Guedes Soares C, Martorell S (2009) Reliability, risk and safety. Theory and applications. CRC Press, London, 2362 pp
5. Cepin M, Bris R (2017) Safety and reliability—theory and applications. Taylor & Francis Group, London, 3627 pp
6. de Jong J, Claas E, Osterhaus A et al (1997) A pandemic warning? Nature 389:554. https://doi.org/10.1038/39218
7. EU (2006) The seventh frame research programme 2007–2013. EU, Brussels
8. Haugen S, Vinnem J, Barros A, Kongsvik T, Van Gulijk A (2018) Safe societies in a changing world. Taylor & Francis Group, London, 3234 pp. https://www.ntnu.edu/esrel2018
9. IAPSAM (2012) Probabilistic safety assessment and management conference. IPSAM & ESRA, Helsinki, 6889 pp
10. Nowakowski T, Mlyńczak M, Jodejko-Pietruczuk A, Werbińska-Wojciechowska S (2014) Safety and reliability: methodology and application. Taylor & Francis Group, London, 2453 pp
11. OECD (1992) OECD environmental indicators towards sustainable development. OECD, Paris, p 152. ISBN 92-64-18718-9
12. OECD (2002) Guidance on safety performance indicators. guidance for industry, public authorities and communities for developing SPI programmes related to chemical accident prevention, preparedness and response. OECD, Paris, 191 pp
13. Podofillini L, Sudret B, Stojadinovic B, Zio E, Kröger W (2015) Safety and reliability of complex engineered systems: ESREL 2015. CRC Press, London, 4560 pp
14. Porada V et al (2019) Security science. Aleš Čeněk, Pilsen, 780 pp. ISBN 978-80-7380-758-0
15. Procházková D (2011) Strategic management of the safety of the territory and organization. CTU, Prague, 483 pp. ISBN 978-80-01-04844-3
16. Procházková D (2014) Fight against terrorism. EU project: improving security by democratic participation—ISDEP. CTU, Prague, 200 pp. ISBN 978-80-0105568-7
17. Procházková D (2017) Principles of management of risks of complex technological facilities. ČVUT, Praha, 364 pp. http://hdl.handle.net/10467/72582
18. Procházková D (2018) Analysis and coping with risks connected with technical facilities. ČVUT, Praha, 222 pp. http://hdl.handle.net/10467/78442
19. Steenbergen R, Van Gelder P, Miraglia S, Ton Vrouwenvelder A (2013) Safety reliability and risk analysis: beyond the horizon. Taylor & Francis Group, London, 3387 pp
20. Tuser I, Hoskova-Mayerova S (2020) Emergency management in resolving an emergency situation. J Risk Financ Manag 13(11):262. https://doi.org/10.3390/jrfm13110262
21. UN (1994) Human development report. New York. www.un.org
22. US ISDR (2006) World conference on disaster reduction declaration. Kobe, 25 pp. www.un.org
23. Walls I, Revie M, Bedford T (2016) Risk, reliability and safety: innovating theory and practice: proceedings of ESREL 2016. CRC Press, London, 2942 pp
24. Wever J (2000) Integral safety in Netherland. Paper presented at the Australian Institute of Criminology in 2000. www.aic.gov.au/conference

Theoretical Concepts Related to the Local Public Order Affairs

Oldřich Krulík and Josef Hrudka

Abstract The topic of local public order affairs (with special emphasis on the functioning of units, known as municipal police and their position in the security system of an individual state) is one of the current or emerging issues of security discourse. As with all outputs of scientific research, the question arises as to whether or to what extent this issue is methodically anchored, or how it is covered by secondary studies from renowned sources (including possible theoretical concepts used here). The chapter summarizes this topic and attempts to create a springboard for other researchers who are interested in the issue. Terms such as pluralistic policing, the extended police family or multi-level policing at local level will be mentioned.

Keywords Local public order affairs · Municipal police · Pluralistic policing · Extended police family · Multi-level policing · Theoretical attitudes

1 Introduction

The topic of local public order affairs (for example, the functioning of municipal police forces) is one of the emerging issues of security discourse today. As for all scientific research outputs, even here the question arises how this issue is methodologically anchored, or how it is covered by secondary studies in reputable sources (including possible theoretical concepts that are applied here). The chapter attempts to describe the related situation and create a springboard for other researchers involved in the issue. In addition to monitoring the frequency of occurrence of related ideas in relevant studies, the authors declare an effort to take their own, constructively-critical position regarding the topic, based on emerging robust empirical data.

O. Krulík (✉) · J. Hrudka
AMBIS College, Lindnerova 575/1, 180 00 Praha, Czech Republic
e-mail: oldrich.krulik@ambis.cz

J. Hrudka
e-mail: josef.hrudka@ambis.cz

© The Author(s), under exclusive license to Springer Nature Switzerland AG 2022
I. Tušer and Š. Hošková-Mayerová (eds.), *Trends and Future Directions in Security and Emergency Management*, Lecture Notes in Networks and Systems 257,
https://doi.org/10.1007/978-3-030-88907-4_3

2 Historical Studies

At first, attention should be brought to two studies, originating from the environment of the United States of America, dealing with some, at the time, rather insignificant findings.

The first text, a 1929 study by Bruce SMITH (National Institute of Public Administration, New York), states that the role of the police in modern society is increasingly complicated [118].

If we look at the situation from more than a century ago, the police stood against a relatively specific group of perpetrators. The urgent need for protection made the ordinary citizen a natural ally of the police, even though the police officer did not enjoy any high degree of public trust.

An important feature of the police administration in the United States is its local character. Although the federal government runs certain specialized law enforcement agencies and some individual states (of the Union) have invested in the creation of a *regional/state police force*, most of the burden in the area lies directly with the municipalities. The related agenda is sometimes quite various and can divert attention from the basic police mission (protection of life, health and property against crime and anti-social behavior). An example offers an indicative list of possible tasks of municipal police forces:

1. Official agenda (licensing of private detectives, pawnshops, bazaars, street sales, dance halls and public exhibitions—including film and stage censorship).
2. Regulation of road transport and parking (agenda very unpopular with the public).
3. Probation service (supervision of released convicts).
4. Social assistance for the poor or homeless.
5. Inspection of requirements for storage of hazardous substances, including filling stations.
6. Census, voter registration and verification of voter lists.
7. Control of examination of prostitutes for venereal diseases.
8. *Dog agenda*, including the sale of dog tags.

The freedom that municipalities enjoy in the management of their police forces leads to a variety of forms of police organization. Although there are some framework federal or state regulations in this area, the number of organizational models is almost the same as the number of municipalities.

The most common type is now largely limited to smaller towns and villages. Relevant committee of a city council, which instructs officials who are generally known as *police managers*, supervises police officers. Police committees are inclined to use their powers to regulate the functioning of the police department in as much detail as possible. **The police force is often under the control of a large number of superiors and ceases to be a cohesive public agency**.

It seems that the European police forces, some of which operate in regimes as democratic as the United States, do not seem to be facing the same difficulties. There are probably a considerable number of reasons for this difference, but they all serve to emphasize the usual influence of local guerrilla politics on municipal administration in the United States. Although the remedies of this situation are outside the scope of the police administration, there is no reason to prevent the introduction of methods that will lead to more successful police work.

The internal organization of the police force is in each individual case conditioned by the scope of its responsibilities. If the unit is authorized to a number of inspection duties of a technical nature, which must be performed by specialized personnel, the operation of the organization is very complicated.

In some larger cities, which include one or more districts, there is often a position of a sheriff, who can act independently of the city's police force. Where state police forces are in operation, new difficulties arise. The local police understand the intervention of the state police as a disregard for their ability to handle the situation, as an attack on what is known as the *home rule*. In addition, the relationship between the police and the public prosecutor can be complex. The public prosecutor can deposit specific detectives from the police force to his or her own office, where they work as a separate agency, or can even set up a completely independent body of investigators. As a result, the municipal police forces are without information about the commission of certain crimes, the solution of which could shed some light.

It is inevitable for the police to take the initiative in gaining public confidence. Developments in police education can also contribute to this change. At the time the study was undertaken, the police training institutions were still relatively small. There were many larger cities that did not provide sufficient police training courses. There was a lack of a sufficient number of study texts adapted to the local situation. Systematic teaching not only of new recruits, but also of veterans and management was (and still is) one of the most promising means of restoring public confidence.

The uniformed foot patrol is a vanishing institution. The repressive influence of a uniformed police officer disappears—and the preventive influence grows. The crime is rarely committed in the presence of a police officer, and if so, it is usually because the police appeared on the scene unexpectedly. No one can measure the repressive effect of a uniformed patrol, but there is no doubt that its impact on deterrence is great. The density of police stations can thus have an important factor in crime prevention. However, each station requires from twelve to fifteen police officers to be able to operate 24 h a day and respond to emergency calls, for example. Around the year 1900, the police districts were very small. In the age of radio communications and patrol cars, there are growing voices to reduce the number of police stations and the number of patrols. However, when a car patrol is entrusted with 20 km of streets, all responsibility collapses. The effect of the patrol routine will inevitably lead to limit the attention to some streets and to neglect the rest.

We also go deep into the past with the study of Weldon COOPER of the University of Texas [26]. He stated that one of the most important features of the city administration is the tendency to experiment with different organizational patterns. An institution is generally defined as the organization of personnel to facilitate the fulfillment of a particular purpose through the allocation of functions and responsibilities. It is a matter of combining efforts and capacities in relation to a common task so that the desired goal is achieved as satisfactorily as possible.

Municipal reformers want to remove structures that fail to cope effectively with the growing burden on the community, and install a model that performs the necessary services more efficiently. When there are general reorganizations, they do not avoid the police structure either. Steps to reorganize the police structure are most often motivated by efforts to better control the police and prevent corruption.

In this respect, cities in the United States differ significantly from cities in England, which use the same general features of a police organization since the creation of modern police forces.

The activities of a police organization can be divided into those that are primary (crime suppression, detection and detention of perpetrators and prosecution, road traffic regulation) and tasks that are ancillary but necessary for the primary purposes (e.g. licensing, provision of public ambulance services, probation and mediation activities, checking, registering and verifying voter lists, censuses, temporary care of the poor, or issuing of dog tags). The growing interest in **crime prevention** has been accepted as a *proper duty* of the police in many jurisdictions.

When the New Scotland Yard was formed, police duties were more straightforward than today. There were no motor vehicles and no traffic laws; crime prevention and detection activities were less extensive; fingerprints have not been introduced and person registers have only just begun to develop; communication systems were in their infancy. There is no part of the police work that has not become extremely complicated and *more scientific* in the last forty years. Each department must employ a number of professionals. At this time, the first operations centers were established. Radiocommunications are being used and the police are being motorized. This development raises serious doubts about the effectiveness of tools such as the foot police patrol.

3   Defining the Concept of Community Police; Variables in Relation to the Role of the Public

Ercan BALCIOĞLU (an expert in forensic and social sciences, Gazi Ankara University, Turkey) and Erkan PALA (a Ph.D. student in police studies, Loughborough University, United Kingdom) state in a 2016 treatise that the philosophy of performance of municipal police activities is an attractive area for authors seeking answers to crime-related questions, from public-police relations to petty crime and anti-social behavior [97].

Deniz KOCAK (a political scientist, Helmut Schmidt Universität; Universität der Bundeswehr, Hamburg, Germany) in a text from 2018 describes the development of relations between the police and the public in the context of opposition to *police brutality* or alleged and real discriminatory practices of police officers since 1970s.

At least since then, police managers and politicians in the United States and Western Europe have been *"desperately looking for new approaches"* in this area [66].

Community policing, or what is sometimes called community-oriented police service (COPS), is a multilateral package of police reforms that have changed the police forces of many states since the 1990s. In evaluating the way in which the concept was institutionalized in the United States of America in the late 1990s, it was possible to identify twelve specific typical community policing practices [144, p. 80]:

1. strong/fixed connection of the police officers to a specific location (*know your police officer*);
2. the use of a denser network of police stations *in the front line*, in a specific location;
3. specialization of police officers to perform specific tasks in target locations;
4. emphasis on foot (cycling, riding) police patrols;
5. the use of unpaid civilian volunteers;
6. neighborhood watch programs;
7. patrol programs in business districts;
8. contact programs for victims of crime;
9. work with *community correspondents*, building channels of communication with the public;
10. meetings between police officers and the public within the community;
11. educational programs for the public, with an emphasis on crime prevention;
12. opinion polls to obtain information on local priorities.

Police officers and politicians have often identified community policing as the *only alternative* to improving police-public relations in ethnically or socially heterogeneous environment, including a positive effect on reducing the detected crime [98].

Ideally, the principles, labeled as the community policing, guide each police officer. Despite frequent mention, however, community policing appears to be a term that is difficult to define. A well-clarified and generally accepted definition of the term is still being sought. However, it would be a mistake to give the impression that there is no consensus on what constitutes the basic elements of this model: police relations and the public are vital; the police and the public need to act as partners; the police and the public constantly consult each other on their demands and expectations. In its most peculiar form, community police include three basic elements: problem solving, public involvement, and decentralization [116].

Problem solving is related to the importance of proactive crime prevention, and the reduction of traditional forms of coercive activity. Community policing therefore emphasize efforts to address issues such as empty, unmaintained houses, graffiti or minors on the streets—but different from the traditional reactive model of policing [143].

Public involvement and strengthening the relationship between the police and the community at the neighborhood level aims to proper identification of the needs of both actors. Citizens can become the *eyes and ears* of police forces (not only in terms of reporting suspected crimes, but also in terms of providing information on neighborhood issues that may turn into violations). In this respect, we are talking about the formation of so-called **social capital** [106].

Decentralization in this context reinforces the first two elements by recognizing the links between the public and the police. Problems can only be solved if police officers are given incentives, training and flexibility to focus on community-based activities (for example, changed schedule of operations and strategy, decentralization of the police structure and strengthening the powers of patrol officers).

The foot patrol, a typical old-new practice (reintroduced since the 1980s), brings police officers into closer contact with those they serve—and it has become an iconic example of the community policing concept [144, p. 75].

Dozens of departments in a number of major cities have implemented specific elements across the United States. This widespread adoption of the concept was encouraged by financial support program during the Clinton term (the prospect of hiring 100,000 new police officers in departments able to introduce to community police support tools) [51, pp. 295–296; 144, p. 75].

There is also another look at community policing that points to two different styles that community policing can adopt. The difference is related to the extent to which police forces use coercion. In the so-called **partnership model** (partnership policing), the police emphasize that they work with the community to solve long-term problems, look for their causes (using situational crime prevention tools) and use other methods far from the actual use of force. It emphasizes the development of alternative activities for young people (sports, recreation and employment opportunities). The community is not considered to be passive audience, but a partner in ensuring security. The success of the police in ensuring security depends on a wide range of partners [126]. It is argued that it would be an unrealistic attempt by the police force to tackle crime solely on its own [113, p. 9; 119]. Intensive two-way communication can not only reduce public fears of crime, but also improve public image, increase public awareness, and enable citizens to increase their resilience on their own (investing in increasing the security of a house or apartment, neighborhood patrols) [112, 128]. It is notable, that functioning partnerships (what can be understood as a standard that must work nearly everywhere) is presented as some excellent above standard.

The paradigm, referred to as *problem oriented policing*, requires the police to know more about victims and perpetrators and to find out why the act took place the way it did [1]. The concept addresses specific issues that have been identified and that could be addressed through civic engagement. Such an approach is also likely to reduce crime and other forms of anti-social behavior [3].

Police forces are seeking feedback from the community on police work priorities. Police forces themselves often see as a priority to tackle the most serious crimes (such as terrorism), while residents of a particular locality are plagued by noisy young people or burglaries. By focusing on these priorities, police forces can easily improve the quality of life in the community. Police officers have to contend with the image of professionals who are not interested in *too common* or non-priority crime in the locality [114, 148].

4 *Soft Policing* Versus *Hard Policing*

One of the main ideas of the concept of community policing is to reduce purely reactive or *traditional* approaches to policing (hard policing, real policing), and to strengthen proactive tools, often based on *mere* interpersonal communication (soft policing). However, this approach did not find its way into the hearts of all police officers. Although there are many reasons to implement elements identifiable as community policing in specific localities or police activities, this phenomenon faces resistance inside and outside the police. The *traditionalists* appreciates the fight against crime, supports aggressive police tactics, and is skeptical about public relations and the soft approach as a whole. They argues that *soft* approaches are not *real* police work.

The view of the traditionalists is closely linked to one of the most controversial and most doubted concepts, based on the so-called **broken windows theory** [10]. This theory has convinced many that the wave of crime plaguing some neighborhoods was set in motion by empty houses, graffiti, and vandalism, as signals of the collapse of informal social control in the locality. If these phenomena are not addressed, they provide a visual signal that law-abiding forces do not control the site—and this will encourage criminal elements [141]. From this perspective, the term *proactivity* is understood in a different way, and involves *"an aggressive approach to fulfilling the desires of the majority of the public"* (for example, the situation around *zero tolerance* in New York). This includes the use of coercive measures against drunks, noisy teenagers, prostitutes and other street people, even if their behavior does not violate the law, only neighborhood norms [123, p. 112]. On the background of rising crime, politicians have not contradicted or even supported this trend, fearing that they might be labeled like someone who is *lenient with crime.*

The idea of a greater emphasis on *real* policing (persecution of criminals, use of force against perpetrators, etc.) dates back to the 1960s. Police forces have become more specialized and professional, and during this process they have *disappeared from the community*, and foot patrols have become a rarity in many locations. Although the increased mobility of criminals and the changing crime patterns provide some support for a stronger law enforcement style, reducing the ability to interact with partners, including the general public, is an objective loss for the police [120, 136, 151].

Some members of the research community have questioned the broken windows theory, or rather the view that vandalism and loitering are a prelude to crime increase. It is said that *hard* police measures and *zero tolerance* means benefit for *richer* parts of society, while ethnic minorities, adolescents and socially disadvantaged members of society find themselves in the role of those targeted by police measures. For this reason, those groups tend to lose the confidence in police impartiality [71]. The paradox is that some police forces (not only in the United States) draw funds for community policing projects, but as a result, they purchase armored vehicles and other repressive or controversial techniques [67].

Proponents of a *soft* approach are of the opinion that policing is not (in the best sense of the word) so different from the work of a field social worker. Like social security and other social programs, police may be instructed to disseminate the expectations of public officials to the respective target population. The message of the community police is likely to vary depending on the nature of such target population (racial and social composition of the community) and the nature of the relationship of this target population with the police [105].

Although we could assume that the benefits of community policing are considerable, they are very difficult to measure, and positive effects can take a long time. If specific research focuses on whether the implementation of community policing tools contributes to reducing the reported crime—the results are unclear [65, p. 49; 83]. The same applies to the evaluation of the introduction of foot patrols—the impact on the detected crime rate is limited (but the results of car patrols are clearly even worse), but there is a visible change in reducing public fears of crime, improving police perception and increasing the willingness of citizens to engage in local security initiatives. After all, public concerns often relate to more forms of anti-social behavior than to crime as such [14, 63, 114, 124, p. 72; 127, p. 86; 133, p. 47; 4, p. 340; 64, 148, p. 8].

5   Pluraristic Policing; Building of the *Wider Police Families*

Geography expert Megan O'NEILL and sociologist Nicholas R. FYFE (both University of Dundee, Scotland) in a 2017 study seek to decipher the complex nature of the very concept of 'plural policing' at local and national levels. The study provides information on how the procedures concerning the police system reflect the influence of history and geography. The text focuses mainly on the relations that have emerged in the public sector through **pluralization** processes, in particular the introduction of institutes of various *police assistants* or municipal police forces. They show that experiences with pluralized police vary widely across Europe and call into question the role of neoliberal forces in this regard [95, pp. 1–5].

They also refer to a number of other authors, such as the sociologists and criminologists Trevor JONES (Cardiff University, Wales) and Tim NEWBURN (London School of Economics and Political Science), who opened the topic at the end of the XX century [57, 60]. The same applies to legal expert and criminologist Adam CRAWFORD (University of Leeds). References to *plural policing, policing beyond the police* and *extended policing family* are now a common part of discussions about policing in modern society [31, 32].

Police work is now provided by various networks of public police, commercial bodies, voluntary and community groups, individual citizens, national and local regulatory agencies etc. [59, p. 1].

The authors first telegraphically mention the studies from which they themselves benefit:

David GARLAND (lawyer and sociologist at New York University) notes in a 2001 study that the key factors in this development are seemingly well established (ranging from fundamental shifts from a *culture of control* as well as the "*myth of a sovereign state monopoly to the fight against crime*" towards recognition of the role of private, municipal and voluntary actors in the control and prevention of illegal activities [47].

Cynthia LUM (criminologist from George Mason University, Fairfax, Virginia, United States) together with the already mentioned Nicholas R. FYFE open the topic of the nature of the built-up area (closed communities, expanding night economy, etc.) in a 2015 study. They believe that there is a danger that generalized claims about the dynamic and changing nature of pluralistic police will become a new orthodoxy, losing their sense of recognition of the importance of local specifics [80, pp. 219–222].

Ian LOADER (lawyer, political scientist and sociologist, Oxford University) states in a 2000 paper that relations that have emerged in the public sector through its pluralization (creation of auxiliary or municipal police services; policing 'below' the police) represent a problem for the *proper* police forces. There are difficulties in defining one's professional identity and building relationships with the community (voices are heard that this or that is not a priority for the *real* police, and this task can be taken over by municipal police officers, however differently defined [77]. And we return again to the study from 2017. The text deals, selectively, with six countries that were more familiar to the authors [95, pp. 1–5].

Scotland, citing the findings of Les JOHNSTON (Police and Public Administration Expert, University of Portsmouth, United Kingdom) in 2003 and Donny Marie BROWN (sociologist, Durham University, United Kingdom) in 2015. It is noted that Scotland community wardens are not members of the police as such, but they are considered a part of an *extended police family* or *multilevel policing*. They work in *high-risk areas* where the anti-social behavior is more likely to occur, usually as a part of community security initiatives. The task of wardens is also to act as mediators, ready to solve community problems. Relations between wardens and police officers have not always been ideal, but over time the situation has improved, because wardens tend to be included in information exchange models with locally relevant police forces [55, p. 3]. Perceptions of wardens have changed over time, and today they are reportedly seen by a large part of the public as a credible force, able to gather knowledge of local criminal activities, that a particular informant would probably not disclose to police officers [13].

England and Wales. The process of finding a professional identity is also a problem for Police Community Support Officers (PCSOs). Unlike wardens in Scotland, they are people who are a direct part of the police family. This complicates finding its acceptable role in the communities it serves. The ambition of these units was originally to conduct highly visible infantry patrols, fulfill the role of *eyes and ears* of full-fledged police officers and field communication with the public. These people *are and are not* police officers, depending on the specific location. Their marginal status within the police family can be interpreted as an advantage, only because they are not considered *real* police officers and can therefore obtain information from the public that no one would tell *real* police officers. This is what PCSOs often consider the most valuable aspect of their work [95].

The Netherlands[1] is a country where Municipal Law Enforcement Officers (MLEOs) operate, a kind of *quasi-police*, employed by municipalities. This model follows the former wardens, who, however, did not enjoy a positive reputation (because they lacked powers and their recruitment standards were low). The new concept has a much higher recruitment standard and the powers to deal with anti-social phenomena, including the possibility of arresting *problematic persons*. The primary task of the officers is to gather information and to oversee behavior in *problematic areas*; reporting problems, and responding to suggestions from the public. Although the police are theoretically entitled to direct these officials, in practice this rarely happens. These practitioners thus represent an example of the erosion of the monopoly of the national police force [140, 54, 99, 130].

Paris agglomeration, France (mentioning the findings of a pair of authors, Jacques DE MAILLARD—a political scientist and sociologist in relation to police activities, and Mathieu ZAGRODZKI, a sociologist and criminal law expert, both from the University of Versailles-Saint-Quentin, France, and Adam CRAWFOR, a criminologist, University of Leeds, United Kingdom). This case represents a unique configuration of pluralized supervision, which does not occur in Anglo-American models and in the rest of France. Despite France's traditional tendency to centralize, security measures in Paris are highly pluralized. Two city agencies and one private platform were involved in addressing parking, security and order issues in the city. Municipal inspectors, city employees, have certain powers and impose fines for less serious offenses, especially in terms of violations of the cleanliness of public spaces. Night municipal officials focus their work on *sensitive* localities. The private company Groupement parisien intrabailleurs de surveillance (GPIS) is also involved in the areas of *sensitive* social housing and its employees can evict problematic tenants from real estate. All three services do not avoid direct contact with the National Police, they are not afraid of *contamination*, but are looking for greater legitimacy of their work in this interaction [34].

Austria (the authors are mentioning the 2003 findings of Clifford SHEARING, a criminologist and an expert in security aspects of public administration—Griffith University, Queensland, Australia, and Jennifer WOOD, an expert in criminal law—Temple University, Philadelphia, United States of America) is an example of limited plural policing. In Austria, there is a national police force, which has a large monopoly on police work. Smaller municipalities in Austria can create their own municipal police forces, but at present less than 2% of them have done so. However, statutory municipalities cannot do this. As a compromise, the government allowed these cities to set up *city guards* (sometimes based on a private basis), guarding public spaces and dealing with certain traffic offenses. The private sector has very little penetration

---

[1] The authors are preparing a separate study on "experiments", regarding the local public order affairs in the Netherlands.

of the police. Neoliberal *security privatization* is not developing as much here as in other countries [68, 109].

Belgium (citing the findings of Elke DEVROE, an expert on internal and international security issues, Leiden University) is an example of a country with a very complex structure of public administration and self-government. Because of corruption scandals involving police officers and politicians, a major reform was carried out in 1998 to restructure police work, taking into account different local, regional and national structures. In Belgium, about a third of police officers are now systematized in the national Federal Police and about two thirds in the ranks of the Local Police, which is part of the national police structure but performing tasks predominantly at the level of micro-regions. In the past, there were also municipal police forces in Belgium, but their powers and resources were absorbed into the national police [146].

6 Institutional Aspects of Police Work

Robert E. WORDEN (a political scientist and criminologist, University of Albany, New York) and Sarah J. McLEAN (specializes in the effectiveness of crime prevention programs, John F. Finn Institute for Public Safety, Albany, New York) in a publication from 2017 shows that the police department (here predominantly in the United States of America) can be described from many points of view, with one possible perspective being institutional. **Institutional theory** is generally used in the study of private and public sector organizations—here in order to understand the dynamics of police activities [142].

Front-line policing often takes place on the street and is carried out by *the lowest rank police officers*. The working environment of police officers is heterogeneous and turbulent, with many unpredictable aspects. Police officers' decisions are subject to great uncertainty about their consequences. Not all situations can be embedded in manuals [125].

Teachers, social workers, police officers and other professions, identifiable as *street-level bureaucracy*, deal with a number of similar variables: chronically insufficient resources, including time and information; vague, ambiguous and sometimes conflicting goals of the organization; and the fact, that it is not easy to control the work efficiency or results of individual oirdinary workers/officers. Workers/officers/officials also often reconceptualize their work, *refining* their interpretation of internal norms to reduce the discrepancy between the ideal and the achievable reality [12, 75, 135].

How individual *ordinary* police officers understand, for example, certain reform pressures will not always be in line with the interpretations of police managers. Some police subcultures will support reforms; others will resist them and promote their own values and attitudes.

The institutional environment formulates expectations regarding the structure of the organization and reward efficient production. Some types of organizations, such as manufac-turing companies, operate in an environment where the market determines the value of products.

Measuring the performance of the police is ambiguous, with variables playing a role in reducing the crime occurence and victimization; detention of offenders; reducing fear of crimne and increasing the sense of security of the public, etc. In addition, the level of expenditure on service and/or the efficient use of public resources can be also monitored [93]. Improving performance in one area (for example, greater emphasis on respect for civil liberties) can be at the expense of performance in another area (for example, crime control). And not all people inside and outside the organization have to resonate with such development.

Despite a number of technological innovations, policing is still a profession that interacts directly with people and relies mainly on communication. Use of the force is only the last option, how to solve a certain situation, the last resort. Enforcement is a professional prerogative for police officers, making police force a unique organization. However, most of the functions that the police perform are dependent on *talking to people* anyway [5].

Some pressures are in conflict with each other. Environmental pressures, for example, lead the police to adopt reforms that are generally considered desirable, although there are no clear links to the outputs or results being assessed. The hearts of not all ordinary police officers are easy to win for any reform. They want to "*change everything so that everything can remain the same*". They will formally support the reform steps, but do nothing in their favor. Any institution or organization tends to resist the change. Police forces are no exception. Changes can be declared with the best of intentions, but it is necessary to remain skeptical about leading to systematic changes (not only in terms of police-public interaction) [137].

7 Socially Disadvantaged Groups of Inhabitants and Community Policing

Without being primarily an academic dissertation, mention should be made of a document, generally referred to as the Kerner Report, which seems to foreshadow the development of the BlackLives Mattter concept, which is so visible today. The text was written in 1968, during the term of President Lyndon Johnson, and at its core is a commission report chaired by Illinois Governor Otto Kerner [153].

The Commission examined deep animosity between police and ghetto residents. Since 1964, virtually every major city in the United States has had tense relationships between the police and minority groups. The police officer in the ghetto is a symbol of the entire law enforcement system. As such, it is becoming a tangible target for complaints about system deficiencies. Activists speak of "*order maintained at the expense of justice*". The police officer alone cannot solve these problems. However, his or her role is one of the most difficult. He or she has to deal daily with a number of problems and people who test his or her patience, ingenuity, character and courage in a way that few other professions are ever tested.

Effective law enforcement requires community support. Such support will not exist if a substantial part of the community considers the police to be an occupying force.

More than 70% of African Americans believe that the police lack respect for them, use offensive language and use unnecessary force during arrests. About 15% of incidents allegedly triggered a *offensive order* by a police officer. In the ghetto, the average age of the population is under 21 and the flats are uninhabitable in the summer due to the heat. Young people spend time on the streets and want to show their courage. They provoke police officers, test their self-control and thus strengthen their hostility towards minorities in general.

The best police officers do not serve in the ghetto, but the worst. Sometimes it is a part of their disciplinary punishment. It is similar to a school in a ghetto with teachers who would not find a job anywhere else. With a few exceptions, the worst in the system gets here. If excellent police officers are assigned to the ghetto, if misconduct is immediately the subject of disciplinary action, then a culture that rejects misconduct may spread in the police.

Ghetto residents are victims of a statistically significant number of burglaries in daylight. However, only a few motorized patrols (and almost no foot patrols) are allocated to the ghetto area. Although motorized patrols are necessary, means must be found to get the patrol *out of the car* and get her to know and understand the locals. A reassuring interview may be the best approach to deal with the situation, but many police officers lack the training or experience to be able to use this tool. If the whole process does not proceed with courtesy and understanding, even the most enlightened proposals may fall apart.

The concept must also be publicly explained so that the inhabitants of the ghettos understand it and know what to expect. It is also necessary not to interpret the whole process as a sign of weakness of the police. It is also proposed to change the concept of police medals. These are traditionally awarded for heroic deeds, for example arresting dangerous perpetrators. Similarly, work that prevents the situation in a closed community from escalating (so intervention was not necessary at all) should be valued.

Minorities' members, especially African Americans, have been kept out of police force due to discrimination. In many larger cities, however, police managers are now actively seeking these candidates. However, African Americans that are more qualified are often attracted to other or better paid positions (for example, in the military). Mention should also be made of the negative image of the police in the community and the limited education of some African Americans.

A compromise could be the creation of a new type of uniformed *civil servant* with limited powers (ideally between the ages of 17 and 21 with "*the skills and integrity necessary to carry out policing*"). A less serious offense in the criminal record alone may not be an obstacle in this career. The person in question would perform various activities, with the primary emphasis on working in a minority environment. At the same time, he or she would continue the education in order to be promoted to the position of a full-fledged police officer as soon as possible.

Neighborhood service centers have opened in some cities. These are usually set up in high-crime areas, in easily accessible places such as public housing projects. Their task is to provide information and connect the citizen with a suitable public agency. The topic of *minority dynamics* and *socially disadvantaged people* is also mentioned in a number of studies from a much more recent period, not only from the United States of America but also from Europe. Conventional attempts to solve problems through consultation meetings, where the public is asked for opinions and suggestions, improve the image of the police force. Nevertheless, it is a largely unrepresentative tool.

The question also remains, who is the *community*? Isn't the *community* understood primarily as *relatively rich* or *relatively law-abiding* (and *white*) part of the population? Then where is the assumption that the police serve everyone? Police support does not mean too much added value for *privileged groups of the population*, which themselves can create informal mechanisms for maintaining order [105].

Measuring *community involvement* is sometimes controversial. The ethnic minorities or unemployed are far from involved in communicating with police structures. They may have a good reason for such an attitude due to the existing profiling or experience with the hostile, rude and inappropriate behavior of police officers [14, 116, 152].

Young people, members of ethnic minorities or people from the margins of society are also far less represented at consultation meetings with the police. The priorities that will be heard here are therefore the attitude of a small minority of people, or a few active individuals [41, 58, 115, 16, 42, 44, pp. 398, 400; 79, 104].

Confidence in the police is lower in larger cities. Older, educated and higher income people, believers and women show a higher level of trust in the police. Members of minorities show much lower confidence in the police than members of the majority, although other socio-economic indicators are taken into account. Marginalized members of the community may hear messages from the police as messages to which they are not invited. Their distrust is then reflected in the form of their limited participation [84]. Too often, these are not community relations programs, but programs designed to improve the image of the department in the community. The citizen can only see what the police want them to see.

Very illustrative is the process that took place in Seattle. Most of the reforms of community police in large cities in the 1990s suggests that this process was initiated and supported by police managers who responded to their own perception of public dissatisfaction with the police. In Seattle, however, there was pressure for reform from the community, and the idea had to overcome strong opposition from the city's police management at the time. First formal police reform was launched in Seattle in 1990 in response to the vision of one of the local crime prevention organizations. However, critics saw the reform as *bastardized*, as a process that catered to only one elite group, whose meetings were *by invitation only* and whose goal was to maintain the state quo. The activities concerned a problematic part of the city, but only limited efforts were made to communicate with the inhabitants of these areas themselves (where 60% were members of minorities and where general sentiment perceived police officers as a repressive force, not collaborates). In 1994, a city ordinance banning sitting or lying on sidewalks in commercial districts from 7:00 to 21:00 aroused emotions. Activists, advocates for socially disadvantaged groups, protested against the alleged anti-social and racist background of the regulation (70% of the transgressions involved African-American, Hispanic or Native American offenders) [82, pp. 136–137; 102, pp. 56–58; 6, 91].

Citizens are asked to engage in certain activities that may be sensitive (such as neighborhood patrols and problem reporting). Activating the public for this form of civic engagement requires a high level of social capital and trust in the police. Selective partnerships with some community organizations may send a message to other segments of the community that they are not part of the *community* in policing [88].

In a 2012 study, Elaine B. SHARP (a political scientist and urban sociologist, University of Kansas) analyzes the possible unintended, demobilizing effects of certain developments or police actions in terms of public involvement in neighborhood associations. The same community police program can send different messages to different types of people. Sometimes there is an effort to reduce crime, obtain information from the public—and sometimes *only* to strengthen trust in the police and build social capital at the local level. It is not necessary to conceive every meeting with the public as dealing with formal complaints—even friendly gossip is to be appreciated [108].

Useful are also studies that directly address the relationship of minorities to national police forces—including their objective or subjective discrimination or share of persons belonging to minorities in national police forces (African Americans in the United States, post-apartheid situation in South Africa, minorities in People's Republic of China, the situation of immigrant communities in some European countries, etc.).

Henry SMART (an African-American public administration expert, City University of New York, United States) in his 2018 paper, describes experiments (modeling, computational simulations) to demonstrate how so called *colorism* can play a role in the local policing context. The author understands this term as a *distribution of advantages and disadvantages* based on the skin color of a person. The concept may help explain some of the trends in the law enforcement system (not only) in the United States of America. For example, emancipation campaigns in favor of African Americans, paradoxically, lead to more police interventions on Latinos [117].

8 Demographic Challenges and *Bifurcation* of the Police

Demographic trends (in Europe and North America) and the related labor market tensions certainly also affected the police activities. However, the situation is even more specific and complicated. The vision of a *bifurcated police* is heard from different directions, with the recruitment of new members going in different directions, looking for *different types of police officers* [50].

The first will be an educated IT specialist, a *sybaritic millennial* who will tackle new challenges in cyberspace or sophisticated economic crime for an above-standard salary; possibly more female forces, performing tasks in contact with the public (clever, intellectual policing) [69]. The second will be the traditional patrol officer, but also members of special order units, gendarmerie (tough, physical, street, trouble policing). A recent example of such an attitude is a project of several European countries, the European Gendarmerie Force (EGF, EuroGendFor) [72]. Developments in some localities may escalate very quickly for socio-economic or ethnic reasons, and brutal responses may become the only solution [35].

At the same time, it is emphasized that the (foot) patrol activity itself, as far as it is rhetorically glorified, represents a relatively unproductive, parochial part of police activities, which does not seem to fit into the globalized world of the XXI. century. In this regard, it is not out of the question to delegate this activities to regions, municipalities (municipal police), volunteers and other similar structures, with the proviso that it is not one of the *really essential police activities.*

9 Social Networks as a Communication Channel

The calls for the greater involvement of police forces in the environment of social networks such as Facebook or Twitter and the Youtube channel can hardly be omited. Traditional methods of leaflets, street interviews and home visits are ineffective, especially in relation to the younger generation [147, p. 3; 154, p. 13]. A large part of young people barely interacting outside cyberspace. After all, crime is also moving into cyberspace [18, pp. 189–228].

Internet social media tools are free and fast. They can quickly reach a large group of people, provide them with information and advice on crime prevention or ask the public for help [44, p. 399; 101]. If you really communicate with the public and answer questions quickly, you will help to create a positive perception of policing and community involvement [49, 27, 23, 25]. These activities can also be carried out by municipal police officers, or variously defined wardens or volunteers, affiliated with the police forces.

10 The *Core-Periphery* Theory and Its Application to the Municipal or Regional Police Forces

A specific view of local public order issues, or the building of regional or municipal police forces, is the use of core-periphery theory (a concept, connected with

Stein ROKKAN, political scientist and sociologist, University of Bergen, Norway)—although only few related studies has been applied to the performance of police activities or municipal police [74].

One of the topics that Rokkan dealt with was the issue of differentiating the continent of Europe and its regions into *core/center* and *peripheries*. Rokkan dealt mainly with the level of individual states. The specific features of their political systems were usually associated with the ranking of certain countries at the core/center or periphery. He understood the core as a privileged place, the center of the political system, embodying the state as such. Ceremonies demonstrating state or national identity take place here. There is a concentration of economic actors (corporate headquarters, banks), cultural and educational institutions. Finally yet importantly, it is the administrative center of the state, the seat of central authorities. Periphery means the territory over which the center exercises variously defined power or control. The horizontal periphery is determined by geography (far from the center). Socio-economic indicators (poverty, backwardness) then determine the vertical periphery. The periphery is also defined in parallel by *difference* from the center, which can best be applied in Europe to the Western Balkans when it comes to ethnic or cultural-civilizational aspects [145].

If this theory is applied to security issues, then it is more about e-government or public administration in general [45, pp. 2–3]. However, this is not a unique view. For example, in relation to developments in Ukraine, the concept of the core/center-periphery and the functioning of the police appears in 2018 in a study that discusses how individual local governors and other actors use the police for their own interests. Not so explicitly, this concept is mentioned in relation to some post-Soviet Central Asian republics, where the *center is far away* and regional leaders are not prevented from abusing the police [85, pp. 144–147]. Somewhat less clear is the approach to the performance of police activities on the border of the United States and Mexico. Police sheriffs are accountable to local elites and cannot exist without their support [81, pp. 8–10].

The periphery is a space where norms are not understood so literally and where written law gives way to unwritten traditions. The various authors, without explicitly talking about the core/center-periphery concept, either describe the unhappy situation on variously defined peripheries (and the need for investment and equality with the center), or directly describe the processes that can be described as the integration of periphery' (involvement of minorities or other disadvantaged groups in the security agenda, multi-ethnic patrols for better local security, community policing, improved communication between police greater confidence of citizens). At the same time, there is a presumption that such an approach is not only possible, but that it sometimes turns to the issue of fears of possible excessive emancipation of the periphery, whether it is an economic peripheral, or a culturally-civilizational and ethnic one.

The view is expressed that in ethnically fragmented countries (Belgium, Northern Macedonia, Bosnia and Herzegovina, and Spain), the building of national police (security) forces can be seen as a unifying element, as something that *holds the state together*. In contrast, the building of regional or municipal security forces is motivated by the effort to emancipate regions and municipalities. If these municipalities and regions are nationally, culturally or religiously specific, this can be seen as a contribution to the *dismantling of the state*.

If we speak specifically in relation to the situation in Bosnia and Herzegovina, then the *excessive decentralization* of the police forces can be seen as a major problem. Police do not intervene against their acquaintances in the locality; defends local interests against the state, against the center; they help to hide internationally wanted people accused of war crimes etc. There are no, or just slowly developing, generally accepted standards for policing. Thus, there is an effort to change the situation by reforms *from inside* and *from outside*, also with the involvement of relevant international organizations [132].

In relation to the Western Balkans, this topic is also historically elaborated, as an illustration of how Nazi Germany, fascist Italy and the puppet states controlled by them between 1941 and 1945 purposefully emancipated the Balkan ethnic group, especially the Kosovo Albanians and Bosniaks (Muslims). Through the military and police forces they formed, other nations of the former Yugoslavia were suppressed. Even the emancipation initiated in this way, after World War II could not be completely erased and these national groups could not be pushed out of national, regional (republican) and local security structures [121].

11 Neopatrimonial Logic of Eastern European Police

In a 2012 study, Stephan HENSELL (a political scientist at the University of Hamburg, Germany) is of the opinion that police forces in Eastern European countries are often too politicized, poorly trained and corrupt. Since the early 1990s, after the fall of socialist regimes, the bureaucratic apparatuses and administrative procedures of these countries are changing only slowly. This also applies to the functioning of the police force [131, 52, 89].

In many countries in South-Eastern Europe and Eurasia, police officers are not seen as public servants, but rather as *uniformed bandits* who primarily pursue private economic interests and are difficult to distinguish from the criminal sphere. Some police administrations are governed by a non-patronage logic derived from the concept of traditional domination by Max WEBER [134, pp. 226–241]. Acquiring a certain position in the public sector is understood as an *investment*, a necessary precondition for unofficial but generally accepted extra income. It can be a specific form of clientelism, a type of governance where the ruling group does not distinguish between the public and its own interests—and uses state resources at its discretion, including forces and resources within the police force. In some settings, it is literally a desperate survival strategy [134, pp. 222–232].

The ideal state is an impersonal organization, based on a clear separation of public and private spheres. However, practice often differs from this ideal. Impersonating a bureaucratic or police position is an ideal that reduces corruption and authoritarianism, but also limits the ability to establish an intimate relationship with the public [48, p. 121].

Local knowledge can thus be as much a positive aspect, as a means of corrupt behavior. Depending on their areas of competence, actors can benefit both from the adoption of legislation and from its implementation and enforcement [107, pp. 21–23]. There is potential to speed up, or to postpone administrative proceedings, or otherwise use a wide range of informal taxes and fees. It is called a *predatory autonomy*. If the police forces are not effectively controlled, they are increasingly independent and predatory [103, 86, pp. 104, 118–119].

The core of the text are two case studies on the situation in Albania and Georgia. The author considers these two countries to be semi-democratic or hybrid—despite international pressure and massive international support for police reform. The police forces in Albania and Georgia are the centralized, corrupt, paramilitary institution that the public likens to robbers rather than law enforcers. The rulers (top public officials) use the police force for their own needs, to maintain power and order at their discretion. Anyone considering a career with the police is aware of this, and either adapts to such a situation or the road is closed to him or her. Max Weber sometimes speaks of the *sultan model* [122, p. 46].

At the same time, the number of police officers in Albania is gradually declining (from around 24,000 in 1991 to 12,000 in 2004). There is a constant rotation of staff. After the change of minister (change of political party in power), the commanders of specialized units or commanders in the territory also change rapidly. Police salaries are low, which makes *extra earnings* necessary. Police officers mainly benefit from drug smuggling, human trafficking, vehicle theft and the import of stolen vehicles from abroad. There is also the withdrawal of charges for payment up to shares in the profits of petty crimes [9, pp. 47–48; 62, pp. 17–18]. Police officers must hurry, because a change in the country's political representation may mean the end of their predatory careers. Post-election politically motivated purges create new opportunities for clientelistic looting for those who have stood in opposition so far. The search for new loot opportunities may also be the reason why municipal police have been set up in a number of cities in Albania.

There is still no clear separation of police, military and civilian agendas in Georgia. Virtually all employees of the Ministry of the Interior, including the Minister, are not civilian employees but officers. Police officers have established themselves as *independent entrepreneurs* and as one of the most corrupt public institutions. The traffic police literally taxed traffic on Georgian roads through countless mobile roadblocks. Additional income is provided by administrative activities, such as the issuance of license plates and driving licenses. The public procurement mechanism (supply of uniforms, armaments etc.) is also burdened by informal *taxation*. Mention should also be made of cigarette smuggling and the theft of humanitarian aid from abroad. At the same time, the mechanism is stable and *new managers honor previous contracts*. The number of police officers in Georgia is still growing.

This practice is more or less possible in other post-socialist countries of Central and Eastern Europe. Towards the east, this aspect is even more visible, especially in Central Asia. This opens up new comparative perspectives, for example between Eastern Europe, Eurasia and Africa, where patrimonial elements in the police force exist to varying degrees [7].

There are also studies directly concerning the possibility of abuse of municipal police officers (municipal guards) by local authoritarian corruption fraternities. In general, however, undemocratic regimes tend to suppress municipal police forces as a whole, especially if they should represent the priorities of municipal authorities (Belarus, some Central Asian post-Soviet republics) [38, p. 109]. For example, in 1996, a team of authors, especially from Slovenia, presented a number of studies on countries in transition (post-Soviet area, Latin America, Africa and South-Eastern Europe). Some chapters generally relate to preventing and combating the violent image of the police [96].

Lousie SHELLEY (George Mason University, Fairfax, United States of America) talks about the *Soviet heritage* in the police force, i.e. low public accountability, corruption, the cult of violence etc. The public in the respective countries is rather passive; the effort to exercise real democratic control of the police forces can only be discussed in fragments. The role of *Western programs* that fund reform processes and training programs for police forces in the post-socialist space is explicitly mentioned [110].

Other studies [20, 33, 43, 61, 96, 110, pp. 76–77; 21, 111, p. 62], which deal at least marginally with the topic of police activities and the provision of local public order affairs in Central, South-Eastern and Eastern Europe, operate with the following information [74]:

As for the *initial situation* in this area, it is stated that the local security forces dealing with (local) public order issues showed similar variables or indicators related to the *transition* after 1989 and 1991, respectively. This effort is characterized by efforts to *democratize* policing (policing oriented towards the service of the public/society more than towards the service of the state; the performance of the service is transparent; the management of police territorial units is to some extent independent; police officers do not have any privileges—in case of failure, they are treated as other citizens etc.). The basis of joining the police is the mastery of relevant skills and their professional implementation—in interaction with the local population [100, pp. 5–13]. The ambition to achieve unspecified parameters common in the police activities of the systems of *Western* European states is also mentioned. Some authors are also talking about *demilitarization* and *decentralization* of policing and the creation of mechanisms for external and internal control of policing. Another common priority, which is at least theoretically mentioned, is the effort to profession-alize and streamline police activities and increase the legitimacy and popularity of police work. Evergreen is also an effort to relieve (de-burdening) the state (national) police from the trivial or misdemeanor agenda, or non-police activities as such. This then represents the space for building and engagement of the municipal or city police forces.

However, the *real state* or *outcome* of the process rather speaks of the fact that the transition, not only in terms of policing but also in terms of socio-economic change, often did not go *according to plan*. What lasted a decades in Western Europe, took place in Eastern Europe during a much shorter period. The loosening of ties in society and literally an explosion of new forms of crime characterized this period. Legal and organizational measures responded only late to this development. Some components have only been renamed, but their performance standards have not changed much. There remained a close link with the state, centralization, relative militarization, and rigid horizontal governance. The rule is relative underfunding of police activities. Personal initiative is not very welcome. A *too ethically* conceived person in the police forces often does not stay for long. Many changes were limited to declarations, action plans, or symbolic changes (change of name—from militia to police, modification of uniforms, patches or rank system). In some countries, there is a boom of privatization of security, including a model where semi-official structures offer *protection* for a fee.

12 Three Studies from Cantenbury Christ Church University

A 2015 study by a pair of authors from Cantenbury Christ Church University (United Kingdom): Bryn CALESS (an author of a series of treatises on police responsibilities, especially in the United Kingdom but also elsewhere in Europe) and Steve TONG (a criminal law expert), contains three chapters, prospectively related to the topic of the theoretical anchoring of the agenda of local public order affairs.

European Policing in Context

According to the authors, until recently, the European police functioned primarily in the local, parochial dimension, which was largely the result of the social and political situation in Europe in the second half of the eighteenth century. Civil militias, introduced since the Middle Ages, have become *professional militarized guards*. Although there have always been exceptions, the crime took place in a territorially limited area; most police officers almost never encountered colleagues from distant jurisdictions. Knowledge was generally based on local experience [2, 18, pp. 27–56; 22, p. 244; 37].

However, the *local dimension* is still part of modern policing, crime has an increasingly international dimension. Even the lowest levels of police forces are handling with the agenda of the fight against terrorism, cybercrime and human trafficking [22, pp. 244–257].

Three hundred years of history and great regional differences have created different models at different times. Six types of police have developed in Europe since 1750 [20]:

- The continental (Napoleonic) model, often involving the armed national police and paramilitary gendarmerie. This model is also introduced in some Latin American countries. Because this concept has persisted in so many countries and for so long, it is not just a matter of external introduction. Napoleon brought intelligibility to previously unorganized systems. Moreover, neither other rulers nor democratic politicians wanted the police force to be too independent. Central coordination and pyramidal bureaucracy, as well as the inhibition of any parish tendencies to change from the norm, is a popular concept. Napoleon wanted an immediate response to his commands, which required a high level of discipline and uniformity, not inconsistency or local autonomy [11, 39, pp. 89–91, 103].

- Anglo-Saxon model, mostly unarmed non-military and decentralized police. This model often sets itself apart from *continental practices* that interfere with the tradition of individual freedoms. Today, however, it is clear that the differences between the Anglo-Saxon and the continental model disappears, and both models are gradually converging [53, 36, 87, p. 18].
- The colonial model of the armed militia (first introduced by the English administration in Ireland and later widespread in British and other European colonies around the world). Police serve the colonists, while the indigenous population is often subject to repression. Colonial forces are effective in maintaining security, but lack effective review mechanisms to enable the public to hold police officers accountable [24, pp. 98–99].
- The repressive model of the secret political police, existing in Nazi Germany and the Soviet bloc, among others. Absolutist governments suppress any form of disobedience, and the police forces are characterized by extensive routine monitoring of the population and a willingness to use repression to pursue political goals (torture, kidnappings, murder, and unlawful imprisonment). These activities take place with the consent of the political leadership of the state and on a large scale.
- Decentralized police model, existing in the Federal Republic of Germany (after 1945 and after 1990, respectively). Germany to some extent dusted off the model that existed before unification in 1871 and which partially functioned until 1933 [40].
- Eastern European model of the transition to democracy in the former socialist countries. The ambition of the reformers is a transparent police model, separate from state power. However, the problem is the absence of a tradition of political culture and the fear of the population, or rather its unpreparedness for the role of a partner of the police force [129, p. 20]. Reform plans have changed frequently (especially with regard to the centralization or decentralization of individual forces) [149, p. 8]. It will take decades to finalize processes of this kind.

Challenges Facing European Policing Today

Different countries or regions in Europe interpret current challenges for policing differently, depending on the socio-economic, political, historical or ethnic perspective they use: what is a problem in the Baltics may not be perceived as such in the Mediterranean. Based on the statements of police managers, the authors identify some of the possible challenges. However, many of them may be time-dependent and it may not soon be a priority [18, pp. 165–188].

Several respondents mentioned *lack of Europeanism*, excessive national interest, or even parochialism as a significant negative phenomenon. Current debates about the aforementioned concept, called plural policing [56, 30], go hand in hand with decentralization or devolution as such [76, 78].

At the same time, however, the variously motivated pressure for increased centralization of police forces in Northern and Western Europe can be seen. This includes the merging of police districts and, in some cases, the creation of new national police forces (Scotland, some cantons of Switzerland) [46]. Certain changes in policing are taking place not because of the search for best practice, but in the context of budget cuts and competition from private security services. It is clear that maintaining an acceptable level of service and at the same time saving costs is a challenge for every police force. Some police forces have responded to this challenge by reducing their commitments. Others were aimed at replacing regular police officers with support staff or volunteer forces (undergoing a short training program). This allows them to withdraw *proper* police officers from tasks that do not require their skills, powers and training. However, this increases the public's sense of insecurity and fear of crime (citizens are affraid that too many security activities are now on the shoulders of non-professional individuals) [138, p. 48; 92, 139].

The police forces of many countries are facing austerity measures that inevitably affect their activities. After all, police forces in many countries expect to achieve savings through the intensifying communication with the community (that will help to reduce the crime and other anti-social behavior occurence) [28].

The number of contracts with private security services is also growing. Governments, as well as municipalities, are seeking the services of private security agencies, even where it was unthinkable until recently. At the same time, private companies are not always subject to the strict regulatory frameworks typical of public police forces. Private security companies also offer community activities, visible patrollings in a specific location, but at a certain price (which the state, municipalities or other actors are willing and ready to pay). The trend was not hindered by the recent failures of some private actors, such as the London Olympics in 2012, when G4S admitted that it had *overestimated its forces* and that some of the contracted tasks had to be taken over by the military [29, 130].

Future of Policing

The authors again gave police managers the opportunity to comment on the future of police in Europe and beyond [18, pp. 189–228]. The perceived crime occurence in the Euro-Atlantic area is declining. The question is whether this is an objective reality or whether crime is rather transformed. In any case, crime statistics are the subject of dramatic developments that will certainly continue into the future. For example, environmental crime could be punished more severely—and conversely, a certain spectrum of drug crime could be decriminalized. There are literal remarks about a possible *redefinition of the enemy* or unwanted street behavior [8].

Side by side are the views that the future (and the security situation in Europe) is unpredictable, optimistic or catastrophic (social upheavals in agglomerations caused by massive economic volatility are inevitable and brutal responses to them are the only choice).

Some respondents believe that police forces can be very shortsighted, parochial, dealing with local issues, focused on the community. At the same time, crime is increasingly international and global. Thus, what many authors perceive as a benefit, as a pillar of the concept of community policing, is criticized here.

The police must balance between competing priorities and new tasks, or respond to competing public sector spending (at the beginning of 2020, coronavirus). Is it more important to be seen on the street, or to fight racism and discrimination against women (inside and outside the force), or to fight cybercrime? Is our priority a universal police officer—or specialization of police officers in relation to some specific activities? Is it possible to expect that the boundary between the police, the gendarmerie, the army, but also social officials will be blurred? Is there a potential for not exercising of a certain agenda, or its transfer to municipalities (municipal police officers, municipal guards) or other actors?

13 Conclusion and Prelude to Subsequent Empirical Effort

In the previous text, the authors tried to trace topics that could form the basis of an emerging comprehensive concept, if not directly of a scientific sub-discipline dealing with local public order affairs. It is, however, uneasy to set some research framework in this regard. Only basic information can be used in the defined text space, without using the full range of possible backgrounds. The issue of relevant concepts and their steady interpretation remains a significant issue not only at national but also at international level.

At present, it is more than premature to say that the study of local public order affairs (let alone the framework and context of the functioning of municipal police forces) would become a separate and generally accepted discipline. After all, even *police science* or *securitology* in any way understood is only slightly better in this respect. Rather, it is just the edge or intersection of publication outputs that are focused on the environment of other scientific disciplines.

The structure of experts producing impacted studies in areas affecting local public order affairs is relatively diverse. We can find here sociologists, political scientists, experts in the field of security or administrative studies. To some extent, also these people from police practice have found a second career in the academic environment. Specifically, William J. BRATTON has held a number of police positions within Boston and New York. In January 1994, Mayor Rudolph Giuliani put him in charge of the New York Police Department and commissioned a comprehensive reform of the force. George L. KELLING had worked in the police environment for many years before he started his career at the John F. Kennedy School of Government. Mention may also be made regarding Daniel DONELLY, a former police expert, now working at the University of the West of Scotland, Hamilton, United Kingdom.

This *academisation* of experts from practice is probably far more common in the case of the army than regarding police veterans. It is very valuable to mention concepts that were referred to, for example, plural policing or multilevel policing.

However, some of these studies are highly selective and address the situation in a single country or in a few countries, depending on the capabilities of the individual authors.

The comparison of the situation in a number of European countries reveals a number of similarities but also differences, concerning not only the framework of the existence or functioning of forces, which can be described as municipal police. At the same time, it must be emphasized that this topic is constantly evolving dramatically and the information obtained may not be valid for some time.

At the same time, it should be emphasized in the context that the concept of *municipality* within different countries represents a differently large and populous unit. The situation in the Czech Republic is relatively specific: the average number of inhabitants of the municipality is about 1,680 people. Municipalities in many European countries are much more populous, consisting of many settlements [73, 94, 70].

The relationship between the private and public sectors can also be potentially very sensitive, precisely with regard to the possible privatization of local public affairs. Opinions may be heard about the erosion of the state monopoly on the legitimate use of physical force [17, pp. 19–20].

Numerous national interpretations of the central concept of *police* demonstrate the diversity of police practice in Europe. The dynamic development is characterized, for example, by the regional polices in Spain [150].

In some countries, the possibility of setting up municipal police is already in place, primarily on a voluntary basis (Czech Republic, Slovakia, Greece, Italy, Liechtenstein, Portugal, Albania, Montenegro, Serbia, etc.). There are countries where municipalities want to build forces, identifiable as municipal police, but the central government is holding back the situation in this area (Austria, Croatia). In certain countries, activity in the area of ensuring local public order affairs is de facto ordered to municipalities (the Netherlands, Slovenia). In some countries, instead of municipal police forces, there are bodies more or less focused on the traffic and parking agenda (Cyprus, Ireland, and Finland).

In some countries, transfers of competencies between the center, regions and municipalities take place simultaneously, in different directions (Germany, Switzerland, Bosnia and Herzegovina, and Spain). *Centralization* and decentralization trends can sometimes run in parallel, as illustrated by the case of Switzerland:

Centralization: Expanding the activities of multicantonal concordats (joint shopping, forensic activities, research and education for several cantons at once). Abolishing of the municipal or micro-regional forces and incorporating their staff and resources into cantonal forces—on the basis of referendums.

Decentralization: Delegation of specific activities from the level of cantons to micro-regions and municipalities. Efforts to involve private security agencies or volunteer groups in individual tasks (especially at night or on weekends). Replacement of abolished municipal forces by otherwise called local forces.

There is yet no comprehensive secondary study, trying to describe the situation, for example, in all Member States of the European Union. Rather, there are parts of the puzzle that can provide a clearer picture only from a distance. The transferability of experience between countries or regions is limited. Some publications describe the situation in the United States or elsewhere outside Europe, which is of limited use to European needs. There are case studies describing the situation in one city or agglomeration, but which do not allow comparisons among more cities.

The relatively usable is the publications of Marina CAPARINI (Stockholm International Peace Research Institute) and Otwin MARENIN (Geneva Center for the Democratic Control of Armed Forces; Washington State University) from 2004 to 2006. The texts also refer to other usable works. Here too, however, the editors merged relatively different chapters, describing the situation in specific countries [19]. Tensions between external recommendations and pressures, and national resistance to change and efforts to preserve police forces, as was the case before, including all their weaknesses, are often described. The involvement of minorities and women in police structures is also painful. The study, as one of the few, explicitly describes the role of municipalities. It reiterates the need to act in the interests of the public and municipalities, as well as the importance of communication and cooperation between the police and municipalities, described as community policing (or even democratic policing) [20, 21].

Also worth mentioning, are the colective publications, mostly from the University of Maribor, led by Gorazd MEŠKO, describing the situation in Austria, Croatia, the Czech Republic, Estonia, Germany, Hungary, Kosovo, Northern Macedonia, Montenegro, the Russian Federation, Serbia, Slovakia, Slovenia and Republika Srpska in Bosnia and Herzegovina. In addition, the structure of the chapters differs, it is almost possible to state *supply determines demand* and some comprehensive overview in this area is rather missing [90].

According to the authors of the chapter, a constructive-critical attitude, based on as robust empirical data as possible, is the only starting point. The search for a theoretical framework can complete, not precede, the most comprehensive inquiry into specific national or regional approaches and examples. All this ideally through monitoring similar variables (existence or non-existence of municipal, micro-regional, regional or other subnational police forces; competences of these bodies; attitude of the state and national police forces towards these forces; the role of volunteering or the perspective of *privatization of security*, the role of municipalities as such—and variables related to the role of municipalities in the security system of the state, etc.).

The authors of the chapter explicitly announce their goal to map whether there is room in individual European countries for the municipality to participate in ensuring local public order affairs on its territory, or how much the state intervenes in this process (for example, some form of permitting, certification, control, restrictions etc.), and if so, for what reasons? It is already possible to trace certain trends and findings, based on a more or less completed analysis of the situation in about 25 countries on the continent. The authors intend to publish their conclusions during 2021 [15, 74].

References

1. (2004) Effectiveness of police activity in reducing crime, disorder and fear. In: Skogan WG, Frydl K (eds) Fairness and effectiveness in policing: the evidence. National Research Council, Washington DC
2. Anderson M, Den Boer M, Cullen P, Gilmore W, Raab C, Walker N (1995) Policing the European Union: theory, law and practice. Clarendon Press, Oxford. ISBN 978-0-198-25965-4
3. Anderson JM (2005) Community policing-working together to prevent crime. Australian Institute of Criminology, Canberra
4. Bahn C (1974) The reassurance factor in police patrol. Criminology 3:340. [online 2. IX. 2020] https://1url.cz/uz7q3
5. Barker T (1999) Jak být laskavým a efektivním policistou: Pětiminutový policista. Pragma, Praha. ISBN 978-8-072-05689-1
6. Beckett K, Herbert S (2008) Dealing with disorder: social control in the post-industrial city. Theor Criminol 1: 5–30. [online 2. IX. 2020] https://1url.cz/4z7qR
7. Beissinger MR, Young C (2002) Beyond state crisis? Postcolonial Africa and Post-Soviet Eurasia in comparative perspective. Woodrow Wilson Center Press, Washington DC
8. Berkowitz L (1972) Frustrations, comparisons, and other sources of emotion arousal as contributors to social unrest. Soc Iss 1:77–91. [online 3. IX. 2020] https://1url.cz/ez7qF
9. Bogdani M, Loughlin J (2007) Albania and the European Union: the tumultuous journey towards integration and accession. Bloomsbury Academic, London. ISBN 978-1-845-11308-7
10. Bratton WJ, Knobler P (1998) The turnaround: how America's top cop reversed the crime epidemic, Chapter 14. Random House, New York. ISBN 978-0-679-45251-5
11. Broers M, Hicks P, Guimera A (eds) (2012) The Napoleonic Empire and the New European Political Culture. Palgrave Macmillan, London. ISBN 978-1-349-31703-5
12. Brown MK (1981) Working the street: police discretion and the dilemmas of reform. Russell Sage Foundation, New York. [online 29. VIII. 2020] https://doi.org/10.1177/027507408101500313
13. Brown DM (2017) Beyond the thin blue line? A critical analysis of Scotland's community warden scheme. Polic Soc 1:6–20. [online 9. VIII. 2020] https://1url.cz/5z7qp
14. Budd T, Sims L (2001) Antisocial behaviour and disorder. Home Office, Londo. [online 2. IX. 2020] https://1url.cz/xz72n
15. Bukačová B, Krédl J, Krulík O (2021) Obecní policie a privatizace bezpečnosti v evropských zemích, 2nd ed. Aleš Čeněk, Praha (details to be confirmed)
16. Bullock K, Leeney D (2013) Participation, 'responsivity' and accountability in neighbourhood policing. Criminol Crim Justice 2:199–214. [online 5. IX. 2020] https://1url.cz/oz7qB
17. Bureš O (2013) Vymezení základních pojmů. Privatizace bezpečnosti: České a zahra-niční zkušenosti. Bureš O (ed). Grada, Praha, pp 19–20. ISBN 978-8-024-78704-6
18. Caless B, Tong S (2015) Leading policing in Europe book. Bristol University Press, Bristol. ISBN 978-1-447-31572-8
19. Caparini M, Fluri P, Molnár F (2006) Civil society and the security sector: concepts and practices in new democracies. LIT Verlag, Münster
20. Caparini M, Marenin O et al (2004) Transforming police in Central and Eastern Europe: process and progress. Lit Verlag, Transaction Publishers, Münster, New Jersey
21. Caparini M, Marenin O (2005) Crime, insecurity and police reform in post-socialist Central and Eastern Europe. J Power Inst Post-Soviet Soc 2. [online 2. IV. 2019] http://pipss.revues.org/330
22. Casey J (2007) International policing. In: Mitchell M, Casey J (eds) Police leadership and management in Australia. Federation Press, Sydney, pp 244–257
23. Cheurprakobkit S (2002) Community policing: training, definitions and policy implications. Policing 4:709–725. [online 3. IX. 2020] https://1url.cz/iz7qo
24. Cole B (1999) Post-colonial systems. In: Mawby RI (ed) Policing across the world: issues for the twenty-first century. University College London Press, London, pp 98–99. ISBN 978-1-857-28488-1

25. Colvin CA, Goh A (2006) Elements underlying community policing: validation of the construct. Police Pract Res 1:19–33. [online 5. IX. 2020] https://1url.cz/sz7qq
26. Coopper W (1938) Municipal police organization. Southwest Soc Sci Q 4:333–342. [online 19. VIII. 2020] https://www.jstor.org/stable/42879463
27. Cordner G (2000) Community policing—elements and effects. In: Alpert GP, Piquero AR (eds) Community policing—contemporary readings. Waveland Press, Prospect Heights, pp 401–418. [online 5. IX. 2020] http://secure.expertsmind.com/attn_files/2303_chp-24.pdf
28. Cosgrove F, Ramshaw P (2015) The value and deployment of police community support officers in achieving public engagement. Polic Soc 1:77–96. [online 1. IX. 2020] https://1url.cz/1z7qi
29. Crawford A (2002) The police, policing and the future of the "Extended policing family." In: Brown JM (ed) Future of policing. Routledge, Abingdon, pp 173–190. ISBN 978-0-415-82162-9
30. Crawford A (2006) Networked governance and the post-regulatory state? Theor Criminol 4:449–479. [online 2. IX. 2020] https://1url.cz/iz7qJ
31. Crawford A et al (2005) Plural policing. Policy Press, Bristol. [online 2. IX. 2020] https://1url.cz/jz723
32. Crawford A (2008) The pattern of policing in the United Kingdom: policing beyond the Police. In: Newburn T (ed) The handbook of policing. Willan, Cullompton, pp 147–182.
33. Das D, Marenin O et al (2000) Challenges of policing democracies. A world perspective. Gordon and Breach Publishers, Reading
34. De Maillard J, Zagrodzki M (2015) Plural policing in Paris: variations and pitfalls of cooperation between national and municipal police forces. Polic Soc. [online 1. IX. 2020] https://1url.cz/Uz7qe
35. Della Porta D (2013) Clandestine political violence. Cambridge University Press, Cambridge. ISBN 978-0-521-14616-6
36. Den Boer M (1999) Internationalization: a challenge to police organizations in Europe. In: Mawby RI (ed) Policing across the world: issues for the twenty-first century. Routledge, London, pp 59–74. ISBN 978-0-203-50084-2. [online 3. IX. 2020] https://1url.cz/Gz7qd
37. Denys C (2010) The development of police forces in urban Europe in the eighteenth century. J Urban Hist 3:332–344. [online 2. IX. 2020] https://journals.sagepub.com/doi/abs/10.1177/0096144209359144
38. Donelly D (2013) Municipal policing in the European Union. Springer, London. ISBN 978-1-137-29061-8
39. Emsley C (2007) Crime, police, & penal policy: European experiences 1750–1940. Oxford University Press, Oxford. ISBN 978-0-199-20285-0
40. Feltes T (2002) Community-oriented policing in germany: training and education. Policing 1:48–59. [online 2. IX. 2020] https://1url.cz/Jz7qu
41. Fielding NG (1995) Community policing. Clarendon Press, Oxford
42. Fielding NG (2005) Concepts and theory in community policing. Howard J Crim Just 5:460–472. [online 3. IX. 2020] https://onlinelibrary.wiley.com/doi/abs/10.1111/j.1468-2311.2005.00391.x
43. Fogel D (1994) Policing in Central and Eastern Europe: report on a study tour. European Institute for Crime Prevention and Control, Helsinki
44. Foster J, Jones C (2010) 'Nice to do' and Essential: improving neighbour-hood policing in an English police force. Policing 4:398–400. [online 2. IX. 2020] https://academic.oup.com/policing/article-abstract/4/4/395/1503912
45. Frank L (2002) Pojmy centrum a periferie v teorii Steina Rokkana a relevance jejich vymezení ve věku informačních technologií. Středoevropské politické studie 2/3. [online 15. V. 2020] https://1url.cz/Fz7q2
46. Fyfe NR, Terpstra J, Tops P (eds) (2013) Centralizing forces? Comparative perspectives on contemporary police reform in Northern and Western Europe. Eleven International Publishing, The Hague. ISBN 978-9-462-36059-4

47. Garland D (2001) The culture of control. Oxford University Press, Oxford. [online 9. VIII. 2020] https://1url.cz/rz72V
48. Geddes B (1999) What do we know about democratization after twenty years? Ann Rev Polit Sci 121
49. Greene JR, Mastrofski SD (1998) Community policing: rhetoric or reality. Prager Publishers, New York. ISBN 978-0-275-92952-7
50. Hainsworth P (2008) The extreme right in Western Europe. Routledge, London. ISBN 978-0-203-96505-4
51. He N, Zhao J, Lovrich NP (2005) Community policing. Crime Delinq 3:295–296. [online 2. IX. 2020] https://1url.cz/Yz728
52. Hensell S (2012) The patrimonial logic of the police in Eastern Europe. Eur Asia Stud 5:811–833. [online 29. VIII. 2020] https://www.tandfonline.com/doi/abs/10.1080/09668136.2012.681244
53. Hurd D (2007) Robert Peel: a biography. Weidenfeld, London. ISBN 978-0-297-84844-8
54. Jammers V (2004) Crime prevention in the Netherlands. In: Kerner H-J, Marks E (eds) Internetdokumentation Deutscher Präventionstag. Deutschen Stiftung für Verbrechensverhütung und Straffälligenhilfe, Hannover. [online 9. VIII. 2020] https://1url.cz/sz7q4
55. Johnston L (2003) From 'pluralisation' to 'the police extended family'. Int J Sociol Law 3:185–204. [online 9. VIII. 2020] https://1url.cz/hz7qz
56. Johnston L, Shearing C (2003) Governing security. Explorations in policing and justice. Routledge, London. ISBN 978-0-415-14962-2
57. Jones TT, Newburn T (1998) Private security and public policing. Clarendon Press, Oxford
58. Jones TT, Newburn T (2006) Policy transfer and criminal justice. Open University Press, Maidenhead. ISBN 978-0-335-21669-7
59. Jones TT, Newburn T (eds) (2006) Plural policing: a comparative perspective. Routledge, London
60. Jones TT, Newburn T (2000) Private security and public policing. Br J Criminol 1:170–173. https://1url.cz/iz7qM
61. Kadar A et al (2001) Police in transition. Central European University Press, Budapest
62. Kajsiu B et al (2002) Albania, a weak democracy, a weak state: report on the state of democracy in Albania. Albanian Institute for International Studies, Tirana, pp 17–18
63. Kelling GL, Pate A, Dieckman D, Brown CE (1974) The Kansas City preventive patrol experiment: technical report. Police Foundation, Washington DC
64. Kelling GL, Moore MH (1998) The evolving strategy of policing. Perspect Polic 5(8). [online 2. IX. 2020] https://pdfs.semanticscholar.org/a614/21a27a6c4fa0e25962ef30e95a2 2371c1b9c.pdf
65. Kerley KR, Benson ML (2000) Does community-oriented policing help build stronger communities? Police Q 1(49). [online 2. IX. 2020] https://www.researchgate.net/publication/255658928_Does_Community-Oriented_Policing_Help_Build_Stronger_Communities
66. Kocak D (2018) Rethinking community policing in international police reform: examples from Asia. Ubiquity Press, London, pp 11–16. [online 29. VIII. 2020] https://www.jstor.org/stable/j.ctv6zdc
67. Kraska P (2007) Militarization and policing—its relevance to 21st century police. Policing 4. [online 2. IX. 2020] https://1url.cz/3z7qt
68. Krulík O (2014) Specifický model zajišťování bezpečnosti v některých statutárních městech Rakouska. Bezpečnostní teorie a praxe 3:81–100
69. Krulík O (2017) VIII. mezinárodní týden v rámci Polizeiakademie Niedersachsen. Bezpečnostní teorie a praxe 2:139–143
70. Krulík O (2011) Ideální rozloha policejního mezičlánku: mýty a evropská realita (part 1). Policista 10, annex, pp I–XII
71. Krulík O, Nováková J (2011) Image of the place and perception of security. Sci Popul Prot 1:59–68. http://www.population-protection.eu/attachments/038_vol3n1_krulik_novakova.pdf

72. Krulík O, Tvrdek T (2013) Evropské četnické síly – mnoho povyku pro nic? Vojenské rozhledy 2:182–193. https://1url.cz/Ez727
73. Krulík O (2015) Obecní policie a privatizace bezpečnosti v evropských zemích. Aleš Čeněk, Plzeň. ISBN 978-8-087-95614-4
74. Krédl J (2016) Obecní policie ve vybraných zemích postsovětského prostoru. Ochrana a Bezpečnost 2. [online 2. IV. 2019] https://1url.cz/Dz7qL
75. Lipsky M (1980) Street-level bureaucracy: dilemmas of the individual in public services. Russell Sage Foundation, New York
76. Loader I, Sparks R (2013) Public criminology. Routledge, London. ISBN 978-0-203-84604-9
77. Loader, I. (2000) Plural policing and democratic governance. Soc Leg Stud 3:323–345. [online 9. VIII. 2020] https://1url.cz/Bz7qr
78. Loader I, Walker N (2001) Policing as a public good: reconstituting the connections between policing and the state. Theor Criminol 1:9–35. [online 1. IX. 2020] https://1url.cz/tz72Y
79. Long J, Wells W, De Leon-Granados W (2002) Implementation issues in a co-mmunity and police partnership in law enforcement space. Police Pract Res 3:231–246. [online 1. X. 2020] https://1url.cz/Cz72y
80. Lum C, Fyfe NR (2015) Space, place and policing. Policing 3:219–222. [online 9. VIII. 2020] https://1url.cz/9z72c
81. Lusk M, Staudt K, Moya E (2012) Social justice in the United States—Mexico Border Region. Springer, New York. ISBN 978-9-400-74150-8
82. Lyons W (1999) The politics of community policing: rearranging the power to punish. University of Michigan Press, Ann Arbor. [online 2. IX. 2020] https://muse.jhu.edu/book/7198
83. Macdonald JM (2002) The effectiveness of community policing in reducing urban violence. Crime Delinq 4:592–618. [online 3. IX. 2020] https://psycnet.apa.org/record/2002-04552-002
84. Maguire ER, Mastrofski SD (2000) Paterns of community policing in the United States. Police Q 1:4–45. [online 2. IX. 2020] https://1url.cz/nz72P
85. Marat E (2018) The politics of police reform: society against the state in Post-Soviet Countries. Oxford University Press, Oxford. ISBN 978-0-190-86150-6
86. Marenin O (1985) Review essay: police performance and state rule: control and autonomy in the exercise of coercion. Comp Polit 1:104–119
87. Mawby RI (ed) (1999) Policing across the world: issues for the twenty-first century. University College London Press, London. ISBN 978-1-857-28488-1
88. Mettler S (1998) Dividing citizens: gender and federalism in new deal public policy. Cornell University Press, Ithaca. [online 2. IX. 2020] https://www.jstor.org/stable/10.7591/j.ctv5rdzmh
89. Meyer-Sahling J-H (2004) Civil service reform in post-communist Europe: the bumpy road to depoliticisation. West Eur Polit 1:71–103
90. Meško G, Fields CB, Lobnikar B, Sotlar A (eds) (2013) Handbook on policing in Central and Eastern Europe. Springer Publishing, Berlin. ISBN 978-4-614-6719-9
91. Miller J, Bland N, Quinton P (2001) A challenge for police-community relations: rethinking stop and search in England and Wales. Eur J Crim Policy Res 1:71–93
92. Mills H, Silvestri A, Grimshaw R (2010) Police expenditure 1999–2009. Hadley Trust, Centre for Crime and Justice Studies, London, p 52. [online 1. IX. 2020] https://1url.cz/Oz72b
93. Moore MH, Thacher D, Dodge A, Moore T (2002) Recognizing value in policing: the challenge of measuring police performance. Police Executive Research Forum, Washington DC. ISBN 978-1-878-73476-1
94. Místní správa v zemích střední a východní Evropy a ve Společenství nezávislých států, (1994) 1995. Institut for Local Government and Public Service, Budapest
95. O'Neill M, Fyfe NR (2017) Plural policing in Europe: relationships and governance in contemporary security systems. Polic Soc 1:1–5. [online 9. I. 2020] https://1url.cz/Wz725
96. Pagon M et al (1996) Policing in Central and Eastern Europe. College of Police and Security Studies, Ljubljana

97. Pala E, Balcioğlu E (2016) Community policing in England, Wales, and European Union: past, present and future. Ankara Avrupa Çalışmaları Dergisi 1:173–199. [online 29. VIII. 2020] https://1url.cz/Ez72X

98. Paoline EA (2004) Shedding light on police culture: an examination of officers' occupational attitudes. Police Q 2:205–236. [online 2. IX. 2020] https://1url.cz/uz72O

99. Polder W (1997) Notes on criminology research. Dutch Penal Law Policy 8. [online 9. VIII. 2020] https://1url.cz/2z72x

100. Pomsaers P (2013) Policing in Central and Eastern Europe as an epiphenomenon of geopolitical events. In: Meško G, Fields CB, Lobnikar B, Sotlar A (eds) Handbook on policing in Central and Eastern Europe. Springer Publishing, Berlin, pp 5–13. ISBN 978-4-614-6719-9

101. Procter R, Crump J, Karstedt S, Voss A, Cantijoch M (2013) Reading the Riots: what were the police doing on Twitter? Polic Soc 4:413–436. [online 2. IX. 2020] https://1url.cz/oz72S

102. Reed WE (1999) The politics of community policing: the case of Seattle. Taylor and Francis, Abingdon-on-Thames. ISBN 978-0-815-33029-5

103. Rose-Ackerman S (1999) Corruption and government: causes, consequences, and reform. Cambridge University Press, Cambridge, pp 9–38. ISBN 978-0-521-65912-3

104. Schneider SR (1998) Overcoming barriers to communication between police and socially disadvantaged neighbourhoods. Crime Law Soc Change 4:347–377. [online 5. IX. 2020] https://1url.cz/Nz72p

105. Schneider AI, Ingram H (1997) Policy design for democracy. University of Kansas Press, Lawrence. ISBN: 978-0-700-60843-5. [online 2. IX. 2020] https://1url.cz/0z72T

106. Scott JD (2002) Assessing the relationship between police-community coproduction and neighborhood-level social capital. J Contemp Crim Just 2. [online 2. IX. 2020] https://journals.sagepub.com/doi/10.1177/1043986202018002003

107. Scott P (1972) The spatial analysis of crime and delinquency. Aust Geogr Stud 1:1–23. [online 2. IX. 2020] https://onlinelibrary.wiley.com/doi/abs/10.1111/j.1467-8470.1972.tb00126.x

108. Sharp EB (2012) Does local government matter? How urban policies shape civic engagement. University of Minnesota Press, Minneapolis, pp 77–114. [online 29. VIII. 2020] https://1url.cz/Ez72j

109. Shearing C, Wood J (2003) Governing security for common goods. Int J Sociol Law 3:205–225. [online 2. IX. 2020] https://1url.cz/Fz72B

110. Shelley L (1999) Post-socialist policing: limitations on institutional change. In: Mawby RI (ed) Policing across the world: issues for the twenty-first century. Routledge, London, pp 75–87. ISBN 978-1-857-28489-8

111. Shelley L (1994) The sources of Soviet Policing. Police Stud 2

112. Sherman LW, Gottfredson DC, Mackenzie DL, Eck J, Reuter P, Bushway SD (1998) Preventing crime: what works, what doesn't, what's promising. National Institute of Justice, Washington DC. [online 2. IX. 2020] https://www.ncjrs.gov/pdffiles/171676.pdf

113. Shotland RL, Goodstein LI (1984) The role of bystanders in crime control. J Soc Iss 1:9. [online 29. VIII. 2020] https://spssi.onlinelibrary.wiley.com/doi/abs/10.1111/j.1540-4560.1984.tb01079.x

114. Skogan GW (1992) Disorder and decline: crime and the spiral of decay in American neighbourhood, 2nd edn. University of California Press, Los Angeles. ISBN 978-0-309-16932-5

115. Skogan GW (1995) Community participation and community policing. University of Montreal, Montreal, pp 2–3

116. Skogan GW (2006) Asymmetry in the impact of encounters with the police. Polic Soc 2:99–126. [online 2. IX. 2020] https://www.tandfonline.com/doi/abs/10.1080/10439460600662098

117. Smart H (2019) Operationalizing a conceptual model of colorism in local policing. Soc Justice Res 1:72–115. [online 29. VIII. 2020] https://1url.cz/tz72I

118. Smith B (1929) Municipal police administration. Ann Am Acad Polit Soc Sci 1–27. [online 5. IX. 2020] https://www.jstor.org/stable/1017543

119. Smith DA, Visher CA (1981) Street level justice: situational determinants of police arrest decisions. Soc Probl 2:167–178. [online 2. IX. 2020] https://1url.cz/4z72q

120. Stephens M (1988) Policing: the critical issues. Harvester Wheatsheaf, London. ISBN 978-0-745-00409-9
121. Stojković L, Martić M (1952) National minorities in Yugoslavia. Jugoslavia, Beograd
122. Taylor SL (2012) 30 sekúnd politiky. Fortuna Libri, Bratislava, p 46. ISBN 978-8-081-42013-9
123. Terrill W, Mastrofski SD (2001) Working the street: does community policing matter? In: Skogan WG (ed) Community policing: can it work? Wadsworth Publishing, Belmont, pp 109–135
124. The Newark Foot Patrol Experiment (1981) Police Foundation, Washington DC. [online 2. IX. 2020] https://1url.cz/Tz72d
125. Thompson J (1967) Organizations in action: social science bases of administrative theory. McGraw-Hill, New York
126. Tiedke K, Freeman W, Sower C, Holland J (1957) Community involvement. The Free Press, Glencoe
127. Trojanowicz RC (1982) Evaluation of the neighbourhood foot patrol program in Flint. Michigan State University, East Lansing, Michigan
128. Trojanowicz RC (1983) An evaluation of a neighbourhood foot patrol program. J Police Sci Admin 4:410–419
129. Tupman B, Tupman A (1999) Policing in Europe: uniform in diversity. Intellect Books, Bristol. ISBN 978-1-871-51690-6
130. Van Steden R, Sarre R (2007) The growth of privatized policing. Int J Comp Appl Crim Justice 1:51–71. [online 3. IX. 2020] https://1url.cz/mz72f
131. Verheijen T (2007) Public administration in post-communist states. In: Peters BG, Pierre J (eds) Handbook of public administration. Sage Publications, Los Angeles, pp 311–319
132. Visoka G (2016) Peace figuration after international intervention. Routledge, London. ISBN 978-131738275-1
133. Wakefield A (2006) The value of foot patrol. The Police Foundation, London. ISBN 978-0-947-69239-8. [online 2. IX. 2020] https://1url.cz/Xz724
134. Weber M (1978) Economy and society: an outline of interpretive sociology. University of California Press, Berkeley, pp 222–241. ISBN 978-0-520-03500-3
135. Weick KE (1995) Sensemaking in organizations. Sage, Thousand Oaks
136. Weinberger B (1995) The best police in the world: an oral history of British policing. Scolar Press, Aldershot
137. Weisburd DL, Mastrofski SD, Willis JJ, Greenspan R (2006) Changing everything so that everything can stay the same: compstat and American policing. In: Weisburd DL, Braga AA (eds) Police innovation: contrasting perspectives. Cambridge University Press, New York
138. What price policing? A study of efficiency and value for money in the police service. (1998) Her Majesty's Inspector of Constabulary, London. ISBN 978-1-840-82065-2.
139. Whitehead T (2010) Lives at risk from civilianised police service, rank and file warn. Telegraph 17. [online 9. I. 2020] https://1url.cz/3z72g
140. Willemse HM (1994) Developments in Dutch crime prevention. Directorate for Crime Prevention, Netherlands Ministry of Justice, Den Haag. [online 9. VIII. 2020] https://1url.cz/gz7qs
141. Wilson JQ, Kelling GL (1982) Broken windows: the police and neighborhood safety. Atl Monthly. [online 2. IX. 2020] https://www.theatlantic.com/ideastour/archive/windows.html
142. Worden RE, McLean SJ (2017) Mirage of police reform: procedural justice and police legitimacy, Chapter II. University of California Press, Berkeley, pp 14–41. [online 29. VIII. 2020] https://doi.org/10.1525/j.ctt1w8h1r1.6
143. Xu Y, Fiedler ML, Flaming KH (2005) Discovering the impact of community policing. J Res Crime Delinq 2. [online 2. IX. 2020] https://journals.sagepub.com/doi/10.1177/0022427804266544
144. Zhao J, Lovrich NP, Thurman Q (1999) The status of community policing in American cities. Policing [online 10. III. 2021] https://www.emerald.com/insight/content/doi/10.1108/13639519910256893/full/html
145. Říchová B (2000) Přehled moderních politologických teorií. Praha, Portál. ISBN 978-8-071-78461-3

Internet Source

146. Devroe E (2015) Plural policing in comparative perspective. Leiden Security and Global Affairs Blog, Universiteit Leiden, 11. V. 2015. [online 2. IX. 2020] https://1url.cz/3z72Z

147. Engage (2010) Digital and social media engagement for the police service. National Policing Improvement Agency, London. [online 5. IX. 2020] http://connectedcops.net/wp-content/uploads/2010/04/engage.pdf

148. Goldstein H (1979) Improving policing: a problem-oriented approach. Crime and Delinquency, University of Wisconsin Legal Studies Research Paper No. 1336, pp 236–258. [online 2. IX. 2020] https://1url.cz/uz72U

149. Paun C (2007) Democratization and police reform (Master of arts thesis). Freie Universität Berlin, Humboldt-Universität zu Berlin and Universität Potsdam, Joint Master's Degree Program in International Relations, Berlin, Potsdam, p 8. [online 2. IX. 2020] https://1url.cz/Uz72D

150. Pearls in Policing. [online 2. IX. 2020] https://www.pearlsinpolicing.com/

151. Policing in the 21st century: reconnecting police and the people (2010). Home Office, 20. IX. 2010, London. [online 2. IX. 2020] https://1url.cz/Uz72H

152. Survey Results. Crime Survey for England and Wales. [online 3. IX. 2020] https://www.crimesurvey.co.uk/en/SurveyResults.html

153. The Kerner Report (2016) Princeton University Press, Princeton. ISBN: 978-0-691-17424-2. [online 2. IX. 2020] https://www.jstor.org/stable/j.ctvcszz6s

154. United States Digital Future in Focus (2012) ComScore, 9. II. 2012. [online 3. IX. 2020] https://1url.cz/az72N

Formal Definition of Scholarly Books in the Czech Republic and Their Evaluation Mainly in the Context of the Social Sciences and Humanities

Petr Kolman and Jiří Kolman

Abstract The purpose of this chapter is to examine the rules formulating scholarly books (monographs) in the Czech Republic and the Czech academic world in general. Yet there is analysed, how formal rules are able to influence the quality and quantity of the scholarly books production. Moreover, formal regulation improvements are properly proposed. The paper aims to target these issues, by means of a content legal analysis of the national rules of the Czech Republic defining scholarly books. This paper also analyses how externalities—formal legal and Research Funding Organisations (RFOs) regulation of the scholarly books have an impact on knowledge production presented by the above-mentioned publication "genre" for Research Performing Organisations (RPOs) in the Czech Republic. The design of the formal (legal) definition of a "scholarly book" based on the analysis is proposed as well.

This study offers a novel perspective on how carefully designed formal rules regulate the knowledgeable production via scholarly book. Therefore, book chapters can improve scientific outcomes. In addition, it also shows how the codified rules can efficiently contribute to research results dissemination with respect to various scientific disciplines that are at stake. These dissemination practices are usually based on a traditional way of scientific practices rooted at the end of the eighteenth century.

Keywords Research quality · Scholarly books · Monographs · Research evaluation · ISREDI · Legal regulation · Academic writing · Knowledgeable creation

P. Kolman (✉)
Department of Security and Law, AMBIS College, Lindnerova 1, Prague, Czech Republic
e-mail: petr.kolman@ambis.cz

J. Kolman
Global Change Research Institute of the Czech Academy of Sciences, Bělidla 986/4a, Brno, Czech Republic
e-mail: kolman.j@czechglobe.cz

I. Tušer and Š. Hošková-Mayerová (eds.), *Trends and Future Directions in Security and Emergency Management*, Lecture Notes in Networks and Systems 257,
https://doi.org/10.1007/978-3-030-88907-4_4

1   Introduction

It is remarkable that even though traditional positivist science is based on Enlightenment in the eighteenth century, the scientific work including the dissemination of scientific results by their publication has been based on traditional unwritten approaches and customs. However, this is slowly changing. Traditional unwritten approaches can be illustrated e.g., by peer reviews of submitted scientific publications (including monographs), missing legal regulations of the scientific publication (e.g. what is "monography", scientific article, conference), scientific career assessment based on assessment by the academic peers, or code of research behaviour interpreted and used by the academic peers. Definitely a scientific freedom and academic autonomy should be respected. Obviously, too much regulation and too many administrative approaches kill the creativity vital for scientific life, new paths, and innovative findings. According to Hammarfelt [16] and Whitley [33, p. 48f], the reputation of an academic is dependent on their recognition among a wider community of peers, mainly from the same scientific field, which means that the peer community, rather than the institutional framework, is the venue where careers are valued. In this sense, research fields are what Whitley [33, p. 48f] calls "reputational work organisations", where labour market position is determined by reputation among colleagues. However, unclear rules based on the word of mouth and unwritten traditions and customs can often create unfair environment. Environment favour, rather unclear methods and processes, provide the advantages which have the power of the last word (e.g. senior researchers living from the past successes or having the power to influence the informal standards of research or evaluations of their team colleagues or scientific peers). In this study, there is the current situation deeply analysed and crucially compared it with the past approaches and it is in comparison with the international scene as well.

Nevertheless, the goal of this chapter is not to kill science by strict positivist legal regulation. On the contrary, it is an attempt to provide formal regulations of the scholarly publication meeting the needs of the whole scientific community and all stakeholders (individual scientist, Research Performing Organisation, Research Funding Organisation, a student using scientific literature and textbooks, wide public awareness interested in or using the scientific results published in the scholarly book, applied in the sphere such are companies, public service or international scientific partners). Compared with the other countries (see e.g., Kulczycki et al. [19] and especially the Netherlands: Van Boom and Van Gestel [31]), the Czech experience with the formal regulation defining monographs has a long tradition and has become a very complex try to grasp all the peculiarities of the academic publication.

Structure of the chapter:

1.   In the beginning, the methodological approach of this legal analysis it properly described.
2.   In the finding parts, there are identified the impacts of the formal rules regulating the scholarly books on the Social Sciences and Humanities. Afterwards, as follows the legal implications and a design of the formal (legal) definition.

3. In the last discussion part, there is a focus on the question of urgency of the reform or a change of the governmental approach related to research assessment that is used for public national research funding.

Why is this text focused in the area of the Czech Republic? Apart from the fact that the Czech scientific world is for the author the most familiar and based on his own personal experience, there are the other interesting features often related with historical milestones that have an impact on the current situation. Hence such historical periods and moments related to scholarly books, formal public (state, national) regulations can be identified as follows: A communist system regulating the researchers by tight administrative rules and a state control, opposite development after the Velvet Revolution of 1989, a division of the Czechoslovakia in 1993, pre—EU access period (till 2004) and the period after the EU accession. As Good and Vermeulen [11] point out (in the context of a growing body of literature on the research evaluation and performance based on research funding systems, discussing the impacts of introducing such systems in the countries including the UK, Spain, Slovakia, Hong Kong, Australia, Poland, Italy, New Zealand, Flanders, Norway, Denmark, and Finland) the Czech case of the Czech performance-based research funding system provides new insights in the interactions between politico-economic regimes and research policy, while directing the attention of research policy scholars to significant developments in Central and Eastern European countries as well.

In 2017 the new science performance evaluation system started to be implemented by the Czech Government. This new system tries to leave the mechanistic scientometric approach by the more use of peer review evaluations. Therefore, his chapter tries to reflect these new rules that have an impact on formulation and interpretation. What is a scholarly book shaped during the peer review evaluations of the scholar books carried on within the science performance evaluations? The issues related with science performance evaluations rule and their impact on scientific work and publications strategies are also apparent in the other countries (e.g. in case of the UK see Mcculloch [22] or Sweden see Hammarfelt [16]).

Another Czech specific feature is linguistics. Czech is a minor language and a dominant scientific language (especially in STEM sciences) is English even at the national level of the scientific publications (mainly scholarly journals). Therefore, a low number of the well referenced international journal publishers' force a Czech scientist to publish in English abroad (mainly US publishers). In case of legal scholarly journals, the situation is very illustrative: Existing rankings of legal journals are relevant for the United States of America. Even the most prestigious English-language journals published in the United Kingdom or in Europe have a lower impact factor. Few German and French legal journals are ranked and journals in other languages are absent [20, p. 148]. Moreover, considering the monographs the situation is worse, because in the Czech Republic there is no globally accepted publishing editor. Additionally, there is no global scholar publisher issuing scientific results in Czech language. Furthermore, the majority of the scholarly books dedicated for the international community is published mainly in English in international publishing houses seated abroad. From this point of view, it brings the effect that the

Czech national official research policy has a very low power how to influence the situation.

Like in the other countries, a new challenge but also the opportunity is the phenomenon of the open access to the scientific results and open scientific data.

With the open access, the challenge is the predatory publishers, where several of the Czech researchers are regrettably actively involved by deliberate publications as well. See the detailed analysis and case study from the Czech Republic published by Stöcklová and Vostal [28].

2 Methods

For the purpose of the analysis, the normative method focusing on balance of rights and obligations under the framework defined by law and formal rules were also used. Additionally, the method is drawn from moral, legal and political philosophy and it aims to provide the analysis around the questions of what ought to be Tyler [30, p. 130]. Such related tools are historical and comparative approaches (in comparison with time and different legal systems).

Writing of scholarly books is considered, for the purposes of this study, in analogy to generally scientific writing used in study published by Mcculloch [22, p. 505] as a set of practices that are embedded in social contexts historically located in time and space [5, 15, 29]. This approach acknowledges that academic writing entails what Van Leeuwen [32, p. 6] calls "socially regulated ways of doing things". And it involves analysing elements of the everyday writing experience and activities of participants. Hence in this study legal and formal regulations are qualified via socially regulated ways. From this point of view, there is applied a related methodical system of theory that is a general approach to distinguish Czech legal system from other normative systems. Moreover, the Czech Republic is, as a member state, a part of the European Union legal system and a member of the Organisation for Economic Co-operation and Development (OECD). OECD provides documents created in collaboration with OECD member states. Those documents are implemented by the Czech Republic as well. Such an OECD document, the most relevant one for the purpose of this study and therefore is related to research policy, is mainly Frascati Manual 2015: Guidelines for Collecting and Reporting Data on Research and Experimental Development, The Measurement of Scientific, Technological and Innovation Activities.

In the context of this analysis, there is a crucial pointing out the terms: "scholarly book(s)", "academic book(s)", "monograph" or "book" used with the same meaning otherwise, it is not explicitly written and specified.

3 Results

3.1 What Are the Scholarly Books and Book Chapters According to the Czech Government?

Currently the scholarly book description is formally defined in the "Czech research area" by the Methodology for Evaluating Research Organisations and RD&I Purpose-tied Aid Programmes (valid for years 2017+). This regulation (Methodology 2017+) is based on the obligation of the governmental body Research, Development and Innovation Council (R&D&I Council) stipulated by the Act No. 130/2002 Coll., on the Support of Research, Experimental Development and Innovation from Public Funds and on the Amendment to Some Related Acts (the Act on the Support of Research and Development), as amended, to provide the evaluation of methodology. Pursuant to the §35 par. 2(c), the R&D&I Council has to ensure the preparation of the national Methodology of Evaluation of the Results of Research Organizations and Results of Finished Programme. Therefore its presentation to the government and further pursuant to the §35 par. 2(d) ensure the evaluation of results of research organizations and also finish programmes according to the Methodology of Evaluation of the Results of Research Organizations and Results of Finished Programmes approved by the Czech Government.

Generally the data about the research results are collected into the national Information System for Research, Experimental Development and Innovations (ISREDI) by means of a process that involves the institutions and the funders of research activities [3]. Records in the ISREDI are mostly locally created in the current research information systems of the Czech Academy of Sciences and moreover such educational institutions are universities. Additionally, the authors of the publications working in the Czech academia and educational institutions are responsible for the data of quality. However, the final records are verified at the national level. The publication records are collected annually from the research organizations via funding bodies to the ISREDI. The collected data are verified in a specific timeframe that includes subject classification in two aspects, algorithmically e.g. the validity of the whole publication record and manually via verification panels e.g. for monographs, chapters, and articles published in Czech journals not indexed in Scopus or the WoS [19].

The main scope and tasks of the Methodology 2017+ are based on the consensus articulated during the implementation of the national Project Effective System of Research Financing, Development and Innovation supported by the Ministry of Education, Youth and Sports of the Czech Republic implemented in the period from February 1st 2012 till June 30th 2015. The main objective of the project was to design a proposal for the new R&D&I evaluation and financing system by means of public funding, which could also become the main information source for the strategic R&D&I system management on civil service authorities, granting authorities, programs and research institutions level. The main outcomes of the project and recommendations providing an "Olympic 'benchmark" underlining the international scientific evaluation context [4] became as a solid base for the purpose of the Methodology 2017+ as follows:

(1) Collecting information for quality management of R&D&I at all levels (the formative aspects);
(2) Enhancing the efficiency of spending public funds (the comprehensive aspects);
(3) Supporting the quality and international competitiveness of Czech R&D&I;
(4) Distributing and adding to the accountability of the stakeholders in the R&D&I system;
(5) Gaining information for granting subsidies for the long-term conceptual development of the research organisation (LCDRO).

The above-described purpose of the Methodology 2017+ is transformed via this document into corresponding so called five modules that research evaluation processes usually tend to focus on four outputs: volume, quality, impact, and utility [7]. In case of the Methodology 2017+ the following modules were articulated:

Module 1—Quality of Selected Results.
Module 2—Research Performance.
Module 3—Social Relevance.
Module 4—Viability.
Module 5—Strategy and Policies.

The characterisation of the types of result, the criteria for their capacity to be verified, and the method of entering data in the national Information System for Research, Experimental Development and Innovations (ISREDI) are updated in connection with the Methodology 17+. This process facilitates to keep national records of the scientific results up to date.

The definitions of the types of result are mainly interrelated with the Module 1—Quality of Selected Results and the Module 2—Research Performance and must be, according to the Methodology 17+, updated and implemented while taking account of the continuity with the existing definitions and preventing retrospective impacts. In order to facilitate updating without the need to open the source document, these issues are also addressed in a separate documentary subject to the governmental approval by the annex Methodology 17+ amendment. The definitions provided in the previous Methodology 2013–2016 had continued applying until the new ones received the governmental approval. The new definitions of the types of the scientific results were approved after more than nine months by the Definition of Type of Results Separate Annex no. 4 of the Methodology 17+ of the Resolution No. 837 of 29 November 2017 by the Czech Government. Furthermore, the new definitions of the scientific results including the scholarly book type have been in force since 1 January 2018.

The Methodology 17+ stipulates itself that the Information System for Research, Experimental Development and Innovations (ISREDI) will be the main data source for the disciplines, in which results are usually books or articles published elsewhere than in international databases for the results of the applied research and development. Apart from the explicit information considering scholarly books in the

Methodology 17+ mentioned in previous sentence, it is the statement contemplating the authorship. In a wide range of disciplines, the results for evaluation are books or articles with many co-authors and this problem will not be addressed by determining the mathematical share of each author of the research organisation in the result in a mechanical way.

Due to the fact that currently there is still missing an official translation of the Annex no. 4, the definition itself published in above mentioned Definition of Type of Results Separate Annex no. 4 of the Methodology 17+ can be described and translated into English for the purpose of this article in the following way:

The "book" presents the original research results that were created by the author of the book or by the author's team whose author was a member as well. The book is a non-periodical professional publication, however, covering a wide range of at least 50 printed pages of its own text without photographs, images, maps or other attachments, printed or electronically, and reviewed by at least one generally recognized expert from the field in the form of a lecturer's report. The expert ought not to be from the author's workplace. Additionally, the book refers, pursuant to the Annex no. 4 definition, to a well-defined problem of a particular field of science. It contains the formulation of an identifiable and scientifically recognized methodology. There is an explicitly formulated methodological basis in monographs for application. Furthermore, the formulation of the new methodology based on the theoretical research is found in the given field. The formal attributes of a professional book are the references to literature in the text, a list of used literature (possibly footnotes and a bibliography of sources) and a summary at least in one world language as well. Besides, the book has its International Standard Book Number (ISBN) or International Standard Music Number (ISMN, used for printed music).

Annex no. 4 points out that the whole book is created by a unified (single) creative team. It is no matter what is the share of the individual team members in the content of the book. Even if the individual chapters of the books have a separate authorship, such as a monograph, scientific encyclopaedia and lexicon, a critical edition of the sources, a critical edition of the artistic (musical, visual, etc.) material diploma, doctoral, habilitation and dissertation work (which are not based on the works already published in journal listed in Web of Science registry, journal listed in SCOPUS registry or in other reviewed scientific journal), critically commented translation of demanding philosophical, historical or philological and similar texts, scientifically conceived language vocabulary and expert interpretation dictionary, critical exhibition catalogue, etc., the formal criteria are met. For multi-volume scholarly books, it is feasible to include each volume in the ISREDI whether every required criterion is individually fulfilled. In addition, it was published as a separate publication with its own ISBN. In case a professional book is included in ISREDI as a result of the type "book" and in the condition of the same submitter of the result, its chapters cannot be classified as a result of the type "chapter".

Annex no. 4. includes the negative definition, which is not for the evaluation purposes considered as a book. Books that do not have an ISBN or even ISMN are not considered as a book. Textbooks (i. e. textbooks, scripts) are not scholarly books unless they are the result of original pedagogical research. Other texts that are not deemed to be a book: expert opinions and assessments, studies, translations, manuals, information and science popularisation publications, yearbooks (except from those which meet the requirements of a professional book), annual or similar periodic reports. Furthermore, published diplomas, doctoral, habilitation and dissertation work that are based on publications in journals listed in Web of Science registry, journals listed in SCOPUS registry or in other reviewed scientific journals with commentary and ISBN. Moreover, other documents that are not esteemed, pursuant to Annex no. 4, to be books are as follows: common linguistic dictionaries; purposefully published summaries of professional work (e. g. within a workplace), print or electronically published summaries of abstracts, or oral communications from conferences, methodical guidelines, catalogues and standards, proceedings (individual contributions in the proceedings are results of a type "article in proceedings"), fiction, popular educational literature, such as travel books, theatre plays; selective bibliographies, annual reports, speeches, reports, student competition chapters, tourist guides; commercial translations from foreign languages; memoirs, information materials; popularizing monographs, biographies and autobiographies; monographically issued final project reports.

Fully in line with the Czech legislation, the Annex no. 4 poses the following condition. Whether a book is published in the Czech Republic, the compulsory copy must be registered with the National Library of the Czech Republic. For results of a "book" published abroad, the verification includes the Reference to Digital Object Identifier (DOI) or Open Access (OA), traceability in an internationally recognized catalogue and verification by a returning loan from the reporting institution certified by the provider as well.

Closely related to scholarly books are book chapters. These are defined by the Annex no. 4 in the following way. "Chapter or chapters in a scholarly book", in case the book meets the definition for the resulting type book, are applied whether the whole book has only an editor or the author is listed as the title of the book co-author (though with minority content) and is a member of the author's team with clearly stated authors. However, the chapter must include a separate author or creator. Whether a professional book is included in ISREDI as a scholarly book, its chapters cannot be classified as a type chapter in the case of the same submitter of the result.

Before analysing the above-mentioned definition of the book and book chapter, it is essential to compare formal definitions of the scholarly book and book chapter that were applied by the Czech government for the period from 2013 until 2017. Nevertheless, the differences will be pointed out further ahead. The definitions are used from the English translated version of the Methodology of Evaluation of Research Organizations and Evaluation of Finished Programmes (valid for the period from 2013 until 2017) that have been published in official governmental websites www.vyzkum.cz since then and are also dedicated to the national research evaluation and research management. The previous definition of the scholarly book was only slightly different. The previous version did not deemed to be a scholarly book (even fulfilling a formal "book" criteria) published diploma, doctoral, habilitation and dissertation work that are not based on the works already published in a journal listed in Web of Science registry, journal listed in SCOPUS registry or in another reviewed scientific journal. Additionally, other similar critically commented translation of demanding philosophical, historical or philological texts was not deemed to be a scholarly book as well. Moreover, compared with current evaluation methodology, whatever textbooks (i.e. textbooks, scripts) were not deemed to be scholarly books even whether they were resulting from original pedagogical research. At present the monographs based on the original pedagogical research are accepted as well. Compared with the current version, printed or electronically published series of research chapters, which were explicitly included in the negative list of results, are not considered as scholarly books. Due to rather pleonastic definition, in the current version of the negative list of the results there was the deleted, printed or electronically published set of extended abstracts. The set of abstracts (that might include also extended abstracts) was retained. The negative list in the current version was fine-tuned by adding "e.g." changing the meaning of the negative list from the closed list to the demonstrative list of the results that are not accepted to be a scholarly book. The previous version of the methodology rendered the obligation in case of published books in the Czech Republic that the obligatory issue must be registered in the Czech National Library.

This obligation is based on the legal requirement of the Act No. 37/1995 Coll. Act on non-periodical publications that in §3 stipulates the obligation of the publisher to provide, free of charge and at his own expense from each issue, a non-periodical publication within 30 days from the date of issue certain number of publications to several libraries that are listed in the act. The non-periodical publication (not only scholarly books but other types of books such as e.g. fictions) shall include as follows:

(a)   a name of the work,
(b)   names of the authors whose works are included in the publication, if known, and the authors have not explicitly forbidden their introduction, or the author's pseudonyms,
(c)   a business name and registered office or name, surname and permanent address of the person who published the non-periodical publication (hereinafter referred to as "the publisher"),
(d)   a year of first issue, if known,
(e)   for the works transposed the original title of the work and the indication of the issue from which the translation was made,
(f)   a designation of the copyright holder,
(g)   International Standard Numbering Number of Books, if issued,
(h)   a business name and registered office or name, surname and permanent address of the person who produced the non-periodical publication and the year in which the publication was published.

The previous version of the methodology did not mention any measures for the results of the "book" published abroad. Furthermore, the positive aspect of the new scholarly book definition adds the ways of proving of the books published in a foreign country. Thus, the international reach is another aspect that is more and more considered in the scientific evaluations. This is not only in case of individual careers but international reach and impacts are deemed in case of evaluations of the research organisations (this is also part of the evaluations carried on within other evaluation modules of the Methodology 17+). The international reach gained in importance in comparison with the situation in the past [16, p. 617, 17]. It is important to point out in case of SSH, readers are more locally oriented (e.g., the Czech researchers who focus mainly in Czech readership). Books are published by national publishers and present in non-commercial databases in Czech governmental ISREDI. The commercial databases of scholar books are the Web of Science´s Book Citation Index (BCI) and registry of Scopus that include books from a selection of scholarly publishers and they seem to focus on highly selective procedures. These procedural coverage concentrates on prestigious international publishers. These publishers are mainly based in the USA or UK and in the servers of the natural sciences, engineering, and medicine [8].

3.2   Features, Specifics and Trends of Scholarly Books, Publication and Evaluation in the Czech Republic

Step by step internationalisation of the Czech research "arena" is visible and is also reflected in the national evaluation methodology described above. Hand in hand with internationalisation of the research that also includes SSH field, there are also changes in publication patterns in SSH (e.g. the preference of publications in journals in comparison with monographs). Additionally, this observation is well shown in the following study [19] Publication patterns in the social sciences and humanities: Evidence from eight European countries. This recent study investigates patterns in the language and type of SSH publications in non-English speaking European countries (the Czech Republic, Denmark, Finland, Flanders ((Belgium)), Norway, Poland, Slovakia, and Slovenia). Furthermore, it demonstrates such patterns related not only to the scientific discipline but also to each cultural and historic heritage of the country. The findings of the study show that publication patterns vary across fields (e.g. patterns in law differ from those in economics and business in Flanders and Finland in the same way) and within fields (e.g. patterns in law in the Czech Republic differ from patterns in law in Finland). Thus it is observed that the publication patterns are stable and quite similar in the West European and Nordic countries, whereas in the Central and Eastern European countries the publication patterns demonstrate considerable changes. Nevertheless, in all countries, the share of articles and the share of publications in English are on the rise. For instance, in Finland the share of foreign academic staff at universities increased in SSH fields from 8% of the full-time equivalent in 2010 to 12% in 2016 [23]. In the Czech Republic in 2011, less than 10% of researchers overall were foreign, and half of those were Slovaks [3], whereas the share of foreign academic staff at Polish universities was 2.5% in 2016 [27].

The Czech Republic compared with above 7 mentioned non-English speaking countries has the highest share of published monographs of the total volume of the all type of published results. In Denmark, Finland, Flanders, Norway, Poland, and

Slovenia monographs did not constitute more than 6% of the total volume in 2014, whereas in the Czech Republic the share of monographs was 12.83% [19].

It is obvious that the Czech evaluation system deems the length of the scholarly books (50 printed pages). The use of numbers for measuring the length of publications is noteworthy in case of scholarly books not only in the Czech Republic but also in other countries as well. For instance, in case of Sweden, Hammerfest based on a content analysis of individual careers´ assessment reports in three disciplines– biomedicine, economics and history showcases that especially as referees in history tend to rely on narrative accounts, focus on their assessments of reports in measuring the length of publications, however, which do not make another use of quantitative data or metrics. Hence the length of the publication is a clearly necessary factor when evaluating publications in history discipline [16, p. 615]. It is crucial to point out that the length of printed pages of the scholarly books in the Czech Republic is the same for all types of research discipline, no matter whether it is STEM or SSH monograph, even though SSH results are usually more narrative compared with STEM publications. Often even, for instance, juristic journal scientific articles include more than 50 pages (see e.g. well reputable juristic journals such as Stanford Law Review or Yale Law Journal).

The Czech evaluation methodology does not esteem the reputation of the publisher, however, this is not challenging because, the reputation might be deemed during the peer review process used in the research performance assessment. The reputation of the monograph publisher is an essential aspect not only in the Czech Republic but also, for instance, in Spain [10]. In Spain, the methodological proposal aimed at the evaluation of the prestige of publishers had already been developed as well [9]. In addition, the Spanish survey, which was launched to 11,647 lecturers and Spanish researchers in social sciences and humanities, with a response rate of 26%, proves that the top-rated publishers determine the assessment of the quality of their published monograph.

The current Czech evaluation methodological definition reflects also the situation especially typical in SSH field where publications of dissertation works have become more and more appreciated as a monograph. This trend is interrelated with a concept of an ideal trajectory of the academic career [6] that is often applied in case of individual career assessments.

In the Czech Republic it shall be concluded that main features, specifics and trends in the field of scholarly books and evaluations, which should be deemed when we talk about a reform of the formal rules defining books and book chapters, are as follows:

(1) Internationalisation of the Czech research arena. Share of publications in English and foreign staff working in the Czech Republic is on the rise.
(2) Compared with other 7 above mentioned countries, the publication patterns in substantial differences between fields are observed.
(3) The publication patterns are stable and quite similar in the West European and Nordic countries, whereas in the Central and Eastern European countries the publication patterns demonstrate considerable changes.

(4) The Czech Republic compared with above 7 mentioned non-English speaking countries represents the highest share of published monographs of the total volume of all the types of published results.

(5) Minimum page length of monographs is stipulated even though this 50-paged minimum can be identified as a standard for scholarly articles published in journals in SSH.

(6) Governmental reflection of the trend in SSH where has become more and more appreciated to publish dissertation work as a monograph.

3.3 Implications and Design of the Formal (Legal) Regulation

As the publication patterns demonstrate considerable changes and trends in internationalisation of the Czech research arena, these processes should be fully reflected in research evaluation methodology documents and in the evaluation, research results (including scholarly books and chapters) process as well. For example, all methodological evaluation documents should be published in English immediately (as currently the English translation of the Annex no. 4. was not published for a long time). Also joint terminology of the crucial words should be preserved by official documents avoiding misleading meanings (e.g. a terminological specialist book is applied either for scholarly books or from time to time used a monograph) or government should explain in the documents that synonyms are applied for certain words. The new stipulation of one of the ways of the verification of the scholarly books published abroad by traceability in an internationally recognized catalogue should be more specified by the exact naming of recognised catalogues such as Web of Science´s Book Citation Index (BCI) and Registry of Scopus. This specification of recognised catalogues would avoid the risks of the cheats of publishing books by predatory publishers (see the case described in Stöcklová and Vostal [28]).

Due to the fact that the publication patterns in substantial differences between fields are observed, it should be explicitly emphasized in the part defining the scholarly books and chapters that during the peer review of the results submitted in the ISREDI, the definition of the scholarly books and chapters might be adopted (e.g. a question of minimum page numbers of the books) by the review panel according to the standards generally accepted in the certain scientific discipline.

Another aspect related to sensitivity of the peculiarities of the scientific disciplines is the need of distinction or specification of the "handbook" by law community as a crucial outcome of the academic work as this is recognised e.g. in cases of handbooks commenting and interpreting various legal acts. The necessity of handbooks for legal community is not specific only in case of the Czech Republic but also generally as this is supported by the survey carried out in the Netherlands [31]. The Dutch survey, among the others, highlights that "lawyers, especially when compared with non-lawyers, still attach a considerable value to the writing of handbooks, since these are commonly excluded from the category of academic publications."

As there is a high number of monographs published in the Czech Republic, the quantity should be questioned. Is this due to predatory publishing? To avoid this suspicion, the lists of Web of Science´s Book Citation Index (BCI) and registry of Scopus should be only deemed in case of books published abroad as it is in case of articles published in foreign journals. In addition to the books published in the Czech Republic, the publishers´ quality should be considered. For instance, Czech scientific publishers might be defined as universities and the Czech Academy of Sciences and its institutes. This narrowing of the monographs´ publishers would assure that the review process is preserved according to standards of the scientific disciplines and there would be the avoidance of the publishing in dubious publishing houses. This would also impede cross financing as e.g. researchers employed in public research organisation or university to publish results paid by public funding in private publishing houses.

Universities should be promoted, among the others, to support the publication of the final works such as dissertations and habilitations in English. Perhaps it would be seminal to put it as an indicator, which is represented by amount of dissertations and habilitations published in English, in the relevant Methodology 2017+ module (probably in the Module 4—Viability) to see the viability of the research organisation.

When focusing on the national regulation in the Czech Republic, it comes apparent the question about the fact whether each country should have its own definition and standards of scholarly books. On one hand, there is one "scientific global culture" of each scientific discipline. On the other hand, there is within the scientific community is a notion what scholarly book is. Hence somebody tries (e.g., a funding agency, senior researcher, national research evaluation institutions) put the scholarly book definition in a chapter, however, the views differ. Thus, the definition or standardisation of the scholarly books is dependent, e.g. on the evaluation and funding systems of the countries. For instance, in various countries, there might be posed the less emphasis on the obligation that each scholarly book should be peer reviewed, because the peer review is part of the research result assessment (as this was the case in the Czech Republic in the period from 2013 until 2016). Another aspect is the open access policy of the funding national agencies that might vary and thus bring peculiarities in the definition what a scholarly book is (e.g. scholarly books definitions might be restricted only to those books that are published openly without any restrictions). The different aspect is that the national research policy might be imposed in scientific community, then it is broadly interconnected with the university education and it imposes other tasks then only to provide excellent science. One of these tasks might be the emphasis on education and appreciation of textbooks as part of scientific knowledge production. We can conclude that when there is no single international research policy and scientific assessment, then we cannot expect to have one single formal definition of the scholarly book. Formal definitions of the monographs are one of the tools of research performance assessment created by the public body for its research policy purposes and definitely it includes its limits.

The Frascati manual, which is dedicated for OECD countries research performance reporting, provides guidelines what the definition of research and experimental development is. The manual also defines R&D main components: basic research, applied research and experimental development research. Obviously Frascati manual does not provide standardisation of the scientific resulting types (journal articles, monographs…) as still the international consensus among the OECD governmental members have not been reached. The same situation is in the case of EU research framework programmes, where are accepted the scientific results being in line with the national regulations. Currently particular European Commission´s specifications of the existing types of the scientific results still miss.

It can be summarised that proper rules defining the scholarly books is important, however, it should be fine-tuned by its interpretation within the peer review process carried on during the research assessment. One fit all approach is not sensitive neither to individual scientific disciplines nor national frameworks, because the funding (governmental) agencies research and research evaluation policies vary from a country to a country.

4   Conclusion

The vast majority of research assessments, which is the case of the Methodology 2017+ and its previous periods, however, are implemented in a top-down manner by either governments or university administrators. In addition, research assessment procedures usually apply bibliometric and scientometric methods developed for the natural and life sciences that do not reflect SSH research and disseminations practices. Bibliometric research shows that these methods cannot readily be applied for the SSH [18, 21, 24]. Therefore, research assessment procedures (and oftentimes research evaluation in general) meet the strong opposition in the scholarly communities of the SSH [25].

However, in case of the Czech Republic, since 2017 this negative aspect of pure scientometrics approach has changed thanks to the inclusion of other aspects and tools of research assessments (see above mentioned five modules). This change also includes the peer review of the results reported annually in the ISREDI. Scientomertics is applied now as a supportive tool. Let us see in the future from the long-term perspective, how this switch from quantitative mechanistic approach to more qualitative sensitive way of evaluation will bring more reliable and useful outcomes.

Clear rules are substantial perquisite for successful assesments, however, they should not be applied mechanically. As recent research surveys underline, quality management tools should not be imposed upon a discipline without the understanding and support of the scholarly community [31, 31]. A sensitive approach shall be utilized, for instance, the application of the peer review method and considerations of other non-scientometric aspects, reflecting peculiarities of the scientific disciplines, expectations and request of the stakeholders and also current trends in the international academic and scientific arena.

References

1. Act No. 37/1995 Coll., Act on non-periodical publications. https://www.zakonyprolidi.cz/cs/1995-37?text=Z%C3%A1kon%20%C4%8D.%2037%2F1995%20Sb. Accessed 9 May 2018
2. Act No. 130/2002 Coll., on the Support of Research, Experimental Development and Innovation from Public Funds and on the Amendment to Some Related Acts (the Act on the Support of Research and Development), the Czech Republic, English unofficial translated version. http://www.vyzkum.cz/storage/att/2D962B39DFEE8904BD6E509A5354FACA/Act%20No130%20_2002.pdf. Accessed 9 May 2018
3. Arnold E (2011) International audit of research, development & innovation in the Czech Republic: Final report: Synthesis report. https://rio.jrc.ec.europa.eu/en/file/8082/download?token=l5tMSCU7. Accessed 9 May 2018
4. Arnold E, Mahieu B (2015) The new evaluation methodology. Conference presentation, Olomouc, 14.5.2015. http://metodika.reformy-msmt.cz/last-conference-with-technopolis. Accessed 9 May 2018
5. Barton D (2007) Literacy: an introduction to the ecology of written language, 2 edn. Blackwell, Oxford
6. Felt U (2017) Under the shadow of time: where indicators and academic values meet. Engag Sci Technol Soc 3:53–63
7. Geuna A, Martin BR (2003) University research evaluation and funding: an international comparison. Minerva 41(4):277–304. https://doi.org/10.1023/B:MINE.0000005155.70870.bd
8. Giménez-Toledo E, Mañana-Rodríguez J, Sivertsen G (2017) Scholarly book publishing: its information sources for evaluation in the social sciences and humanities. Res Eval 26(2):91–101. https://doi.org/10.1093/reseval/rvx007
9. Giménez-Toledo E, Román-Román A (2009) Assessment of humanities and social sciences monographs through their publishers: a review and a study towards a model of evaluation. Res Eval 18(3):201–213. https://doi.org/10.3152/095820209X471986
10. Giménez-Toledo E, Tejada-Artigas C, Mañana-Rodriguez J (2013) Evaluation of scientific books' publishers in social sciences and humanities: Results of a survey. Res Eval 22(1):64–77. https://doi.org/10.1093/reseval/rvs036
11. Good B, Vermeulen N, Tiefenthaler B, Arnold E (2015) Counting quality? The Czech performance-based research funding system. Res Eval 24(2):91–105. https://doi.org/10.1093/reseval/rvu035
12. Government of the Czech Republic. Czech Government Resolution no. 475 of 19 June 2013 on methodology of evaluation of research organizations and evaluation of finished programmes (valid for years 2013–2017). http://www.vyzkum.cz/storage/att/A7FE6F4477F5064B57B683C62C4A4CFD/Methodology.pdf. Accessed 9 May 2018
13. Government of the Czech Republic. Czech Government Resolution no. 107 of 8 February 2017 on Methodology for evaluating research organisations and RD&I purpose-tied aid programmes (valid for years 2017+). http://vyzkum.cz/FrontClanek.aspx?idsekce=695512. Accessed 9 May 2018.
14. Government of the Czech Republic. Czech Government Resolution no. 837 of 29 November 2017 Annex no. 4 of the methodology for evaluating research organisations and RD&I purpose-tied Aid programmes (valid for years 2017+). http://vyzkum.cz/FrontClanek.aspx?idsekce=799796. Accessed 9 May 2018
15. Hamilton M (2012) Literacy and the politics of representation. Routledge, London
16. Hammarfelt B (2017) Recognition and reward in the academy: Valuing publication oeuvres in biomedicine, economics and history. Aslib J Inf Manag 69(5):607–623. https://doi.org/10.1108/AJIM-01-2017-0006
17. Hemlin S, Montgomery H (1993) Peer judgements of scientific quality: a cross-disciplinarydocument analysis of professorship candidates. Sci Technol Stud 28(1):19–27
18. Hicks D (2004) The four literatures of social science. In: Moed H, Glänzel W, Schmoch U (eds) Handbook of quantitative science and technology research. Kluwer Academic Publishers, New York, pp 473–496

19. Kulczycki E, Engels TCE, Pölönen J, Bruun K, Dušková M, Guns R, Nowotniak R, Petr M, Sivertsen G, IsteničStarčič A, Zuccala A (2017) Publication patterns in the social sciencesand humanities: evidence from eight European countries. Scientometrics, Firstonline 26 March 2018, pp 1–24. https://doi.org/10.1007/s11192-018-2711-0

20. Křepelka F (2014) Dominance of English in the European Union and in European Law. In: Sierocka H, Swieczkowska H (eds) Issues in teaching and translating english for special purposes. WydawnictwoUniwersytetu w Bialymstoku, Bialystok, pp 137–150

21. Lariviere V, Gingras Y, Archambault É (2006) Canadian collaboration networks: a comparative analysis of the natural sciences, social sciences and the humanities. Scientometrics 68(3):519–533. https://doi.org/10.1007/s11192-006-0127-8

22. Mcculloch S (2017) Hobson's choice: the effects of research evaluation on academics' writing practices in England. Aslib J Inf Manag 69(5):503–515. https://doi.org/10.1108/AJIM-12-2016-0216

23. Ministry of Education and Culture and the Finnish National Board of Education (2017) Vipunen—Education statistics Finland. Vipunen—Education Statistics Finland. https://vipunen.fi/en-gb/. Aaccessed 9 May 2018

24. Nederhof AJ (2006) Bibliometric monitoring of research performance in the social sciences and the humanities: a review. Scientometrics 66(1):81–100. https://doi.org/10.1007/s11192-006-0007-2

25. Ochsner M, Sven H, Galleron I (2017) The future of research assessment in the humanities: bottom-up assessment procedures. Palgrave Communications volume 3. Article number 17020. https://doi.org/10.1057/palcomms.2017.20

26. Organisation for Economic Co-Operation and Development (OECD) (2017) Frascati Manual 2015: guidelines for collecting and reporting data on research and experimental development, the measurement of scientific, technological and innovation activities. OECD Publishing, pp 44–45. http://www.oecd-ilibrary.org/docserver/download/9215001e.pdf?expires=151 3286734&id=id&accname=guest&checksum=44C0856E1A164785BF3BC694CE3FD0C4. Accessed 9 May 2018

27. PAP Nauka w Polsce (2017) Prawie 700 uczonychrozpoczęłoprace w Polscedziekiprogramowi MarieSkłodowska-Curie Actions, Nauka w Polsce. http://naukawpolsce.pap.pl/aktualnosci/news%2C414316%2Cprawie-700-uczonych-rozpoczelo-prace-w-polsce-dzieki-programowi-marie-sklodowska-curie-actions.html. Accessed 9 May 2018

28. Stöckelová T, Vostal F (2017) Academic stratospheres-cum-underworlds: when highs and lows of publication cultures meet. Aslib J Inf Manag 69(5):516–528. https://doi.org/10.1108/AJIM-01-2017-0013

29. Tusting K (2012) Learning accountability literacies in educational workplaces: situated learning andprocesses of commodification. Lang Educ 26(2):121–138. https://doi.org/10.1080/09500782.2011.642879

30. Tyler TR (2017) Methodology in legal research. Utrecht Law Rev 13(3):130–141. https://doi.org/10.1080/09500782.2011.642879

31. Van Boom WH, Van Gestel R (2017) Evaluating the quality of dutch academic legal publications: results from a survey. Utrecht Law Rev 13(3):9–27. https://doi.org/10.18352/ulr.404

32. Van Leeuwen T (2008) Discourse and practice: new tools for critical discourse analysis. Oxford University Press, Oxford

33. Whitley R (2000) The intellectual and social organization of the sciences. Oxford University Press, Oxford

Security and Emergency Maanagement

The Definition Frame of the Conflict of the Crisis Management in the International Relations

Rastislav Kazanský

Abstract Most theories and experts in the field of security science argues that it is necessary in crisis management in international relations define conflict frame. That is what exactly and precisely determines the type of conflict. This specific definition framework activity is the prerequisite for the theory of conflicts. Thanks to this, we can study them further and determine the causes of conflicts, the development and goal of the parties involved and the progress and a possible solution of conflicts. The first chapter of the publication focuses on the characterization and definition of the term 'conflict'. A conflict, as a multidimensional phenomenon, may be classified into several groups according to its examined properties. The following analysis deals with the various phases of the conflict, from its beginning to its end. The final part of the chapter offers a list of conflict databases and projects, which examine, categorize and divide conflicts.

Keywords International relations · Crisis management · Conflict analysis · Methodology · Typology

1 Introduction Definitions of the Term 'Conflict' in Relation to International Security

The term 'conflict' accompanies the human race and society from their very origins. Conflicts are present in the entirety of human history [6]. We can find a number of different approaches to the definition of conflict in available contemporary professional literature. A conflict is a social phenomenon and its definition is quite complex. During the examination, analysis and creation of specific definition, it is necessary to take into account the structure, diversity and complexity of this concept. In general, we can characterize a conflict (lat. conflictio, ger. der Konflikt) as a dispute, discrepancy, disagreement, armed encounter or war [12]. A situation where people, groups

R. Kazanský (✉)
Faculty of Political Sciences and International Relations, Kuzmanyho 1, 974 01 Banská Bystrica, Slovak Republic
e-mail: rastislav.kazansky@umb.sk

© The Author(s), under exclusive license to Springer Nature Switzerland AG 2022
I. Tušer and Š. Hošková-Mayerová (eds.), *Trends and Future Directions in Security and Emergency Management*, Lecture Notes in Networks and Systems 257,
https://doi.org/10.1007/978-3-030-88907-4_5

or countries enter into serious dispute, may be an alternative definition. Basic meanings of the term 'conflict' include: a situation in which violence is used, a struggle between two countries or a situation in which thoughts, feelings, opinions, ideals, etc. are in contradiction.

Several authors deal with the definition of the term 'conflict'. O. Krejčí defines a conflict as a situation, in which a certain group (tribe, ethnic group, ideological group or state) or an individual is in a purposeful dispute with one or more groups or individuals. A conflict is a struggle for values relating to the maintenance or increase of welfare, status or power. Opponents of these values try to neutralize, hurt or remove their rival or rivals [13].

According to Š. Waisová, a conflict is a social reality, in which at least two parties (individuals, groups, states), with a different outlook on certain facts or different, contradictory interests, stand in opposition (Waisová 2002). According to her, a conflict represents a situation in which, at the same time, a minimum of two parties are striving to obtain the same goods, which are deficient and cannot satisfy the needs of both (all) parties (Waisová 2005).

D. Kusá defines a conflict as a state in which one or two (or more) people and communities feel that their interests are incompatible. They usually have an antagonistic approach towards each other, which they show by trying to cause the other party harm. They seek to assert their own interests by influencing the other party (Kusá 2006).

F. Glasl provides a more specific definition of a conflict, as an interaction between agents (individuals, groups or organizations), where at least one agent understands that their thoughts, ideas, perceptions and/or feelings are incompatible with the will, thoughts, feelings, etc. of another agent (or agents) and s/he feels limited by their activities [16].

According to L. Hofreiter, a conflict represents a certain quality of relations between units of a social environment (parties involved, which may be individuals, social groups, states or a coalition of states), which are manifested in the efforts of certain parties involved to promote their own needs, achieve their own interests and objectives at the expense of and against the wishes of their opponents, or which are contradictory to the interests of their opponents [2].

The number of definitions of the term 'conflict' depends on the complexity of its concept. Some definitions define a conflict in general and only create basic starting points from which we may further explore this concept. On the other hand, other definitions deal with certain, particular and specific, characteristics of conflicts based on their typology. To analyse a conflict from the point of view of international security, it is necessary to lay down those properties and elements which are, in general, common for all conflicts, regardless of their specificities. While exploring this concept, we may use two basic models, which occur during conflicts in natural environment as well as those in social environment. The static and dynamic model [21, 22].

The static model examines a conflict as a social complex, which consists of two elements. Parties participating in the conflict (people, animals, objects, theories, etc.) and the relations between them.

The dynamic model draws on the behaviourist theory of psychology. According to this theory, the 'stimulus–response' principle affects the behaviour of a person. A person reacts to the stimuli from the external environment.

If these impulses are in contradiction with the interests of the object (person or group), their reaction to the situation is adequate and a conflict arises [2]. It is these relationships that have an impact on the dynamic aspect of the conflict.

However, the concept of 'conflict' cannot be put into contrast with 'competition'. During a conflict, the parties involved seek to strengthen their position and status at the expense of the others. One of the parties may even attempt to remove or destroy their opponents. Compared to this, competition means that even though the parties involved are trying to achieve the same goal at the expense of other parties, their mutual relationship is not as critical as to warrant a fear of elimination of one of them [13].

Equally, we cannot identify 'conflict' with 'tension' because tension means a hidden hostility, fear, suspicion, a perception of the divergence of interests and perhaps also a wish of superiority, or the wish of gaining independence. In this case, fear does not usually escalate from attitudes and perceptions to mutual hostile acts [13].

When characterising and examining political conflicts, it is necessary to define the term 'political crisis', which represents the beginning of military solving of conflicts of interests and powers. The word crisis, in post-modern terms, expresses restlessness and chaos within international politics and it also defines the concept of disorder and non-compliance on a global scale.

Not every conflict is a political crisis, but every crisis includes the state of conflict. The crisis is usually a conflict which stems from a dispute about a certain issue. In this phase, stability transforms into instability, or certainties into uncertainties in certain processes of development. A crisis is a particular moment or a period of time, after which a significant twist in the evolution or a change of the system may occur.

A crisis is more than simple tension, restlessness or separation within international relations. O. Krejčí defines a crisis as a type of conflict which is typically represented by a sudden outburst of unexpected events and hostilities, caused by existing conflicts [13]. A crisis in international politics is characterised by unforeseen and unexpected reactions from opposing parties, a feeling of great danger, a sense of a lack of time to come to conclusions and decisions and by the feeling that inactive will have horrific consequences.

Conflicts and crises, which the human civilization faces today, become increasingly complex and harder to resolve, as a result of growing globalization (Ivančík and Jurčák 2013a, b). Within international relations of the twenty-first century, conflicts are characterized by four basic components: (a) the parties involved, (b) the issues causing the conflict, (c) attitudes, (d) actions.

(a)   the parties involved in conflicts are usually the states. However, international organizations, non-state organizations, revolutionary movements and ethnic groups may also become involved. For illustration, in the period from 1818 until 1996, states participated in 41% of all conflicts. They were some of the decisive factors in the initiation of conflicts and belonged to the most active parties involved in international conflicts [13]. Presently, the number of states involved in conflicts is decreasing while the number of non-state parties involved is rising.

(b)   the issues causing the conflict are the objects and/or the position the persons involved want to achieve. The parties involved in the conflict (e.g. states) attempt to gain assets which, on one hand, strengthen their power and/or their potential to obtain power and, on the other hand, take some of their power away. These are, e.g. territories, safe areas and regions, control over resources, a world revolution or dissolution of certain states, etc. The conflicting behaviour of the parties involved implies their attitudes and actions. Such behaviour is caused by the fact that Party A has or gains a certain status, which opposes the wishes, ideas and interests of Party B.

(c)   attitudes represent the behaviour which may be expected from the persons involved. They are associated with hostility, distrust, stereotyping and a sense of justice. They also represent a source of tension and help the leaders of revolutions and citizens to become committed and to act in the conflict.

(d)   actions that occur during conflicts may be diplomatic, commercial, serve as propaganda or other. The parties involved tend to use them against each other [13].

2   Procedure of Conflict Analysis

Conflict analysis is the examination of the nature, causes, dynamics and parties involved in a conflict. Exploration of these elements allows us to better understand specific conflicts and to provide appropriate and accurately targeted means by which to deal with them. On the other hand, it is necessary to realize that the dynamics of conflicts is extremely complex and we often need to use different processes to analyse them.

An analysis of a conflict happens on multiple levels (e.g. local, regional, national or global). One of the objectives of this analysis is to define the links and relations among the given levels of conflict. It is necessary to correctly identify the point of view for the analysis on different levels. For example, the dynamics and issues of a given dispute may be different on one level than the dynamics and process on another level and they may have a different intensity curve. Understanding these links creates prerequisites for a comprehensive and explicit examination of the intervention and dynamics of conflicts. All of these levels affect one another.

Conflict analysis aims to define how it is possible to transform conflict situations and settle disputes among the parties involved. It is necessary to understand the context of the conflict, as well as the interactions between intervention and context, in order to understand its transformation.

This interaction is the basis for the following process, which is designed to prevent unwanted effects and, conversely, to maximize positive effects on the conflict itself.

The basic points of conflict analysis are:

(a) the profile of the conflict,
(b) the parties involved in the conflict,
(c) the reasons for the conflict,
(d) the dynamics of the conflict.

The conflict profile defines a brief characterization of the context unique for the given conflict. When defining the conflict profile, it is necessary to answer a few basic questions which would help determine the nature of the conflict's environment more precisely. R. Mischnick defines the initial questions:

- What is the geopolitical, economic, political, and socio-cultural context of the security situation? Geographical localization, political, economic and social structure, history, composition of the population, geostrategic location, environment, etc.
- What are the acute social, economic, political and environmental issues in the country? Destruction of social sphere, new infrastructure, decentralization, elections, reforms, issue of refugees, military and civilian victims or presence of armed forces.
- Which dispute-affected areas may be present within this context? Area under the influence of individual parties involved, close proximity of battle fronts to natural sources or strategic infrastructure, population exiled to the edge of society.
- Is the history of the conflict present? Key events, attempts at mediation, external intervention [16].

The term 'causes of conflicts' comes to play here. A cause, in a broader sense, is a phenomenon that gives rise to another phenomenon's appearing. It is necessary to realize that conflicts are multidimensional phenomena without a single explicit cause. They have several causes, the conflict-generating potential of which is combined. On the other hand, we also have to consider the fact that the stimulus (the cause) which causes a conflict in a certain group, may remain without response in a different environment.

Generally, we can divide conflicts into the following basic categories, based on their causes:

- conflicts over identity and self-determination, which are characterized predominantly by ethnonational and ethnocultural conflicts;
- economically motivated conflicts, during which a specific type of conflict emerges solely to gain profit, a so-called "war for profit";
- conflicts based on political basis, due to the poor functioning of the government, the inability to ensure primary function of the state [24].

Some of the main causes of conflicts are: (a) structural causes of conflicts—illegitimate government, lack of government power, low political participation, unequal political and social opportunities, unequal access to natural resources, etc.; (b) events which are the immediate causes of conflicts—e.g. uncontrolled security services, human rights violations, destabilizing situation in neighbouring countries, increase in the ownership of light weapons, etc.; (c) the so-called "conflict defractors" (conflict triggers) which may cause an outbreak of violence and a subsequent escalation of conflicts. Such triggers are, for example, elections in the country, a collapse of local currency, an enormous increase in unemployment, an increase in prices or a shortage of basic commodities, a leak of state capital, the imprisonment or assassination of a key political leader; (d) factors which prolong conflict dynamics—e.g. opposing parties becoming more radical, the development of war economy, the availability of weapons, etc.; (e) factors which contribute to establishing peace—a dialogue between the parties involved, the process of demobilisation, reforms, anti-discrimination measures, the commitment of the civil society to maintain peace.

The parties involved in conflicts may be individuals, groups or institutions, organizations or, in the context of international relations, explicitly defined states, which are immediately (positively or negatively) affected by the conflict, which create a conflict or work with a conflict in the process of managing or transformation of its dynamics. Parties may be directly or indirectly involved in a conflict. Parties directly involved in a conflict are those participants, who are in a direct, immediate dispute— they are the so-called subjects of the conflict. Parties indirectly involved in a conflict are the so-called third parties. They play a secondary role in the course of the dispute. We distinguish parties involved on the basis of their relations towards the opposing parties, their interest, goals, positions and strategies. According to R. Mischnick, the main parties involved may be: national government, political parties, the security sector (police forces, the army), the private sector, local military leaders and armed groups, neighbouring states, donor organizations and foreign embassies, multilateral and regional organizations, political and religious groups, the civil society, peace groups, trade unions, refugees and others.

The parties involved may pursue global interests, political ideologies, political participation, political commitments, economic activities, resources or religious ideals [16].

We can understand the dynamics of a conflict as a result of the interaction among the conflict's profile, the parties involved and its causes [2]. All conflicts and disputes within international relations go through certain developmental stages and levels of intensity, during their course [1]. Research and correct understanding of these development stages are a necessity if we wish to effectively interject, appropriately solve and prevent the escalation of conflicts. Understanding the relations between the parties involved in the dispute is a key factor when determining the way in which to solve a conflict. A change of the code of conduct, goals, interests or the way in which the parties involved negotiate can change the dynamics of the conflict.

Every conflict has certain phases (stages), which follow one after another. Long-term studies have shown that not every conflict necessarily needs to go through all of the stages. These may be interrupted during the course of the conflict (e.g. after negotiations or mediation by a third party), they may be repeated after a certain interruption or return to a stage of lower intensity. In some cases, conflicts may stagnate at certain points for decades. Experts look at phases of a conflict in different ways. They approach the stages of conflicts differently but the evaluation of the level

Chart 1 Phases of conflicts.
Source [16]

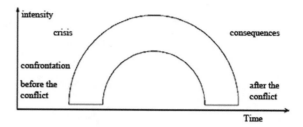

of intensity of violence over a period of time is key. In general, we can summarize these studies into the following phases (stages) of conflicts:

- the pre-conflict phase,
- the confrontation,
- the crisis,
- the consequences,
- the post-conflict phase.

Chart 1 shows the aforementioned phases.

There is always a potential for the existence of a conflict when the parties involved have identical objectives, needs, interests and values, the achievement and satisfaction of which is limited. This latent phase is the phase before the conflict itself, when the dispute is not yet shown openly. It is characterized by tension between the parties involved in the conflict, or by the effort to avoid mutual conflict. The dispute may not occur at all, if there is no "trigger event" or "incident", which leads to the opening of the conflict and then to the second phase—the confrontation. At this point, the opposing parties begin to accumulate resources and, possibly, search for allies in case the dispute will escalate. The crisis is the peak stage of the conflict, in which tension and violence are the most intense. At this stage, the opposing parties usually cease all communication. The next stage of the development of the conflict is the consequences which every crisis inevitably leads to. One of the parties involved may defeat the opponent, back down and accept the terms of the opposing party or surrender. In this stage, there is a possibility to settle the dispute. During the stage after the conflict, a situation which allows a non-violent settling of the dispute, may occur. On one hand, there is a possibility to settle the relations between the parties involved, on the other hand, things may return to the pre-conflict phase, if the causes of the conflict have not been adequately resolved.

A more specific definition of the various stages of a conflict may be found in the study of Š. Waisová, who divides conflicts into seven phases, displayed in Chart 2.

(1) latent conflict,
(2) manifestation of the conflict,
(3) escalation of the conflict,
(4) a stalemate in the conflict,

Chart 2 Dynamics and stages of a conflict. *Source* Waisová (2005)

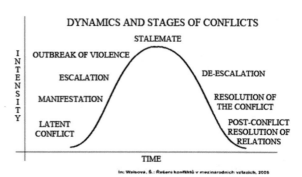

(5)   de-escalation of the conflict,
(6)   resolution of the conflict,
(7)   post-conflict settlement of relations—peace-building.

3   Conflict Typologies

We may find many divisions, typologies and classifications of conflicts in contemporary professional literature. Several factors play a part in the genesis of conflicts and the behaviour of the parties involved. It is, in particular, the history of the parties' mutual relationship, their nature, their perception and explication of the conflict situation. When examining conflicts, it is necessary to delve into their essence and to understand their basic nature. We may then divide conflicts into multiple groups according to multiple criteria. These divisions will depend on the common features and criteria which we will consider essential and crucial to express the main basis for the conflict. According to L. Hofreiter, such features and criteria are:

- the parties involved in the conflict (intrapersonal, interpersonal, between an individual and a group, between groups, between states or groups of states);
- the level of the conflict (horizontally or vertically oriented conflicts);
- the nature of the needs that caused the conflict (material, immaterial, spiritual);
- the duration of the conflict (short-term, quick, long-term, etc.);
- the consequences of the conflict (constructive, destructive) [2].

Social nature affects the investigation of conflicts. According to Š. Waisová, during classification, it is necessary to include (Waisová 2005): 1. research of the background of conflicts (the geopolitical and economic status of the parties involved, the history of their mutual relations and the history of the conflict itself), 2. the type of parties involved (states, non-state organizations, international organizations, movements for independence, revolutionary or insurgent groups, etc.), 3. research of the character and nature of the opponents of the conflict, 4. research of the causes of the conflict (the subject of the dispute), 5. research of the environment and the context of the conflict (who is involved in the conflict, who is supporting the opposing parties, which party is seeking a solution).

By investigating the aforementioned characteristics of conflicts, we can get a comprehensive image of the nature of the conflict, the stages of its future development, the strategies and means of the parties involved, etc. Based on the definition of these properties, it is possible to define the causes of the dispute, the parties involved in the conflict and how we can specify conflicts.

When analysing a conflict, we may use the following classification: (a) natural or physical conflicts, when an individual stands in opposition with nature; (b) social conflicts, when a person (social groups) stands in opposition to another person (social group); (c) internal or psychological conflict, when an individual is in conflict with themselves, their desires are in conflict with their options and their conscience [2].

L. Hofreiter also distinguishes the following types of conflicts [2]: conflict of relations (an aversion toward another person, etc.), conflict of interests (the clash of different interests and needs), conflict of values (the dispute about what is right or wrong, correct or incorrect), a structural conflict (organisational structures with an imbalance of power), a conflict of information (different sources and interpretation of data).

Classification according to interests is another possible division: The interests of the parties involved differ and they depend on several factors (needs, desires, concerns, etc.). Conflicts take place when these factors clash. They relate, in particular, to the areas of (a) resources (territorial, financial, personnel and material) and their distribution, which means the contribution to the process of fund and resource distribution and to the process of political decision-making; (b) identity (of social, religious, cultural and political communities and of communities with which individuals identify); (c) values (specifically those that stem from religion, ideology or the system of government); (d) status (relating to individuals or social groups and their status in society, their compliance with and respect towards values and traditions) (Kusá 2006; [14]).

The current conflict theory within international relations recognizes two types of conflicts, the symmetric and asymmetric conflict. A symmetric conflict is a conflict of interests between relatively similar parties involved, for example, between states, political parties, etc. An asymmetric conflict is a conflict between different groups, for example, between a minority and a majority, employees and employers, the government and rebels, etc. During such a conflict, the dominant party has better conditions to assert its interests, values and needs, because it has the means and resources to do so. The proportionality of power and a change of the status of the parties involved is a solution to an asymmetric conflict [2].

We may also define a symmetric armed conflict as a large-scale armed military confrontation of the armed forces of the participating states, coalitions or other integration groups (alliances, pacts, etc.), the result of which is usually easily measured, e.g. by freeing or occupying a certain territory, by the destruction, defeat or elimination of a known adversary, by achieving set objectives, etc. Unlike the symmetrical armed conflict, an asymmetrical armed conflict is a relatively small-scale and low-intensity military confrontation, in which the parties involved differ by their strength and tactics. It is mostly a conflict, in which a superior external military force, represented by a state (alliance, coalition, group), enters into a military confrontation with an inferior internal military force, represented by a state or non-state party, the territory of which is where the conflict takes place. Since the "weaker" party cannot succeed in an open military confrontation, because its capabilities, capacities and resources are incomparably smaller, it attempts to succeed by using asymmetric operations and forms of struggle [9].

The 2005 Human Security Report used the following division of conflicts for the first time: state-based armed conflicts—if two states are involved in a conflict, or if there is a conflict between a state and a non-state party (rebels, an uprising, etc.); non-state armed conflicts—when non-state groups are involved in a conflict; one-sided violence—violence (genocide,[1] politicide[2] and other violent assaults) committed by the government or political groups against civilians [3].

4 Analytical Levels of Conflict Categorization

We may divide conflicts based on several categories, such as essential or accidental, controlled and uncontrolled, ones that can or cannot be solved. In conflict theory, we most commonly encounter the following categorization: 1. According to the "position" of the conflict within the system of international relations: (a) system conflicts, (b) interstate conflicts, (c) domestic conflicts; 2. According to the means used in the conflict: (a) armed, (b) unarmed; 3. According to the causes of the conflict, in other words, the conflicting interests: (a) power struggles, (b) constitutional, (c) ideological, (d) economic, etc. (Waisová 2005). When defining conflicts according to the "position or location" within the system of international relations, we may use three analytical levels.

(1)   The first level is the international system, the largest set of interacting and interdependent set of particles, which are not under any sort of influence and are not subjected to supremacy and inferiority. It creates space for the persons involved in international relations to communicate, negotiate and act. Their interests, goals, needs, and behaviour are largely influenced by the general distribution of power, structure and rules of the system. System conflicts are those, which change and affect the aforementioned elements and aspects. In principle, system conflicts change the relations and context among most of the parties involved in international relations and establish new standards, norms and institutions of the international system. These mostly historical-power disputes are, essentially, about creating coalitions or alliances, the primary objective of which is to change the system balance and tilt it towards their own

---

[1] Genocide—is an act carried out with the intent to destroy the entirety or a part of a national, ethnic or racial group. It is also a crime against humanity (Tusičisny 2007).

[2] Politicide—is a more neutral term, used for the murdering of groups which are not included in the 1984 Convention on the Prevention and Punishment of the Crime of Genocide (Tusičisny 2007).

side of the dispute and their favour, dividing the profit in the end. Such conflicts were, e.g. the Thirty Years' War, First and Second World Wars and the conflict between the East and the West (Waisová 2002).

(2)   The second level is determined by the specific parties involved. These parties involved are interdependent and, at the same time, independent enough to stand out among other parties. Disputes within international relations, that take place among the parties involved, are referred to as international, or inter-state conflicts. The persons involved in inter-state conflicts are individual states or groups of states, alliances, pacts, transnational corporations, international non-government organizations, etc. An interstate conflict[3] is, basically, a conflict of interests in the sphere of national values and themes (borders, territory, resources, independence, sovereignty and international distribution of power). This type of conflict has a certain length, intensity and depth, and it takes place between at least two parties (states, groups of states, international organizations or organized groups). The goal of the individual parties involved is to advance their interests and to win the dispute. Such a conflict is conditioned by the fact that one of the parties involved in the dispute is the state. Inter-state conflict doesn't need to escalate into war, it transforms into a military conflict as a result of the behaviour of the parties involved.

Inter-state conflicts may be:

- bilateral (two-sided)—between two sovereign states, or where the state is one of the two parties involved in the conflict;
- multilateral (multi-sided)—between multiple parties involved. When this type is concerned, the variety of the combinations of the parties involved depends on the nature of the dispute and on the relations between these parties.

Some of the causes of international conflicts may be, e.g. a wish to claim territory or material goods (resources, assets). Such conflicts usually have a historical background. These conflicts may be expected from states which have, in the past, had strong influence (superpowers and former superpowers) and a dominant role within international relations—conflicts due to losing their influence, status, territory or resources. Another group of causes of interstate conflicts is an effort of the parties involved to gain a dominant status on a global or regional scale, in the form of a unilateral position or a position in the power hierarchy of states. Factors, which are a prerequisite for the creation of this type of conflict are e.g. a sense of grievance or scarcity, a sufficient quantity of resources to initiate and lead the conflict and an idea and vision of a potential victory in the end of the dispute. Examples of conflicts with

---

[3] Examples of international conflicts: The First World War between the Triple Entente and the Central Powers (1914–1918), the Second World War, initiated by Germany (1939–1945), the war between Paraguay and Bolivia in the area of Gran Chaco (1932–1935), the Falklands War between Argentina and Great Britain (1982), disputes between Vietnam and China for the Paracel Islands, the conflict between the Republic of South Africa and Namibia for the border on the Oranje River (since 2000) (Waisová 2005).

enormous consequences were the two world wars, started by Germany with the intention to gain a dominant position in the contemporary international political system. Economic reasons may also be the cause of conflicts among states, for example, the effort to achieve a strong economic influence or to strengthen their economic situation by gaining the resources of another state.

The basic critical factor of international conflicts is the possibility that they'll gradually escalate. The escalation may be vertical (increasing the intensity of the conflict) and horizontal (extending the conflict to other countries of the region). It is also necessary to include inter-state armed conflicts as part of the characteristics of interstate conflicts. Wars among states may arise, provided that there are interests which explicitly need military intervention to be achieved. The conditions for the emerging of an armed conflict include: (a) a favourable ratio of the armed forces, (b) the belief of military-political elites that the armed conflict is needed, (c) a good degree of support from the population for the conduct of an armed conflict, (d) the existence of potential allies and a favourable attitude of the international community, (e) sufficient economic potential, (f) building the infrastructure necessary to lead an armed conflict [2]. The potential for a military conflict can proportionally increase with the concentration of risk factors, deepening of crises and disputes, tension between the parties involved or with a cross-border transfer of the conflict from a neighbouring country. The presence of war or an armed conflict in a neighbouring country largely increases the likelihood that the state in question will get involved in the military conflict. This fact is not even affected by the degree of the involvement of the state in the globalization processes (Kahler, Walter, 2006). In the recent period, the number of inter-state conflicts is decreasing and they rarely reach their peak stage—a military conflict.

(3)   The third analytical level is determined by subunits. Political parties, opposition groups, lobby groups, the mafia, rebel movement, etc. may all be subunits. They are organized groups of individuals inside of an organisational unit, which have the ability to affect the existence and behaviour of the units themselves. We define conflicts which take place inside these units as internal[4] (or domestic). They are currently the most common and dominant form of conflicts. Internal conflicts take place inside of the territory of a certain state between certain parties, which may be e.g. social groups, political parties, ethnic, national or religious groups, interest groups, etc. These parties have irreconcilable, contradictory objectives and interests. The parties in conflict wish to purposefully remove the opposing party from the political struggle, even if it means using violence. In addition to this objective (eliminating the opponent), they also have the potential to do so. Internal conflicts may be violent or non-violent.

---

[4] Examples of domestic conflicts are the wars for independence (Algeria—France, Ireland—United Kingdom, Basque Country—Spain, Chechnya—the Russian Federation, East Timor—Indonesia, Katanga—Zaire/Congo), autonomistic conflicts (Uyghur—China, Abkhazia—Georgia, Adjara—Georgia, Corsica—France, Guerrero and Chiapas—Mexico) and civil wars (Somalia, Sierra Leone, Cambodia, Sri Lanka, Bosnia and Herzegovina, Colombia) (Waisová 2005).

Internal (especially violent) conflicts are characterized by human casualties, material losses, disruption of state infrastructure, enormous violations of basic human rights and civil liberties. The consequences of this type of conflict may also have a negative impact on the environment in the state, the disruption of the social and political system of the state or the moral degradation of its population. These conflicts also cause a number of secondary problematic factors, such as poverty, diseases, hunger and health care which is insufficient, of poor quality or non-existent. States with internal conflicts tend to decline economically, because of hard or non-existent economic activity and economic development, suspended production or an economic isolation of the area of the conflict. The decline is also influenced by the loss of human resources, the so-called "intellectual genocide". It occurs when labourers, experts in economy and state management, educators, doctors and medical staff leave the country in conflict. Due to conflicts, a large number of citizens is in the position of refugees, or are displaced to live in different locations within the country [2].

The threat of domestic conflicts exists, in particular, in the so-called dysfunctional states. This type of conflict occurs mostly in poorer countries with weak or undemocratic governments. As in the case of international conflicts, domestic conflicts also pose a threat to the surrounding states [19]. The occurrence of conflicts within states has a devastating impact not only on the states themselves, but it also contributes to an overall regional instability. In the case of such states, governments are unable to, or do not address the real problems in their countries, do not ensure the basic human needs or guarantee the basic human rights and civil liberties. The majority of these countries do not have the relevant political institutions and lack legal elements of the political system. The course of the conflict is worsened by the absence of a leader, or a central authority, and it allows an extreme escalation and increases the number of human victims of domestic conflicts. Not only horizontal and vertical escalation, but also its long duration, intensity, mobilization of forces and resources and great emotional exertion, are some of the dangerous aspects of domestic conflicts.

The causes and the trigger mechanism of national conflicts may be based on various factors, or may be a combination of more than one of them. Such factors may be, e.g.: (1) restoration of an old conflict which happened in the past, (2) low income of the population tends to be a result of poverty (poor citizens are more prone to illegal activities and often become members of terrorist organisations and organised criminal groups), (3) dependence on the export of a single kind of material (the fluctuation of its price or problems with export may greatly destabilise the state's economy, increase tension and provoke conflicts), (4) high dependence on the import of materials and energy), (5) disproportionate allocation may result in the deepening of disparities within regions, (6) resistance of the population against reigning elites (authoritarian or dictatorial rulers and undemocratic governments), (7) limitation of fundamental human rights and civil liberties, (8) enormous militarization of the state and society, (9) migration, which can disrupt the homogeneity of the indigenous population (ethnic, national, religious or cultural), (10) repression of the rights of ethnic, national and religious groups in the country, (11) separatist and irredentist tendencies, which may evolve into breaking away from the original parent state, (12) disturbance of the stability of a state, or its entering into conflict because of a conflict taking place in the neighbouring state, (13) radicalization of youth, due to high unemployment and impossible personal realization, (14) penetration of organised crime into government structures [2]. Most of these examples may be applied to current conflicts in North Africa and Middle Eastern countries.

A civil war[5] is the most extreme form of domestic conflict. It remains the most radical form of such conflict even despite the fact that, in modern times, fewer and fewer disputes escalate into an armed conflict.

Currently, the following workplaces lead complex studies, for the purposes of a deeper analysis of database creation and a comparison of the methodologies of conflicts within international relations:

- Bonn International Center for Conversion
- Freedom House
- Heidelberg Institute for International Conflict Research
- Internal Displacement Monitoring Centre
- International Crisis Group
- International Energy Agency
- International Institute for Democracy and Electoral Assistance
- International Relations and Security Network
- Political Terror Scale
- Reporters without Borders
- Stockholm International Peace Research Institute
- Swedish Defence Research Agency
- The Fund for Peace
- Transparency International
- United Nations Development Programme
- United Nations Statistics Division
- Uppsala Conflict Data Program
- Vision of Humanity
- The World Health Organization (Ramsbotham 2010).

5   Examples of Supranational Parties Involved in Conflict Resolution

International crisis management operations, led by transnational parties, represent one of the most important instruments of the international community, used to defuse crises and centres of tension, consolidate or stabilize the situation in crisis areas and to

---

[5] There are several definitions of a civil war. We can, generally, define it as the result of a domestic conflict [2]. The US military uses characterisation of the civil war, based on the following criteria (Patten 2007): (a) the challengers in the conflict must have control over (at least some) territory, (b) there must exist a functioning government, (c) each of the parties involved must have some recognition from abroad, (d) the parties must have identifiable and proper armed forces, (e) the parties involved meet in major armed operations.

help countries recover, especially after armed conflicts (Ivančík 2013c). The United Nations Organization, the North Atlantic Treaty Organization and the European Union are among the most important supranational parties involved in the field of conflict resolution.

6 The United Nations Organization

The United Nations Organisation is, in terms of the nature of its members and the range of its competences, a universal and versatile organization. In fact, it is the only organisation of its kind in the contemporary world. It has 193 member countries from all continents, with the exception of Antarctica, and a virtually unlimited range and scope of competences, which include both economic and social development and the protection of human rights, peace and security [7].

The Charter of the United Nations was signed at the United Nations Conference on International Organization in San Francisco on the 26th of June 1945 and came into force on the 24th of October 1945. The Slovak Republic entered the UNO on the 19th of January 1993. According to the Charter of the United Nations, the main goals of the organization is the preservation of international peace and security by means of collective peace-keeping measures in accordance with international law, the development of friendly relations among nations based on their mutual equality, dealing with international issues with cooperation and the endorsement of human rights and freedoms, as well as being a place where these objectives are achieved. The main bodies of the UNO are the General Assembly, the Security Council, the Economic and Social Council, Trusteeship Council, International Court of Justice and the Secretariat (Charter of the United Nations... 1945).

The UNO's peace-building activities affect a wide range of areas, such as military security, in which it involves the disarmament, demobilisation and reintegration of combatants and the demining of territories. The UNO's humanitarian activities involve caring for and repatriation of refugees affected by conflicts, providing health care and food. Activities such as the support and rehabilitation of public institutions and public administration, reforms and elections are related to the area of politics. Activities such as the ensuring of human rights, reforms of the judiciary, security and the investigation of crimes, fall under the field of human rights. The economic and social areas include activities such as helping with rebuilding destroyed infrastructure, creating conditions for public administration and economic growth, as well as the elimination of social and economic inequalities [11].

These and many other peace-building activities are conducted by the UNO's peace-keeping missions. According to the UNO's main doctrine from 2008, peace-keeping missions are divided according to their objectives and the means used to prevent conflicts, keep the peace, enforce peace (the UNO may designate regional organizations to enforce peace) and build peace, on the principles of consent of the parties involved, impartiality and by not using force, except in self-defence or to protect the mandate of the mission (UNO 2008).

The legal basis for the conduct of missions is anchored in Chaps. 6 and 7 of the Charter of the United Nations. Based on it, the UN Security Council may decide on the conduct of activities necessary to maintain and restore international peace and security [7]. As is apparent from the aforementioned facts, the relevant bodies of the UNO for the resolution of conflicts are the Security Council and the Secretariat, in particular its specialized branches:

The Department of Peacekeeping Operations, the Department of the Promotion of Peace and the UN Peace-building Commission, a subsidiary body.

The missions are made of voluntary military, police and civil contributions from the member countries (because the UNO does not have its own army and police forces), which are controlled by the Secretary-General and the relevant departments of the Secretariat. Peacekeeping missions, according to their type, usually begin within 30 days (a traditional peacekeeping operation), or within 90 days (a multi-dimensional mission) after they have been approved by the General Assembly. The request to begin a peacekeeping mission is the resolution of the Security Council, which is based on the initiative of the Secretary-General. This initiative comes in the form of information analysis [11].

The overly long and cumbersome decision-making process of the UN, which applies not only to conflict resolution, is a problem in this respect. Therefore, the discussion about the need of institutional and procedural reforms of the UNO is absolutely justified. The same applies to the efforts for uniform, commonly acceptable and respected conflict and crisis resolution procedures.

The Slovak Republic is engaged in the following UNO missions: UNPROFOR, UNATES, UNAVEM II, UNOMIL, UNOMUR, UNAVEM III, MONUA, UNAMSIL, UNMEE, UNTSO, UNDOF, UNGCI, UNFICYP, UNTAET and UNMISET.

7 North Atlantic Treaty Organization—NATO

The North Atlantic Treaty Organisation is a regional military-political intergovernmental organisation. It was established on the 4th of April 1949 with the signing of the North Atlantic (Washington) Treaty in Washington and, to this day, has 28 members. It has its headquarters in Brussels. The Slovak Republic became a member state of NATO on the 29th of March 2004.

NATO's main objectives, according to the North Atlantic Treaty, are the collective protection of the safety and the freedom of their members by political and military means, in accordance with the Charter of the United Nations (NATO 1949). The collective and individual defence of the members of the UNO may be found in Article 51 of the Charter of the United Nations (Charter of the United Nations... 1945).

The North Atlantic Council, Secretary General, Defence Planning Committee, Nuclear Planning Group and Military Committee form the basic organizational structure of NATO. The authorities relevant to conflict resolution are embodied within the structure of NATO crisis management (Otřísal et al. 2011), which includes the North Atlantic Council, Secretary General, Military Committee, Political Committee, Civil Emergency Planning Committee, Group of Policies Coordination, Situation Centre and, possibly, other committees [24]. The process of crisis management is divided into five phases. In the first phase, indication and warning, the potential crisis area is monitored by the Situation Centre, which reports its findings to the North Atlantic Council. The Council may react by sending relevant organs to assess the situation. This is the second phase—assessment of the crisis situation. Based on the results of the assessment, the North Atlantic Council may entrust the authorities of crisis management with the task of drawing up possible variations of crisis response. This is the third phase—development of crisis-response variations. Individual variations are the subject of the Committee's consultations. Their consolidated form is, eventually, discussed by the North Atlantic Council. The output of this process is a preliminary proposal for a directive of NATO, which is handed out to all of the relevant authorities in the fourth phase—planning and implementation. Its content includes a response strategy, goals and a vision of the final outcome. An operational plan is created on its basis. The headquarters and Office of Crisis Management carries out the supervision of the implementation of the operational plan. The fifth phase is a return to stability and it means that the required state has been achieved [11].

Apart from the threats that relate directly to the collective security of NATO members, there are also other relevant threats and risks, to which NATO should be able to respond, in order to maintain international peace. Responses to such threats, however, do not fall under Article 5 of the North Atlantic Treaty, which deals with the collective defence of its members (NATO, 1949). Threats connected with CBRN problems are a typical example, These threats are mainly typical with weapons of mass destruction proliferation [17]. Nowadays, they are very often mentioned in connection with toxic industrial materials leakage (Otřísal 2014, Štěpánek 2012). All these aspects are challenge for a scientific community to cooperate each other and to develop new methods and technologies for protection against their affects [20, 25].

The operations of crisis-management, which are not included in Article 5 of the Treaty, are meant to respond to crises… which may threaten the security and stability of the Alliance's member states and lead to a conflict on NATO's periphery [11].

Peace-supporting operations also belong to the operations outside of Article 5. By means of such operations, NATO wants to contribute to international peace and security, to the strengthening of stability in the world, to the prevention of conflicts and, in the event of a crisis, to its efficient and effective solving, in accordance with international law [8]. Peace-support operations serve to prevent conflicts, create peace, keep peace, enforce peace, build peace and also as a humanitarian aid. Except for the humanitarian aid, the characteristic of these operations may be found in a separate chapter. Humanitarian aid is an operation aimed to eliminate human suffering in places where the state fails to care for its citizens. It may occur in the form of dropping down food and other useful materials and it may occur together with other operations. The main principles of peace-support operations are: impartiality, consensus and limiting the use of force (NATO 2001).

Impartiality is both the behaviour of units and also the perception of both sides of the conflict. A consensus should occur both between the units taking part in the operation and also between the parties involved in the conflict. The success of the operation largely depends on it. In the case of the restriction of the use of force, we can speak about an adequate, but reasonable use of force.

The Slovak Republic has engaged in the following NATO operations: IFOR, Operation Allied Harbour (AFOR), participation in the SFOR and KFOR operations.

8 Organization for Security and Cooperation in Europe—OSCE

The OSCE is a regional security organization which initially functioned as a series of conferences, under the name Conference for Security and Cooperation in Europe. The first conference was held on the 3rd of June 1973 in Helsinki, Finland. The name and status was changed from conference to an organization during a summit in Budapest at the end of 1994. The Secretariat and the Secretary General of OSCE are based in Vienna. Today, OSCE has 56 members. The Slovak Republic became a member of OSCE on the 1st of January 1993.

The Final Act of CSCE is the most important document. It was adopted on the 1st of July 1975 in Helsinki, Finland. The main principles of the activities of the CSCE Member States are sovereign equality and respect for the rights stemming from sovereignty, not using force or threats of force, inviolability of borders, territorial integrity of states, a peaceful settlement of disputes, not interfering within internal affairs, respect for human rights and freedoms, equality and the right of nations to self-determination, the cooperation among states and the fulfilment of commitments in accordance with international law.

The OSCE institutions are divided into political and executive. The political ones are the OSCE Summit, the Ministerial Council, the Permanent Council, the Forum for Security Co-operation and the OSCE Parliamentary Assembly. The executive ones are the OSCE Chairmanship, the acting Chairman, the Secretary-General, the Office for Democratic Institutions and Human Rights, the High Commissioner on National Minorities, the Representative on Freedom of the Media, the Secretariat and the High-Level Planning Group. The OSCE Conflict Prevention Centre is the relevant authority for dealing with conflicts. It is a part of the Secretariat and is managed by the Secretary-General [11].

Conflict resolution within OSCE takes place not only by means of diplomatic measures and negotiations, but also through field operations, the main agenda of which is early warning, conflict prevention and post-conflict rehabilitation of areas. These missions are led by the Permanent Council. Civilian experts from different regions of the OSCE Member States constitute the personnel of OSCE field operations. Cooperation with other international organisations, as well as the ability and willingness to quickly and adequately respond, are an important part of OSCE field operations. Rapid Expert Assistance and Cooperation Team(s) (REACT) have been formed for this purpose. They should be able to intervene before conflicts escalate. Among the tasks that these operations should fulfil are, help, advice and recommendations for the hosting country, monitoring of the implementation of OSCE commitments, observations of elections and assistance with their organization, promoting and maintaining legal order by democratic institutions, peaceful resolution of conflicts by means of preparation of the appropriate conditions for negotiations, verification and support of the implementation of peace agreements, assistance in the post-conflict reconstruction of society in various areas (Charter for European Security).

The Slovak Republic participated in the OSCE Kosovo Verification Missions and the OSCE Mission to Georgia.

9 The Position of European Union—EU

The European Union is an economic and political partnership of democratic European countries, which have voluntarily joined into a political and economic alliance, in order to achieve common objectives, using a single foreign and domestic policy for its sovereign Member States [4]. The Slovak Republic became a Member State of the EU the 1st of May 2004. The European Council, the European Commission, the European Parliament, the Council of the European Union and the European Court of Justice are the main institutions of the EU. Three main pillars formed the basis of the EU during the validity of the Treaty on European Union (the so-called Maastricht Treaty, which entered into force on the 11th of November 1993): 1. the Economic and Monetary Union, 2. the Common Foreign and Security Policy (CFSP), 3. cooperation in the field of justice and internal security. The Lisbon Treaty entered into force on the 1st of December 2009. It amended the Treaty on the European Union and the Treaty Establishing the European Community, which simplified the overall structure of the European Union and cancelled the three aforementioned pillars, though the CFSP retained its specific nature.

The reason why the CFSP was created had been, in particular, the necessity to represent the Union within the sphere of international relations because, even though the EU is not subject to international law, it is still an important institution which aims to defend its interests within these relations [11].

The aim of the CFSP, covering all areas of the EU's foreign and security policy, is:

- to protect the common values, fundamental interests, independence and integrity of the EU, in accordance with the principles of the Charter of the United Nations,
- to strengthen the security of the EU in all regards,
- to preserve peace and strengthen international security, in accordance with the principles of the Charter of the UNO, as well as the principles of the Helsinki

Final Act and the objectives of the Paris Charter, including those at its external borders,
– to promote international cooperation,
– to develop and consolidate democracy and the rule of law, the respect for human rights and fundamental freedoms.[6]

The Member States play a crucial role in the formulation of the CFSP, while the institutions of the EU only enforce the policies agreed upon by the Member States, which do not give up the right to pursue their own independent foreign policies. They do, however, promise to take into account the jointly agreed approaches and actions of the EU in their policies. This is why they are able to speak as a single voice at international conferences or in the institutions of international organisations. Another advantage is the ability to benefit from the joined political, economic and defensive weight of all of the EU Member States during negotiations [4].

The CFSP covers all of the questions concerning the safety of the EU, including the progressive definition of a common defence policy, which may lead to a common European defence, if decided by the European Council. In that case, it shall recommend that the Member States adopt a resolution, in accordance with their respective constitutional requirements. The EU's policy, in accordance with this article, does not affect the specific characteristics of the security and defence policy of certain Member States and it respects the obligations of certain Member States, which share a common defence within NATO and which are in accordance with the CFSP (Treaty on EU, Article 17). The questions referred to in this Article of the Treaty include humanitarian and rescue roles and missions aimed to keep the peace and tasks of the combat forces when dealing with crisis situations, including the establishment of peace [11].

The Common Security and Defence Policy of the EU (hereinafter referred to as "CSDP") is an integral part of the CFSP and, together with other instruments, forms a part of the EU's external relations. It is a means of support of the CFSP, which grants the EU the tools and capacity needed to carry out crisis-management operations (hereinafter referred to as "OKM EU") outside the territory of the EU.

Through the CSDP, the EU has the ambition to strengthen its ability to respond to world crises without geographical limitations and, thus, fulfil the key requirement defined in the European Security Strategy—to strengthen the EU's role in ensuring global security in accordance with its potential.

Through the CSDP, the EU is becoming an important tool in ensuring global security and stability, together with NATO. According to the statement of the European Council of December 2008, over the past ten years, the EU has established itself as a global political entity. It has adopted an increasing responsibility, as proven by its

---

[6] Title V, Article 11, Treaty on European Union, Official Journal of the EU, C321 E/13 of 29th December 2006. Available at: http://eur-lex.europa.eu/LexUriServ/LexUriServ.do?uri=OJC:2006: 321E:0001:0331:SK:PDF.

increasingly ambitious and diverse civilian and military operations in the name of effective multilateralism and peace.[7]

The main purpose of the CSDP is to provide military and other (police and civilian) means of preventing and resolving international conflicts and managing crises. During the last few years, the CSDP has seen a shift. This was reflected through the number of military and civilian operations and missions [5].

10   Methodology of Conflict Database Creation

The definition and categorization of conflict is associated with certain methodological problems and, therefore, is constantly the subject of discussions among the professional public. During its research, it is necessary to clearly and precisely determine which conflict may be defined as a crisis, armed conflict or war, and which may not. The differences, which arise among individual categorizations, are caused by specific approaches, different use of definitions, analytical methods and criteria, as well as subjective views of experts. Conflicts are categorized and assigned to specific datasets based on specific criteria.

The political status of the parties involved in a conflict is one of the fundamental criteria of conflict (particularly armed) and war division and typology. It leads to different international-political, geopolitical and legal consequences, as well as different levels of interest of politicians, experts or the general public. All of the following cause different reactions of the (international) public: aggression of one state towards another; international intervention under the UNO or without it; domestic national conflicts (civil wars); conflicts within a certain community [24].

Several projects deal with the division of conflicts. In our work we name a few of them. The most commonly used conflict database is a project created by two institutions, which work in the field of security studies and research conflicts—Peace Research Institute Oslo (PRIO) and the Department of Peace and Conflict Research of Uppsala University with its Uppsala Conflict Data Programme (UCDP). This research studies armed and violent conflicts and uses a quantitative approach. The project specifies factors, which are characteristic for this type of conflict.

Among them are: (a) the government or the territory, which are a source of incompatibilities, (b) at least two parties must participate in such conflicts, (c) one of the parties is represented by state power, the other by opposing organization, (d) the use of armed forces has caused at least 25 combat victims per year; a conflict with at least 1,000 combat victims per year is considered a war [24].

---

[7] Declaration of the European Council on the strengthening of the European security and defence policy. The conclusions of the meeting of the Bureau in Brussels on the 11th and 12th of December 2008. Available at: http://www.rokovania.sk/appl/material.nsf/0/EA3B35FD1F564915C1257544 003DE252/FILE/priloha_1.rtf.

According to this methodology, armed conflicts are divided into three basic groups, according to the intensity of violence, which means according to the number of victims:

- a minor armed conflict is a conflict with at least 1,000 victims overall and, at the same time, at least 25 per year,
- an intermediate armed conflict is a conflict with over 1,000 victims overall,
- a war or armed conflict is a conflict with over 1,000 victims per year during the entire duration of the conflict.

The University of Uppsala in Sweden uses the following division of conflicts: (a) conflicts with state participation, in which the opposing parties are either two states, or a state and a non-state party (rebels, mutineers, rioters, etc.), (b) conflicts without state participation, in which the parties involved are non-state groups, (c) one-sided violence, committed by the government or political groups, against civilians. Genocide, politicide and other violent attacks belong to this category [3]. The dividing of conflicts according to their political status has spread thanks to the aforementioned projects of the PRIO/UCDP.[8] This dividing is easier and it divides conflicts into: (a) inter-state (international), (b) domestic (national)—these can be divided into civil wars and others, and (c) armed separatism, which means conflicts that occur during the establishing of states.

Certain problems arise with this simplified definition.

E.g. according to this division, there were no conflicts in the world during 2004 and 2005 because not even the conflicts in Afghanistan or Iraq fall under the above mentioned categorization.

The HIIK (COSIMO)[9] database is another project, which deals with the classification of political conflicts from the year 1945 until present. It currently works with over 500 conflicts, which are then distributed according to the phases which they happen to be in at the moment. It employs the qualitative approach of categorizing conflicts, which means that it doesn't use the criterion of the amount of victims, but divides conflicts based on the significance or intensity of violence. This division has a five-level scale, which is then divided into two subcategories. The first category is an unarmed conflict (non-violent conflict), in which there is the possibility of using natural violent means, but they are not directly used. Non-violent conflicts may be (a) latent conflicts, (b) manifested conflicts. The second category is an armed conflict (violent), which is divided into (a) a crisis, (b) a serious crisis and (c) a war. During their course armed conflicts occasionally, periodically or systematically and in an organized manner, use arms and other violent means [24].

---

[8] The UCDP program, for example, indicates that in 2009, there were 29 national and 7 international conflicts. Appendix 2 compares the development of conflicts, according to the way the UCDP divides them, in 2009 and 1999.

[9] According to the HIIK method, there are currently 365 active conflicts in the world, 108 of which are latent, 114 are manifested, 112 are crises, 24 are serious crises and 7 are wars. Localization according to region: 113 conflicts take place in Asia and Oceania, 66 in Europe, 85 in Sub-Saharan Africa, 46 in America and 55 in the region of Maghreb and the Middle East.

A latent conflict is defined by the existence of a conflict, dispute and disagreement about certain values, among the parties involved. One of the parties makes certain demands, which the other party radically rejects. It is possible to predict the behaviour of the parties in conflict and an escalation of the conflicts may not happen at all, if the parties are willing to negotiate concerning the issues causing the conflict and find a solution that would be accepted by both. A manifested conflict is characterized by an escalation of the intensity of the conflict and by an increasing hostility between the parties involved. The mutual relations between the parties are based on the use of threats, verbal coercion of the opponent or the implementation of various sanctions and restrictions (diplomatic, political and economic). In such cases, the likelihood of an outbreak of physical violence increases.

A crisis within international relations, as a specific category, is characterized by the fact that there is a strong tension between the parties involved and they mobilize their armed forces. Occasionally, there may be some less serious armed military clashes, or the use of armed violence. During a crisis, in the case of domestic conflicts, revolts and coups orchestrated by the citizens or social groups may break out in the country [2].

A serious crisis is a type of conflict characterized by a reoccurring and organized use of armed violence by the parties involved in a dispute. There may also be serious armed clashes. In this case, extensive and numerous terrorist attacks and guerilla fights are typical.

A war is characterized by a systematic and purposeful use of armed violence by armed military components of national power or other parties or groups. Such violence is used on a large scale and with a great destructive effect. The consequences of war may result in the breakup of society, the destruction of the socio-economic system of the country, the disintegration and change of the government, annexation of the territory, taking over the territory, resources, etc.

The HIIK (COSIMO) database allows for a detailed description of the development of specific conflicts in their specific stages (violent or non-violent). A systematic documentation of individual conflicts, detailed information about the parties involved, etc. is based on this. This information is the basis for follow-up measures and suggestions of conflict resolutions, the means of intervention in specific disputes (Das HIIK erfasst... 2011).

The University of Michigan and their project Correlates of War (COW)[10] systematically deals with databases and conflict distribution. The project is dedicated, in particular, to armed conflicts. The main subject of its interest is states or inter-state conflicts, which is why the project focuses on understanding the root causes of armed conflicts, based on national interests and geopolitical rivalry. The basic criteria for the definition of war are—the participation and status of organised forces and the number of casualties. In the framework of the COW project, we may distinguish three types of war (The COW Typology... 2010):

- interstate war—at least 1,000 casualties from the ranks of armed forces is a criterion,

---

[10] According to the COW database, there were 231 armed conflicts in the period from 1964 to 2005, which were then divided into three basic types: domestic, international and internationalized.

- extra-system war—1,0000 casualties a year from among soldiers on the side of a member of the international system (usually anti-colonial wars or wars for the freedom of countries),
- civil war—1,000 victims of the fighting.
- It can be assumed that surveying casualties (usually soldiers) is easier and more efficient than the demographic methods used to estimate indirect victims (mostly civilians) [24].

Civil wars present a specific category of armed conflicts.

The COW project defines the criteria which must be met, in order for a war to be characterized as a civil war. (1) organised military action, (2) at least 1,000 victims per year, (3) participation of state (government) armed forces, (4) effective resistance from the opposing forces (the proportion of casualties on the side of the stronger party, usually the government's armed forces, must be at least 5% of the total number of victims).

The COW project also deals with the examination of the process of the transformation of conflicts. According to the project, it is impossible to accurately and precisely determine conflicts, because every conflict is specific, has certain characteristics, and its progression may be different from others.

This, in particular, may be the cause of the metamorphosis or transformation of a conflict from one stage to other stages of the phase process.

As an example, we may mention the change of domestic conflicts—civil wars to interstate conflicts, if another state or security organization chooses to intervene [24].

11 Conclusion

Approaches to the definition, analysis and typology of conflicts may differ. There are quite a large number of definitions of a conflict and they all are significantly different. These differences depend on the point of view through which we look at particular conflicts. Generally, we may define a conflict as a dispute among the parties involved in the conflict, which may be individuals or groups. In such a dispute, the parties involved attempt to enforce their objectives, interests and requirements and use certain (violent or non-violent) means to fulfil these objectives.

As is the case with defining conflicts, conflict analysis also has several approaches that depend on specific factors and characteristics of specific conflicts. Every conflict is a specific phenomenon and may have different attributes and properties, which is why the analyses may differ significantly. According to the elements of the analysis, categorisations and conflict databases are created. They either research individual specific properties or compare several properties. Currently, several institutions deal with conflict databases. We have already mentioned them in the previous chapter. Their methodology and findings form an exceptional concept of the use of a multi-disciplinary synergy of natural and social sciences, and their application not only into the theory, but also into the practical aspects of decision-making within the field of international relations. Major differences in the various databases occur when comparing individual studies, because of many different approaches and examined properties. It is, therefore, necessary to take into account the individual aspects, on the basis of which we define, explore or categorize conflicts.

It is necessary to include the fact that conflicts are a multidimensional social phenomenon, which is quite complicated and volatile. Because of this, individual definitions and divisions are very different and must be assessed individually. The result is a simpler overview, which is not always completely accurate and relevant.

Correct research, definition and typology are the key for other issues of conflict typology, such as the causes, resolution and prevention of conflicts. We will deal with these concepts in the following sections of our work.

References

1. Fabián K, Rýsová L, Dobrík M (2019) Urban disasters crisis management scenario design and crisis management simulation; Borseková K, Nijkamp P Resilience and urban disasters: surviving cities. In: Borseková K, Nijkamp P (eds) New horizons in regional science. Edward Elgar Publishing, Cheltenham, pp 199–231. ISBN 978-1-78897-009-9
2. Hofreiter L (2008) Conflict theory and resolution. Liptovský Mikuláš: Academy of Army Forces gen. M. R. Štefánika Liptovský Mikuláš, 206 pp.
3. Human Security Centre (2005) The human security report 2005. War and peace in the 21st century. Human Security Centre, Oxford
4. Ivančík R, Jurčák V (2013a) Peace operations of selected international crisis management organizations. Liptovský Mikuláš: Academy of the Armed Forces gen. M. R. Štefánika, 230 pp.
5. Ivančík R, Jurčák V (2013b) Peace operations of International Crisis Management. Ostrowiec Św. Wyższa Szkoła Biznesu i Przedsiębiorczości, 180 pp.
6. Ivančík R, Nečas P (2012) International security from the view of postmodern conflicts on African continent. Publishing house Amelia, Rzeszów, 168 pp.
7. Ivančík R (2012a) UN peacekeeping operations—an international crisis management tool to ensure international security and peace in the world. In Almanac—current issues of world economy and politics, vol 7, no 3, pp 120–137. http://fmv.euba.sk/files/Almanach_3_2012_FMV.pdf
8. Ivančík R (2012b) UN peacekeeping operations—an international crisis management tool to ensure international security and peace in the world. In: Almanac—current issues of world economy and politics, vol 7, no 3. http://fsi.uniza.sk/kkm/files/admincasopis/KM%202%202 012/08%20Ivancik.pdf
9. Ivančík R (2013a) Vojenské aspekty asymetrie v medzinárodnej bezpečnosti. In: Political sciences, vol 16, n 3, pp 6–37. http://www.politickevedy.fpvmv.umb.sk/userfiles/file/3_2013/IVANCIK.pdf
10. Ivančík R (2013b) Crisis management operations—the European Union's contribution to international security. In: Košická bezpečnostná revue, vol 3, no 1, pp 32–41. http://www.vsbm.sk/data/revue/revue-1-13.pdf
11. Jurčák V a kol (2009) International crisis management organizations. Liptovský Mikuláš: Academy of the Armed Forces of General M. R. Štefánik, 235 pp.
12. Kačala J et al (1997) Krátky slovník slovenského jazyka. Bratislava: Veda, Vydavateľstvo Slovanskej akadémie vied, 943 pp.
13. Krejčí O (2007) International politics. Ekopress, Praha, 743 pp.

14. Kudlák A et al (2020) Determination of the financial minimum in a municipal budget to27.deal with crisis situations. Soft Comput 24(12):8607–8618
15. Kusá D (2005) Conflict resolution. Mirius, Bratislava, 29 pp. http://www.equalslovakia.sk/fil eadmin/user_upload/projekty/27_1.2_Riesenie%20konfliktov%20I.pdf
16. Mischnick R (2007) Nonviolent conflict transformation. Train Train Manual. Don Bosco, Bratislava, 164 pp. http://www.trainingoftrainers.org/img/manual_sk.pdf
17. Otrisal P et al (2018) Preparation of filtration sorptive materials from nanofibers, bicofibers, and textile adsorbents without binders employment. Nanomaterials 8(8):564. https://doi.org/10.3390/nano8080564
18. Otřísal P, Florus S (2014) Present and perspectives of physical and collective protection against the effects of toxic substances. Chem Lett 108(12):1168–1171
19. Potůček R (2020) Life cycle of the crisis situation threat and its various models. Stud Syst, Decis Control 208:443–461. https://doi.org/10.1007/978-3-030-18593-0_32
20. Sarakhman O et al (2019) Voltammetric protocol for reliable determination of a platelet aggregation inhibitor dipyridamole on a bare miniaturized boron-doped diamond electrochemical sensor. J Electrochem Soc 166(4):B219–B226. https://doi.org/10.1149/2.0381904jes
21. Stodola P et al (2013) The real-time control algorithm and control curves for servomotors. Int J Circ, Syst Sig Proc 7(2):118–125
22. Stodola P (2015) Tactical models based on a multi-depot vehicle routing problem using the ant colony optimization algorithm. Int J Math Models Methods Appl Sci 2015(9):330–337
23. Štěpánek B, Otřísal P (2012) The development and establishment process of centres of excellence in North Atlantic organization. Croat J Educ-Hrvatski Casopis za Odgoj i obrazovanje 14(1):169–174
24. Tomeš J et al (2007) Conflict of worlds, world of conflicts. Prague: Nakladateľstvo P3K, 350 pp.; Tušer I et al (2020) Evaluation criteria of preparedness for emergency events within the emergency medical services. Stud Syst, Decis Control 208:463–472. https://doi.org/10.1007/978-3-030-18593-0_33
25. Prikryl R et al (2018) Protective properties of a microstructure composed of barrier nanostructured organics and SiO_x layers deposited on a polymer matrix. Nanomaterials 8(9):679. https://doi.org/10.3390/nano8090679

Building Community Resilience in the Czech Legislation

Eliška Polcarová and Jana Pupíková

Abstract This chapter analyses the Czech legislation, conceptual documents, and methodological manuals dealing with the issue of safety and protection of the population. Ensuring the protection of the community in the conditions of the Czech Republic is mainly related to dealing with emergencies and crisis situations. The chapter, therefore, evaluates the legal enshrine of this issue using the qualitative method SWOT analysis. The chapter focuses explicitly on the prevention and preparedness of the population in the Czech legislation concerning building community resilience and proposing criteria for assessing community preparedness in these situations.

Keywords Community · Resilience · Community preparedness · Security environment · Czech Republic

1 Introduction

Based on the ever-increasing disasters not only in Europe but also in the Czech Republic, it is necessary to deal with the safety of the community. Safety is mostly perceived subjectively. Since the beginning of humanity, the community has been threatened by disease, wildlife, and natural disasters [13, 14, 15]. Over time, threats caused by one's actions have also emerged, such as the struggle for tribal territories, terrorism, or industrial accidents. Throughout the historical development of the community, not only knowledge has changed, but also the requirements for safety and everything that can endanger it. The concept of resilience is one of the tools to increase the safety of this community.

E. Polcarová (✉)
Department of Population Protection, Faculty of Safety Engineering, VŠB – Technical University of Ostrava, Ostrava, Czech Republic
e-mail: eliska.kristlova@vsb.cz

J. Pupíková
Department of Security and Law, Ambis College, Prague 8, Czech Republic
e-mail: jana.pupikova@ambis.cz

© The Author(s), under exclusive license to Springer Nature Switzerland AG 2022
I. Tušer and Š. Hošková-Mayerová (eds.), *Trends and Future Directions in Security and Emergency Management*, Lecture Notes in Networks and Systems 257,
https://doi.org/10.1007/978-3-030-88907-4_6

According to the Sendai Framework for Disaster Risk Reduction 2015–2030 [26], the state is responsible for the safety of its citizens. The essential function of each state should be to ensure the protection and development of protected interests, mainly to protect human lives and health, property, safety, the environment, critical infrastructure, and the existence of the state itself. Furthermore, this framework points to reducing the vulnerability of the community and strengthening its resilience at all stages of risk management (not only response and recovery, but also prevention, mitigation, preparedness, and remedial action). Within the Czech Republic, no suitable strategic approach has yet been developed to assess community resilience and define "vulnerabilities" that may have a negative impact on community functioning and to strengthen the prevention and preparedness of the territory and community for future disasters.

The presented chapter tries to examine the key area of development of community building against security risks in selected Czech legislation focused on crisis management and protection of the population in the conditions of the Czech Republic. Based on the information obtained, a definition of community and community resilience will be developed, which have not yet been implemented in Czech legislation. The second part focuses on the identification of critical sectors that can be the basis for future strengthening of community resilience in the Czech Republic.

2 Methodology

This chapter aims to analyse selected Czech literature (legislation, conceptual documents, and methodological manuals) to assess the development of the view of the population's safety and preparedness for future disasters and to identify shortcomings in the area of community resilience building. Ensuring the protection of the community in the Czech Republic is primarily related to dealing with emergencies and crises. The perception of community safety and protection varies based on threats (external and internal) that may threaten protected interests.

2.1 Development of Security Environment in the Czech Republic (1992–2015)

Crisis legislation began to be prepared after the separation of the Czechoslovak Republic. The fundamental pillar was the Constitution of the Czech Republic (Constitutional Act No. 1/1993 Coll.), which laid down the primary conditions of national security. The Act on the Security of the Czech Republic, despite meeting the foundations of a democratic state, was not passed. The change occurred in 1997 when extensive floods hit the Czech Republic, and the Act on the Security of the Czech Republic (Constitutional Act No. 110/1998 Coll.) was passed without delay. Following the adoption of the Act, the process of creating further legislation related to the safety and protection of the population was initiated.

The Security Strategy of the Czech Republic is a necessary conceptual document of the state security policy. The strategy is based on the Constitution of the Czech Republic (Constitutional Act No. 1/1993 Coll.), the Charter of Fundamental Rights and Freedoms (Constitutional act No. 2/1993 Coll.), and Act No. 110/1998 Coll. On national security (Constitutional Act No. 110/1998 Coll.). Each of the Security Strategies [17–21] expresses, in particular, the current state of the security environment at the time the individual strategy was developed.

In the Security Strategies, threats and risks are defined as phenomena that directly or indirectly adversely affect society, the state function, or citizens of the Czech Republic. A security risk is an event that occurs with some probability and is considered undesirable from a security point of view. Security risks are defined in the areas of civilization and social, political-military, economic, environmental, crime, and organized crime. On the other hand, the external or internal security threat is capable of harming the interests of Czech citizens; they may be of a natural or human nature. Security risks can, in some cases, grow into a security threat.

The Security Strategy distinguishes security interests according to their degree of importance into three categories: vital, strategic, and other significant. The vital interest of the Czech Republic is to ensure the existence of the Czech Republic, its sovereignty, territorial integrity, and political independence; furthermore, the defence of democracy and the rule of law and the protection of the fundamental human rights and freedoms of the population are vital interests. To protect the vital interests, this is necessary to preserve strategic interests, which also serve to ensure the social development and prosperity of the Czech Republic.

Since the establishment of the Czech Republic, the view of security has changed significantly. The Security Strategy 1999 is the first document approved by the Government of the Czech Republic on security at the strategic level. Given the general content of the safety issue, this strategy is instead viewed as a methodological aid. Security risks and interests are defined based on concerns about external military assaults, and internal risks focus more on crime. While the strategy focuses on the external and internal security of the environment, it does not offer any concrete measures. The mention of crisis management and ensuring security in non-military crisis situations is probably a response to the 1997 flood.

Security environment and interests in the Security Strategy 2001 was defined based on the Czech Republic's accession to NATO and the upcoming accession to the European Union (EU). In the area of the security environment, fear of aggression and military assault by another state is retreating. Still, it is necessary to respect the contractual obligation of the North Atlantic treaty of collective defence. The threat to democracy is also no longer defined as one of the security risks. There are indications of possible measures, notably the need for international cooperation and the development of humanitarian aid in regions where there are regional conflicts and low living standards (the Middle East, North Africa, Central Asia, and the Caucasus). This situation may be due to increased migration. The last paragraph of the document mentions that citizens' active involvement is necessary to ensure national security.

On the contrary, the Security Strategy 2003 defines new trends, threats, and the resulting risks, which form the environment in which the Czech Republic protects and promotes vital, strategic, and other vital interests (response to the 2001 New York terrorist attack). This strategy calls for the development of new, better, and more efficient analyses to better assess threats. A direct military attack is no longer foreseen, but a new risk of attack is emerging attacks on electronic, communication, and information networks.

The Security Strategy 2011 has been updated following the signing of the Lisbon Treaty [24] in 2009 and the adoption of the new NATO Strategic Concept [22]. To ensure the security of the Czech Republic, it is not only the active cooperation of citizens but also the public authorities that emphasize it. The risk of direct threat is kept to a minimum, while the fear of asymmetric threats is increasing. The process of globalization is negatively viewed, as the interconnection of financial markets, information, communication technologies and infrastructure can lead to cyber-attacks and misuse of communication and intelligence channels to influence society (a newly defined security risk). Demographic change, climate change and their impact on the environment, availability of natural resources and strategic raw materials, natural disasters, and not only these phenomena can cause local armed conflicts and uncontrolled mass migration, which can significantly undermine the security of the Czech Republic. Newly defined risks are threats to critical infrastructure functionality, and disruptions of strategic raw materials or energy supplies. The method of prevention and measures against individual security risks are specified.

The Security Strategy 2015 has been updated based on the ever-increasing severity of non-military threats. There is a constant focus on collective defence within the North Atlantic Alliance and interest in stability within the EU. Furthermore, we need to start adapting to a rapidly changing security situation, a proactive approach, and early detection of threats. Again, there is mention of the Ukrainian crisis [18]. Emphasis is placed on cybersecurity and the provision of communication and information networks, which are becoming critically endangered by technological growth. There is a shift in the perception of crime, especially in the fight against tax evasion, organized crime, the fight against corruption, and the narcotics trade. Figure 1 presents the development and comparison of security threats and risks in individual security strategies of the Czech Republic.

2.2 Legal Security to Prepare the Population in the Event of a Disaster

The fundamental laws in the area of population protection include Act No. 239/2000 Coll., On the Integrated Rescue System and on Amendments to Certain Acts (the IRS Act) and Act No. 240/2000 Coll., On Crisis Management and Amendments to Certain Acts (the Crisis Act). These laws imply that an individual natural person residing in the Czech Republic has the right to information on (crisis) measures and to ensure the

THREATS AND RISKS	SS 1999	SS 2001	SS 2003	SS 2011	SS 2015
Natural disasters, industrial and environmental accidents, the emergence and spread of epidemics	x	x	x	x	x
Terrorist attacks and organized activities of an international nature	x	x	x	x	x
Large-scale migration waves	x	x	x	x	x
Proliferation and use of weapons of mass destruction	-	x	x	x	x
Economic instability	x	x	x	-	-
Cybercrime (cyber-attacks), misuse of scientific research	-	x	-	x	x
Corruption	-	-	x	x	x
Developments and instability in SE Europe, the Middle East, Russia, and North Africa	-	x	-	-	x
Instability and regional conflicts in and around the Euro-Atlantic area	-	-	-	x	x
Endangering the functionality of critical infrastructure	-	-	-	x	x
Disruption of supply of strategic raw materials or energy	-	-	-	x	x
Threat to the fundamental values of democracy and freedom of citizens in other countries	x	-	-	-	-
Violent actions by foreign entities (state and non-state)	x	-	-	-	-
Extensive and serious diverse activities	x	-	-	-	-
Threat of military attack	x	-	-	-	-
Leak of classified information	-	x	-	-	-
Economic, social, and demographic imbalances in the world	-	x	-	-	-
Global warming	-	-	x	-	-
Competition for resources (water or oil)	-	-	x	-	-
Crime and socio-pathological phenomena in socially excluded localities	-	-	-	-	x

x = is included in the Security Strategies;
− = is not included in the Security Strategies.

Fig. 1 The security threats and risks in the security strategies of the Czech Republic

protection of the population and their health, life, and property. Moreover, providing instruction and training on emergency activities, including preventive educational and promotional activities, while having the duty and responsibility for its protection.

Crisis legislation imposes on the authority's crisis management specific tasks in the area of measures and population protection. The central administrative body for the protection of the population is the Ministry of the Interior. The other bodies are responsible for ensuring that citizens are aware of the threat, the planned measures, and emergency response procedures.

The Concept of population protection [7–9] is a non-legislative document dealing with the issue of population protection. The concepts describe the structure of the population protection system, a detailed description of individual tasks, and the deadlines for completion. The basis for these strategic plans are the Security Strategies of the Czech Republic, and the security threats and interests of the Czech Republic identified therein. The concepts aim to analyse and develop visions and tasks in the field of population protection and crisis management and to ensure the implementation and realization of individual tasks in practice.

During the creation of all concepts, an understanding of the concept of protection of the population develops. All defined strategic goals are aimed at ensuring the essential functions of the state, primarily to ensure the safety of the population and the protection of their lives, health and property. The last concept focuses on the broader involvement of citizens in the system of protection of the population. It is about increasing their self-protection by increasing the information and knowledge gained within the system of education, and by preparing for a disaster, and thus increasing the resilience of the whole community. However, vulnerable members of the community must not be forgotten.

Decree of the Ministry of the Interior No. 380/2002 Coll., On the preparation and implementation of tasks of population protection, addresses the procedure for establishing civil protection facilities, their personnel composition and material resources and the training of their personnel. It also determines the way of informing people about the nature of the threat, measures, and implementations. It deals with the technical, operational, and organizational security of the system of warning and notification and the provision of emergency information. From the historical point of view, the principles for the procedure of providing shelters and collective and individual protection of the population are solved.

Methodological guides advise how residents should behave in selected emergencies. The analysed manuals are freely available on the Internet portal of the Fire Rescue Service of the Czech Republic (https://www.hzscr.cz/clanek/prirucky.aspx) and are intended for the general public, including crisis management bodies.

The first methodical tool Self-protection by Hiding the Population was published in 2001. The handbook is primarily intended for state administration authorities, local self-government, legal entities, and natural persons doing business. The manual aims to provide adequate information to protect the population in the event of an emergency. The methodology was developed in response to security risks arising from the Security Strategies of 1999 and 2001, mainly fear of a military attack and thus preparing the population for an external attack. The proposed measures focus on the use of suitable underground and above-ground parts of residential buildings of houses, operating, manufacturing, and other buildings for their adaptation for improvised shelters. The preparatory work (selection of suitable premises, the capacity of shelters, the proposal of the technical solution of adjustments) are carried out in peacetime. In the case of military attacks, self-help is needed in building improvised shelters. The methodology guides on the use of improvised respiratory and body surface protection devices.

In 2003, two other handbooks were published with necessary information on the protection of the population against emergencies. The first handbook, In case of emergency—the handbook for residents, contains basic instructions and recommendations on how to behave in the event of an emergency. The methodology includes an emergency number list, siren instructions, evacuation policy, and evacuation baggage contents. The second part of the handbook proposes the necessary precautions before, during, and after the flood, in a chemical and nuclear accident with a leak of a dangerous/radioactive substance. General instructions are also described for the use of biological and chemical weapons, anonymous announcements (placing bombs, explosives, use of hazardous materials, etc.), receiving a suspicious shipment (letter, package). Last but not least, instructions for improvised respiratory protection and body surface protection are also given.

The second handbook, Protecting People in Emergency, is intended for primary and secondary school teachers. The methodology serves to integrate the issue of emergency protection into educational programs at schools, especially to prepare pupils for the impact of possible consequences of events caused by natural forces or human activities. The manual uses the terminology corresponding to current laws. It is a recommendation document (the school head decides on the content, manner, time scope and classification of topics).

The following two methodologies are focused on the release of hazardous chemical and radioactive substances in emergency planning zones. The behaviour of the population in the event of an accident involving the release of dangerous substances (2004) is intended primarily for state administration authorities, local authorities, legal entities, natural persons doing business, and the population. The entities mentioned above can find information on the properties of hazardous chemicals, the principles of providing first aid in the event of exposure to these substances, and the principles of the population's behaviour in the event of a leak (hiding, improvised means, individual protection, evacuation procedures).

On the other hand, the Handbook for the Protection of the Population in case of the Temelin Nuclear Accident for the Period 2020–2021 and the Handbook for the Protection of the Population in the case of the Radiation Accident of Dukovany for the period 2020–2021 are directly addressed to people living or working in emergency planning zones. In the beginning, information about nuclear power plants is presented, as well as how to ensure protection against ionizing radiation, principles of behaviour in case of a radiation accident, and an overview of reception centres.

In 2015, another handbook We Live in the Flood Area was published, aimed at preparing and managing floods with the least possible loss. Population protection is dealt with in three phases:

1. preparation (forecast, individual preparation, evacuation luggage),
2. recovery (sanitation and cleaning work, liquidation of damage, errors, psychological assistance),
3. prevention (non-structural, structural, and individual flood-control measures, increasing the price of water in the landscape and the city).

2.3 Partial Conclusion

From the analysis of documents dealing with the issue of population protection, it can be stated that the fulfilment of the tasks of population protection is adequately enshrined in Czech legislation. The Czech Republic seeks to respond to the current situation in the world in the security environment and to implement it into existing legislation. Individual objectives and tasks of population protection are based on the analysis of security risks identified in individual security strategies.

On the other hand, the issue of preparing the population for self-protection and mutual assistance (content, scope, skills, and knowledge) is not yet legally addressed. As a result (although the community has a duty and responsibility for their protection and is provided with information on emergency preparation in various forms), there is a lack of effort on the part of the population to further educate themselves in the field of prevention and preparedness. The community relies on assistance from the state to which it is legally entitled. On the website (Fire Rescue Service of the Czech Republic, websites of municipalities and cities), it is possible to find various manuals, methodological aids, as well as video demonstrations of how to prepare and respond to an emergency event.

The evaluation of the legal basis of this issue and the possibility of developing safety and building resilience is demonstrated using the qualitative method of SWOT analysis in Fig. 2. The results of the analysis are also based on the outputs of the Concept of Population Protection 2020–2030.

INTERNAL ORIGIN	
Strengths	**Weaknesses**
• sufficient anchoring of the issue of protection of the population • a clear division of roles and competencies of individual bodies in ensuring the safety and protection of the population by existing legislation • expertise of members and employees • systematization of preventive educational activities • opportunities to improve education	• the preparation of the population for self-defence and mutual assistance (content, scope, skills, and knowledge) is not legally addressed • current strategic security documents do not take into account all national and global security threats (e.g. climate change and its global impact) • there is no strategic approach to assessing community resilience and defining "vulnerabilities" that may have a negative impact on the functioning of the community • insufficient efforts on the part of the population to learn about prevention and preparedness • lack of funds and staffing for building and developing community resilience
EXTERNAL ORIGIN	
Opportunities	**Threats**
• use of foreign experience in the field of community resilience and their implementation in the Czech legal environment • increasing the importance and needs of development in security and community resilience in the context of global security threats • possibilities of multi-source financing, use of funds from transnational projects • involving the private sector and citizens in building resilience • science and research in the field	• the emergence and development of new natural and anthropogenic threats for which society is not prepared • adverse changes in the security environment mainly due to non-state participants • budget cuts for community prevention and preparedness participants

Fig. 2 SWOT analysis

3   Resilience and Related Terms

Nowadays, in the scientific sphere, the issue of community resilience is a continually evolving concept. The definition of the term is based on the actual use in the field of research. The Czech Republic is also trying to implement elements of resilience into its strategies aimed at ensuring the safety of the population. The purpose of the following chapter is to get acquainted with the concept of resilience and to introduce definitions of terms adapted to the conditions of the Czech Republic.

3.1   Definition of Resilience

According to preserved records [3], the term resilience was first used in ancient Rome in the field of law and literature. At the beginning of the nineteenth century, resilience began to be used in the field of mechanics, where it characterized the behaviour of a pring. The term became more widely known in the middle of the nineteenth century, where it was first used in ecology to describe an ecosystem that tries to function in the same way even after a catastrophe. The term was then integrated into the field of psychology, where an effort was made to clarify how one should adapt to changing environmental conditions. Today, the concept of resilience is most widely used in conjunction with disaster risk reduction and climate change. Figure 3 presents the development of the concept in individual areas from literature/law (antiquity) to disaster management (present).

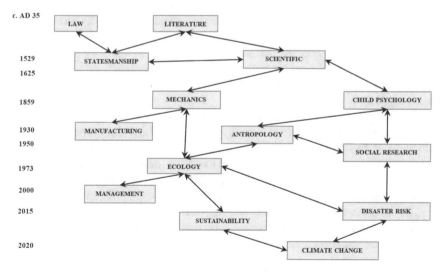

Fig. 3 Development of the concept of resilience from antiquity to the present [3]

In Czech literature, it is possible to meet the concept of resilience most often in the field of psychology [5, 16, 23], where this expression is most associated with a person's ability to cope with adversity and adapt to new situations. It is the ability to survive, not to break our spirit and not to lose the will to continue living.

3.2 Definition of Community

The term community comes from the Latin word "communitas" and can be translated as "human community, society, or community," but can also mean "kindness, sympathy, and companionship" that is associated with giving and receiving service. The Czech Republic does not define the term community in its legislation [6].

Within the solution of the given issue, the community is defined as (from the perspective of crisis management) is defined as a group of people living (temporarily) in a limited area (association, church, business, municipality, the municipality with extended powers, region, state) connected by similar social ties and related interests (religion, culture, customs), and resources. At the same time, members of the community are exposed to the same security risks.

3.3 Community Resilience and Vulnerability

Community resilience is taken as the ability of the community (with an acceptable level of vulnerability) to adequately react with available tools, resources, and information to the security of the risk, thereby managing and mitigating the consequences of these risks. The different skills and resources of community members can be used to effectively respond, prepare, and address the impacts of large-scale security risks. Practical experience of working with disaster-affected communities internationally is presented, for instance, in the book Community Engagement in Post-Disaster Recovery [15].

Community resilience is closely related to vulnerability and plays an essential role in disaster management (how quickly it can respond and recover from a negative situation). Vulnerability can be understood as a set of economic, social, and political conditions that differently affect individual members (or even the entire community) to respond to and recover from security risk. However, a high level of vulnerability does not necessarily mean that the community is not resilient. It only points to the inability of the community to resist and respond to disasters. The acceptable level of community vulnerability is such that the community can function on its own, without the assistance of the state, when a security risk arises.

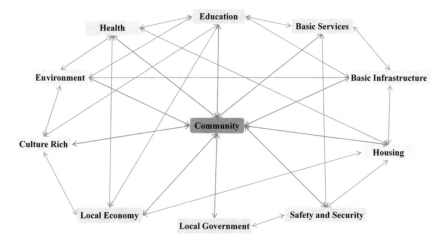

Fig. 4 Vulnerable sectors

4 Proposed Criteria for Evaluating Community Preparedness

This chapter aims to identify vulnerable sectors in the field of prevention and preparedness that will help prepare and strengthen the community in the Czech Republic for future disasters. The proposed vulnerable sectors/criteria are modified for the conditions of the Czech Republic as defined by The Healthy Community [4] and the Resilience Wheel according to the methodology How To Make Cities More Resilient [25]. The individual sectors (Fig. 4) have their specific criteria (Fig. 5). Nevertheless, integration across these sectors can help build the resilience of the community, with the community directly dependent on each proposed sector. The proposed sectors aim to identify vulnerabilities in prevention and preparedness and to strengthen community building through the proposed measures and procedures. An integrated security strategy planning will not only increase community resilience but will also help reduce the cost of rebuilding individual sectors after a disaster.

5 Conclusion

Since the establishment of the Czech Republic, citizens have been constantly threatened by various disasters. To better prepare and respond to the identified dangers, the government has proposed legislation aimed at the safety and protection of the population. The primary purpose of the article was to examine and evaluate this legislative development using SWOT analysis. The analysed legislation aims to ensure the safety and protection of the population, but building the resilience of the community

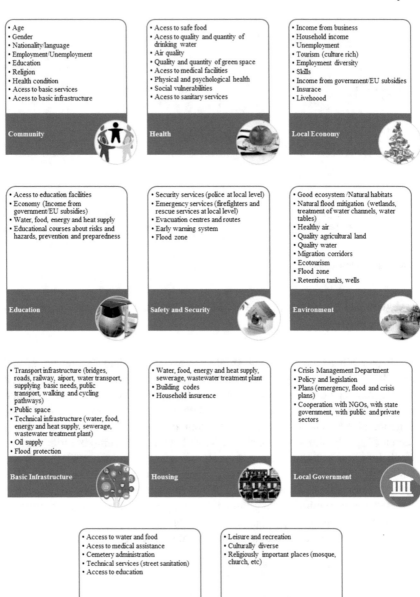

- Age
- Gender
- Nationality/language
- Employment/Unemployment
- Education
- Religion
- Health condition
- Acess to basic services
- Acess to basic infrastructure

Community

- Acess to safe food
- Acess to quality and quantity of drinking water
- Air quality
- Quality and quantity of green space
- Access to medical facilities
- Physical and psychological health
- Social vulnerabilities
- Acess to sanitary services

Health

- Income from business
- Household income
- Unemployment
- Tourism (culture rich)
- Employment diversity
- Skills
- Income from government/EU subsidies
- Insurace
- Livehoood

Local Economy

- Acess to education facilities
- Economy (Income from government/EU subsidies)
- Water, food, energy and heat supply
- Educational courses about risks and hazards, prevention and preparedness

Education

- Security services (police at local level)
- Emergency services (firefighters and rescue services at local level)
- Evacuation centres and routes
- Early warning system
- Flood zone

Safety and Security

- Good ecosystem /Natural habitats
- Natural flood mitigation (wetlands, treatment of water channels, water tables)
- Healthy air
- Quality agricultural land
- Quality water
- Migration corridors
- Ecotourism
- Flood zone
- Retention tanks, wells

Environment

- Transport infrastructure (bridges, roads, railway, aiport, water transport, supplying basic needs, public transport, walking and cycling pathways)
- Public space
- Technical infrastructure (water, food, energy and heat supply, sewerage, wastewater treatment plant)
- Oil supply
- Flood protection

Basic Infrastructure

- Water, food, energy and heat supply, sewerage, wastewater treatment plant
- Building codes
- Household insurence

Housing

- Crisis Management Department
- Policy and legislation
- Plans (emergency, flood and crisis plans)
- Cooperation with NGOs, with state government, with public and private sectors

Local Government

- Access to water and food
- Access to medical assistance
- Cemetery administration
- Technical services (street sanitation)
- Access to education

Basic Services

- Leisure and recreation
- Culturally diverse
- Religiously important places (mosque, church, etc)

Culture Rich

Fig. 5 Criteria for evaluating community preparedness

is currently not primarily included in this legislation, as the Czech literature does not contain a definition of community and community resilience. One of the goals within the solved problem is a theoretical description of the problem of the community of resilience defining these concepts in the Czech Republic. The last part of the article focuses on the identification of vulnerable sectors that can fundamentally help crisis management authorities to build a resilient community for future disasters.

Acknowledgements The work presented in this chapter has been supported by the Student Grant Competition "The proposal (the creation of) the categorization of resources enabling the management of emergency events related to the community" (research project No. SP2017/140).

References

1. Act No. 239/2000 Coll. the integrated rescue system and on amendments to certain acts. Ministry of the Interior of the Czech Republic
2. Act No. 240/2000 Coll. on Crisis management and on amendments to certain acts. Ministry of the Interior of the Czech Republic
3. Alexander DE (2013) Resilience and disaster risk reduction: an etymological journey. Natural Hazards and Earth System Science. London: Institute for Risk and Disaster Reduction, University College London. 13(11), pp. 2707-2716. ISSN 1684–9981. https://doi.org/10.5194/nhess-13-2707-2013
4. Ashby DT, Pharr J (2012) What is a healthy community? 1–8. https://digitalscholarship.unlv.edu/lincy_publications/4
5. Baštecká B (2013) Psychosocial crisis cooperation. ISBN 978-80-247-4195-6
6. Buriánek J (2001) Sociology for secondary schools and colleges. ISBN 80-7168-754-5
7. Concept of population protection up to 2006 with a view to 2015 (2002) Government Resolution No. 416 of 22 April 2002, on the Concept of Population Protection up to 2006 with a view to 2015 General Directorate of the Fire Rescue Service of the Czech Republic
8. Concept of population protection up to 2013 with a view to 2020 (2008) Government Resolution No. 165 of 25 February 2008, Evaluation of the state of implementation Concept of population protection up to 2013 with a view to 2020. General Directorate of the Fire Rescue Service of the Czech Republic
9. Concept of population protection up to 2020 with a view to 2030 (2013) Government Resolution No. 805 of 23 October 2013, on the concept of population protection up to 2020 with a view to 2030. General Directorate of Fire Rescue Service of the Czech Republic
10. Constitutional Act No. 1/1993 Coll. the constitution of the Czech Republic. Ministry of the Interior of the Czech Republic
11. Constitutional Act No. 2/1993 Coll. the charter of fundamental rights and freedoms. Ministry of the Interior of the Czech Republic
12. Constitutional Act No. 110/1998 Coll. on the Security of the Czech Republic. Ministry of the Interior of the Czech Republic
13. Decree No. 380/2002 Coll. on the preparation and implementation of tasks of population protection. Ministry of the Interior of the Czech Republic
14. Maléřová L, Kristlová E, Pokorný J, Wojnarová J (2017) Using of mobile flood protection on the territory of the Moldova as possible protection of the community, vol 92, IOP Publishing, IOP conference series. Earth Environ Sci 1–5. https://doi.org/10.1088/1755-1315/92/1/012039
15. Marsh G, Ahmed I, Mulligan M, Donovan J, Barton S (2017) Community engagement in post-disaster recovery. ISBN 978-1-138-69167-4
16. Paulík K (2017) Psychology of human resilience. ISBN 978-80-247-5646-2

17. Security Strategy of the Czech Republic (1999) Government Resolution No. 123 of 17 February 1999, on the Security Strategy of the Czech Republic, AVIS
18. Security Strategy of the Czech Republic (2001) Government Resolution No. 80 of 21 January 2001 on the security strategy of the Czech Republic. Ministry of Foreign Affairs
19. Security Strategy of the Czech Republic (2003) Resolution of the Government of the Czech Republic No. 1254 of 10 December 2003, amending the Security Strategy of the Czech Republic. Ministry of Foreign Affairs, Institute of International Relations
20. Security Strategy of the Czech Republic (2011) Government resolution No. 665 of 8 September 2011, security strategy of the Czech Republic. Ministry of Foreign Affairs, Institute of International Relations. ISBN 978-80-7441-005-5
21. Security Strategy of the Czech Republic (2015) Government Resolution No. 79 of 04/02/2015, Security Strategy of the Czech Republic 2015. Ministry of Foreign Affairs, Institute of International Relations. ISBN 978-80-7441-005-5
22. Strategic Concept for the Defence and Security of the Members of the North Atlantic Treaty Organization (2010) NATO Public Diplomacy Division
23. Šolcová I (2009) Development of resilience in childhood and adulthood. ISBN 978--80-247-2947-3
24. Treaty of Lisbon amending the Treaty on European Union and the Treaty establishing the European Community (2007) Off J Eur Union
25. UNISDR (United Nations International Strategy for Disaster Reduction) (2012) How to make cities more resilient a handbook for local government leaders. https://www.unisdr.org/files/26462_handbookfinalonlineversion.pdf
26. UNISDR (United Nations International Strategy for Disaster Reduction) (2015) Sendai framework for disaster risk reduction 2015–2030. https://www.preventionweb.net/files/43291_sendaiframeworkfordrren.pdf

Approaches to Soft Targets Protection in the Czech Republic: A Comparison with European Union Guidelines

Aleš Kudlák, Tomáš Zeman, and Rudolf Urban

Abstract The development of guidance materials is one of the key points in both the European Union Action Plan to Support the Protection of Public Spaces and the Framework of Soft Targets Protection approved by the government of the Czech Republic. The goal of this chapter is to assess the degree of compliance by the Czech Framework and guidance materials related to the protection of soft targets with European Union Good Practices to Support the Protection of Public Spaces. The absence of requirement for the assessment of security and physical protection issues in the design process of a new facility in the Czech Framework was identified as being the most significant discrepancy. Subsequently, the state of grant support for the protection of soft targets in the Czech Republic was analysed.

Keywords Crowded places · Funding · Security · Soft targets

1 Introduction

Soft targets or crowded places are generally sites whose properties make them highly vulnerable to terrorist attack [3]. Unfortunately, a more detailed and universally accepted definition is not available. According to the Fourth progress report towards an effective and genuine Security Union of the European Commission (EC) [10],

A. Kudlák
Faculty of Health and Social Sciences, University of South Bohemia, J. Boreckého 1167/27, 370 11 České Budějovice, Czech Republic
e-mail: ales.kudlak@centrum.cz

T. Zeman (✉)
Department of Military Science Theory, Faculty of Military Leadership, University of Defence, Kounicova 65, 662 10 Brno, Czech Republic
e-mail: tomas.zeman2@unob.cz

R. Urban
Department of Security and Law, AMBIS College, Lindnerova 575/1, 180 00 Praha 8, Czech Republic
e-mail: ubn17@centrum.cz

© The Author(s), under exclusive license to Springer Nature Switzerland AG 2022
I. Tušer and Š. Hošková-Mayerová (eds.), *Trends and Future Directions in Security and Emergency Management*, Lecture Notes in Networks and Systems 257,
https://doi.org/10.1007/978-3-030-88907-4_7

soft targets are typically "(…) civilian sites where people gather in large numbers (e.g. public spaces, hospitals, schools, sporting arenas, cultural centers, cafés and restaurants, shopping centres and transportation hubs). By their nature, these locations are vulnerable and difficult to protect and are also characterised by the high likelihood of mass casualties in the event of an attack. For all of these reasons they are favoured by terrorists." The definition of the U.S. Department of Homeland Security differs slightly, since it considers Soft Targets and Crowded Places (ST-CPs) as being locations "that are easily accessible to large numbers of people and that have limited security or protective measures in place making them vulnerable to attack" [32]. Another definition of soft targets, or more precisely crowded places, which highlights the significance of predictability in soft targets identification, was adopted by the Australia-New Zealand Counter-Terrorism Committee (ANZCTC) in Australia's Strategy for Protecting Crowded Places from Terrorism: "Crowded places are locations which are easily accessible by large numbers of people on a predictable basis" [2].

2 Protection of Public Spaces in the European Union

2.1 EU Action Plan

In recent years, the European Union (EU) has made a significant effort to enhance EU resilience against terrorist attacks against soft targets. In October 2017, the European Commission (EC) approved the Action Plan to Support the Protection of Public Spaces (hereinafter referred to as the Action Plan) [11]. According to the Action Plan, "Member States are primarily responsible for the protection of public spaces (…)", however, the EU should support the efforts of Member States to protect soft targets. The support provided by the EU includes:

(1)   funding for support of Member States in the protection of public spaces;
(2)   organisation and support of platforms for the exchange of experiences, expertise and best practices between Member States (e.g. an EU Policy Group on Soft Target Protection or a Practitioners' Forum);
(3)   development of common EU guidance materials and tools for protection of soft targets;
(4)   mediation of the cooperation between stakeholders.

Despite the fact that the security of soft targets is the responsibility of Member States, the demands on ensuring an equal level of security across EU Member States for specific types of soft targets have been emerging in EU legislative documents. In the Fifteenth progress report towards an effective and genuine Security Union of the EC, the need for equal levels of security for EU rail passengers and transport operators across borders was declared [14]. This effort by the EC is based on Article 91(1)(a) of the Treaty on the Functioning of the European Union, which "assigned the

Union and the Member States the task of laying down the common rules applicable to international transport to or from the territory of a Member State or passing across the territory of one or more Member States" [15]. For the purpose of establishing a common transport policy, the EU Rail Passenger Security platform was set up according to Article 1 of the Commission's decision of June 29, 2018. On the basis of open public consultation with citizens, railway stakeholders, national authorities and other related organisations, most respondents support action at the EU level for international trains [17]. On the other hand, the support for EU actions for national and local trains is less than 50%. Some suggested measures include the sharing of information on security incidents, implementation of a common methodology for risk assessment, promotion of passenger security awareness, wider use of security technologies, and the creation of a level of controls or performance of common exercises. The preparation of concrete measures for the increase of rail passenger security, however, is still pending.

In contrast to rail transport, the security of many other soft targets in Member States is not regulated by the European Union. In fact, the security of soft targets is regulated only weakly by the individual Member States. As reviewed by [22], most Member States still have no soft target protection strategy or supporting materials, such as guides, information sheets, assessment tools etc.

2.2 Funding Supports

Although the protection of soft targets remains the responsibility of Member States, the EU provides funding to enhance the protection of soft targets in its Member States [11]. This activity by the EU led to the inclusion of Urban security as one of four topics included in the 4th call for proposals in the Urban Innovative Actions (UIA), which was launched on October 15, 2018 and closed on January 31, 2019, with a total budget of 80–100 million euros as part of the European Regional Development Fund [34]. Under the UIA, only urban authorities of local administrative units with the status of city, town or suburb with a population exceeding 50,000 inhabitants and associations or groupings of urban authorities of cities, towns or suburbs with a total population exceeding 50,000 inhabitant may apply for support [33].

Other funds are also available for the increase of soft targets security, e.g. the Internal Security Fund—Police (ISFP) and security research, e.g. Horizon 2020 [11]. Under ISFP, beneficiaries may be state and federal authorities, local public bodies, non-governmental organisations, as well as private and public law companies [18]. A call for proposals under the ISF for projects on the protection of public spaces that also addresses CBRN-E (chemical, biological, radiological, nuclear and explosives) threats (ISFP-2017-AG-PROTECT), which opened on October 26, 2017 and closed on February 1, 2018, offered funding of 18.5 million euros [12]. In 2018, another call for proposals under the ISF for projects on the protection of public spaces and other soft targets and the protection of soft targets and critical infrastructure against CBRN-E in line with the EU Action Plan [11] (ISFP-2018-AG-CT-PROTECT) was opened on October 10, 2018 and closed on January 16, 2019, with a total budget of 9.5 million euros [16]. In both programmes, the first of four declared priorities was aimed at improving the protection of public spaces and other soft targets in line with the EU Action Plan. Another priority of these programmes is to address emerging security threats to critical infrastructure and public spaces, e.g. CBRN-E or aerial vehicles.

Recently, the call for proposals under Horizon 2020 with the name Protecting the infrastructure of Europe and the people in the European smart cities (SU-INFRA02-2019) was launched on March 14, 2019 and closed on August 22, 2019, with a budget of 16 million euros [13]. The proposals should develop the components of an open platform for sharing and managing information between public and private security operators, including security practitioners in smart cities. This mainly includes methods to detect weapons, systems for video surveillance and methods to identify and neutralise perpetrators with minimal intrusion in crowded areas. Although any legal entity may be a participant of the Horizon 2020 project according to Article 7 of the Regulation No 1290/2018 of the European Parliament and of the Council [9], in the SU-INFRA02-2019 programme, the local governments of at least 2 cities or metropolitan areas from 2 Member States or Associated Countries must be beneficiaries involved in the project [13]. In addition, the participation of industry is also required.

2.3   EU Good Practices

In December 2017, the EU launched a public–private Operators Forum which brought together politics and stakeholders from various sectors (e.g. shopping malls, mass events, hospitality, transportation). Based on the materials provided by the Operators Forum, participants, and the discussion within the Operators Forum, Good Practices for the Protection of Public Spaces (hereinafter referred to as the Good Practices) were identified by the Commission Staff [18]. The Good Practices for the protection of public spaces include:

(1)   performance of vulnerability assessment;
(2)   development and implementation of a security plan, including preparatory, emergency and recovery measures;
(3)   appointment and training of persons responsible for the coordination and implementation of security measures;
(4)   development and implementation of a crisis management plan;
(5)   performance of public awareness campaigns (reporting suspicious behaviour, reaction in the case of terrorist attack);
(6)   development and implementation of a security awareness programme for employees;
(7)   development and implementation of an insider threats awareness programme;
(8)   development of a security training programme for all staff;

(9)   conducting regular security exercises;
(10)   assessing security and physical protection issues from the beginning of the design process of a new facility or event;
(11)   assessing the necessary access controls and barriers;
(12)   utilising the most appropriate detection technology for explosives, firearms, bladed arms, as well as chemical, biological, radiological and nuclear agents;
(13)   appointing contact points and clarifying respective roles and responsibilities in public private cooperation on security matters;
(14)   establishment of trustworthy and timely communication allowing specific risk and threat information exchange with responsible public authorities and local law enforcement;
(15)   coordination of the work on protection of public spaces at local, regional and national levels;
(16)   developing practical recommendations and guidance materials to detect, mitigate or respond to security threats.

3 Soft Targets Protection in the Czech Republic

3.1 Main Documents

Currently, terrorism is considered one of the major security threats to the Czech Republic (for review see [21]). The basic strategic document for protection of soft targets in the Czech Republic is the Framework of Soft Targets Protection for the Years 2017–2020 (hereinafter referred to as the Framework) [28] approved by the Government of the Czech Republic in 2017. The Framework aims to create a functional national system of protection of soft targets in the Czech Republic. As declared in the Framework, this system should be able to respond quickly to the threat of terrorism attack. In 2017, the issue of soft targets protection was also included in the [20] approved by the Government of the Czech Republic (for review see [31]).

In the Framework, soft targets are defined as any object, place or event characterised by the frequent presence of large numbers of people on one hand, and by the absence or low level of security on the other [28]. A possible problem of the Framework lies in the fact that although the Framework speaks about a national system of soft targets protection, it does not address in any way those who are responsible for this system, or from what sources it will be funded. At present, the protection of soft targets is not incorporated in any way in the crisis management system of the Czech Republic, which was established in Act No. 240/2000 Coll., on Crisis Management and amending certain acts (Crisis Act). The Framework [28] does not seek to change this state; it neither suggests changes in laws highlighting the need to realise that the safety of soft targets is a matter for all stakeholders and not just for the state. However, the Framework does emphasize the significance of cooperation and communication between owners or operators and the police.

The Framework presumes that the system of soft targets protection in the Czech Republic should be based on 4 core pillars [28]:

(1)   guidance and training

In accordance with the Framework [28], the Ministry of the Interior of the Czech Republic is tasked with creating a series of training courses for different actors of soft targets protection, i.e. mainly owners and providers of soft targets, but also the general public. In 2018, the Ministry of the Interior of the Czech Republic started a subsidy programme for security enhancement of public places, buildings or events of public administration and school facilities [29]. Two of the seven main areas of this programme are educational and training activities. Only the self-governing regions of the Czech Republic may apply to this subsidy programme. Therefore, until now, performance of the information campaign is primarily in the competence of regional authorities. Additionally, the Ministry of the Interior of the Czech Republic has prepared several guidelines, manuals and booklets aimed at owners or providers of soft targets, e.g. [4, 7, 8, 23].

(2)   grant funding

The Framework [28] has assigned the task of preparing grant programmes for soft targets security enhancement to the Ministry of the Interior of the Czech Republic, the Ministry of Education, Youth and Sports of the Czech Republic, the Ministry of Transport of the Czech Republic, the Ministry of Culture of the Czech Republic, the Ministry of Industry and Trade of the Czech Republic, the Ministry of Defence of the Czech Republic and the Ministry of Health of the Czech Republic.

(3)   communication, cooperation, changing of information and good practices

In order to establish regular communication between stakeholders, a permanent advisory board of the Ministry of the Interior of the Czech Republic for the protection of soft targets was created in 2017, which consists of permanent members, i.e. representatives of state institutions and voluntary members, i.e. experts from the public. The advisory board serves not only to exchange information on the progress of fulfilling the tasks of the Framework, but also to share good practice regarding soft targets security and to provide regular consultation of this issue at the expert level. Within this pillar, the Framework [28] sets tasks to create a standardized security plan form for soft targets. This template was previously published by the Ministry of Interior of the Czech Republic in January, 2019 [6]. The next task in this area is to create a system of early warning and transmission of emergency information about the security situation or incidents to owners and operators of soft targets through either phone calls or SMS warnings. To date, however, this system has not yet been launched.

(4)   proactive approach of the Police of the Czech Republic to soft targets protection

According to the Framework [28], the Police of the Czech Republic should play the role of expert adviser to owners or providers of soft targets. The Police of the Czech Republic should also be prepared to establish direct police protection of some soft targets whenever the security situation requires it. In an extremely serious security situation, the Framework also suggests deployment of military units of the Army of the Czech Republic to enhance soft targets protection.

On the basis of the Framework [28], several guidelines and manuals related to the issue of soft targets protection have been published. The most comprehensive is the Basics of Soft Targets Protection Guidelines (hereinafter referred to as the Guidelines) [23], published in June, 2016 by the non-governmental Soft Targets Protection Institute registered in the Czech Republic. Other significant materials include the Vulnerability Assessment of Soft Target [5], published in June, 2018 by the Centre against Terrorism and Hybrid Threats (CTHT) at the Ministry of the Interior and the Security Plan of Soft Target [6], published by the CTHT in January, 2019.

3.2   Comparison with EU Good Practices

In Table 1, a comparison of measures outlined in the Framework [28] and the Guidelines [23] are compared with the EU Good Practices [19] described above. Although both documents are largely in line with the Good Practices [19], some the good practices are not fulfilled or are only partially fulfilled as can be seen in Table 1.

The first good practice from the EU Good Practices calls for a vulnerability assessment to be carried out in order to identify potential vulnerabilities [19]. A vulnerability assessment is not addressed in either the Framework [28] or the Guidelines [23]. This hole was filled, however, in June 2018, when the Vulnerability Assessment of Soft Targets [5] manual was published by the CTHT. The manual represents a clear guidance intended for owners or operators of a soft target.

The need for the appointment and training of the responsible persons (good practice 3 according to the European Commission [19]) is not addressed in the Framework [28]. This requirement is, however, clearly formulated in the Guidelines [23]. A similar situation may be seen in the good practices, from good practice 6 to good practice 9, as well as in good practices 11 and 12, which relate to awareness programmes and training programmes, exercises performance, access controls, barriers and detection technologies [19].

With regards to the recommendation for the development and implementation of the crisis management plan (good practice 4 according to the European Commission [19]), even though the plan is not explicitly mentioned in both the Framework [28] or the Guidelines [23], this point may be considered accomplished. This is due to the fact that according to the Good Practices, this plan should contain an overview of planned actions in the case of emergency (system of early warning, communication, evacuation etc.) [19]. The guidance [6] published by the CTHT in January 2019, however, recommended that the security plan includes a list of standardized procedures in the case of an incident. Moreover, the Guidelines [23] encourage owners or providers to create a Management Coordination Plan. This recommendation in the Guidelines is accompanied by a brief template of this plan.

Table 1 Comparison of the EU Good Practices [19] with two key soft targets related documents in the Czech Republic: the Framework of Soft Targets Protection (the Framework) [28] and the Basics of Soft Targets Protection: Guidelines (the Guidelines) [23]

Good practices according to the EU good practices	Measures in the framework/the guidelines?	Who is responsible in the Czech Republic according the framework?
(1) Performance of vulnerability assessment	No/no	Owners or operators (vol.)
(2) Development and implementation of security plan	Yes/yes	MI[1] for creation of the template Owners or operators (vol.)
(3) Appointment and training of responsible persons	No/yes	Owners or operators (vol.)
(4) Development and implementation of crisis management plan	No/yes	Owners or operators (vol.)
(5) Performance of public awareness campaigns	Yes/yes	MI[1] Owners or operators (vol.)
(6) Development and implementation of security awareness programme for employees	No/yes	Owners or operators (vol.)
(7) Development and implementation of insider threats awareness programme	No/yes	Owners or operators (vol.)
(8) Development of security training programme for all staff	No/yes	Owners or operators (vol.)
(9) Conducting regular security exercises	No/yes	Owners or operators (vol.)
(10) Assessing security and physical protection issues from the beginning of the design process of a new facility or event	No/no	–
(11) Assessing the necessary access controls and barriers	No/yes	Owners or operators (vol.)
(12) Utilizing the most appropriate detection technology	No/yes	Owners or operators (vol.)
(13) Appointing contact points and clarifying respective roles and responsibilities in public private cooperation	Partially/no	PCR[2] for operating the Hotline for owners or operators of soft targets

(continued)

Table 1 (continued)

Good practices according to the EU good practices	Measures in the framework/the guidelines?	Who is responsible in the Czech Republic according the framework?
(14) Establishment of trustworthy and timely communication allowing specific risk and threat information exchange with responsible public authorities and local law enforcement	Yes/yes	MI[1] for creation of early warning system using phone calls and SMS Owners or operators (vol.)
(15) Coordination of the work on protection at a local, regional and national level	Yes/no	MI[1]
(16) Developing practical recommendations and guidance materials	Yes/no	MI[1]

Note [1]The Ministry of Interior of the Czech Republic (MI), [2]The Police of the Czech Republic (PCR), vol.—voluntarily

The most significant discrepancy may be seen in the fulfilment of the recommendation to assess security and physical protection issues in the design process of a new facility (good practice 10 according to the European Commission [19]). Neither the Framework [28] nor the Guidelines [23] contain any methodical recommendation or reference to them that deal with this issue.

According to the Good Practices [19] "public authorities and operators should establish clear communication channels in case of a security event and update each other on the person(s) responsible for particular tasks (…) so that both sides can clearly and easily know whom to contact". This may be considered only partially fulfilled. Although the Framework [28] declares that contacts between the local departments of the Police of the Czech Republic and owners or operators of soft targets should be established, the Framework does not contain any details about the contact or how this contact should be established (whether it should be established proactively by the Police of the Czech Republic or at the request of owners or operators of soft targets). The only systemic measure addressed in the Framework was the creation of a hotline at the Police Presidium of the Czech Republic in August 2016. The hotline is intended for providing methodological support for owners or operators of soft targets.

In summary, the Czech Framework of Soft Targets Protection for the Years 2017–2020 [28] fully addresses only five of 16 good practices identified in the EU Good Practices [19]. Other Czech guidelines related to soft target protection, especially the Basics of Soft Targets Protection Guidelines, however, address the other 9 good practices. Thus, only one good practice remains completely unresolved and one good practice partially resolved in the Czech Republic.

3.3 Funding Support

As mentioned before, funding support for security enhancement of soft targets is an integral part of the EU Action Plan to Support the Protection of Public Spaces [11]. Accordingly, the Framework assumes the establishment of a system of grant funding for owners and operators of soft targets [28]. As mentioned previously, seven ministries were tasked in the Framework to prepare their own grant programmes for the enhancement of soft targets security. In fact, the subsequent government resolution No. 527 from 2017 (Resolution of the Government of the Czech Republic of 24 July 2017 No. 527 on the proposal to establish grant programs to enhance the protection of soft targets) imposes tasks on four ministries only. The Minister of the Interior was given the task of preparing and launching a grant programme aimed at increasing the protection of public spaces, public administration buildings, schools and school facilities by March 31, 2018. In addition, the Ministers of Transport, Culture and Health were given the task of preparing and launching subsidy programmes by September 30, 2018. The extension, targeting and selection of recipients were in the competence of individual ministries. Until now, however, only three ministries have allocated funds for these grant programmes. As can be seen in Table 2, in the Czech Republic's state budgets for 2019 and 2020, a total of 260 million Czech crown (CZK) was earmarked for grant programmes aimed at increasing the security of soft targets [25].

The Ministry of Culture of the Czech Republic launched the call for a proposal on September 30, 2018 and finished it on December 3, 2018. The grant programme was intended for owners or operators of soft targets in the area of culture [24]. An overview of activities that may be funded from this programme are given in Table 3. Four criteria were used in the process of recipient selection:

(1)   social significance of the soft target;
(2)   location of the soft target;
(3)   number and concentration of people inside the soft targets or in its immediate vicinity;
(4)   openness of the soft target to public.

The Ministry of Health of the Czech Republic launched the call for proposals on November 1, 2018 and completed it on December 31, 2018 [27]. The grant programme was intended for providers of inpatient care with a total number of beds

Table 2 Budgets earmarked by the ministries from the Czech Republic's state budget for 2019 and 2020 for grant programmes aimed at increasing the security of soft targets

Ministry	Budget for 2019	Budget for 2020
Ministry of the Interior	40 million CZK	52 million CZK
Ministry of Health	32 million CZK	32 million CZK
Ministry of Culture	67 million CZK	37 million CZK
Total	139 million CZK	121 million CZK

Note CZK—Czech koruna
Source The Ministry of Finance of the Czech Republic [25, 26]

Table 3 Activities that may be funded from grant programmes established by the ministries in the Czech Republic

Activity	MC	MH	MI
Risk assessment	x	xx	x
Security plan processing	x	x	x
Education and training	x	xx	x
Security services financing	x	x	
Information and learning materials processing	x	x	x
Promotion and public education	x	x	
Exercises organisation	x	xx	x
Operative cards processing	x	x	
Purchase of security equipment	x	x	x
Creation of a physical security standard for soft target	x		
Plan of implementation of security analysis results processing		x	
Presentation and information technologies supporting the security of soft target		x	
Personal costs related to the grant project		x	x
Security documentation processing		x	

Note x—supported activity, xx—prioritized activity, MC—the Ministry of Culture of the Czech Republic, MH—the Ministry of Health of the Czech Republic, MI—the Ministry of the Interior of the Czech Republic
Source the Ministry of Culture of the Czech Republic [24], the Ministry of Health of the Czech Republic [27], the Ministry of the Interior of the Czech Republic [29]

exceeding 350. An overview of activities that may be funded from this programme are given in Table 3.

The Ministry of the Interior of the Czech Republic launched the call for proposals within its grant programme on November 5, 2018 and completed it on February 14, 2019 [29]. An overview of activities that may be funded from this programme are given in Table 3. The recipient of the dotation may only be the self-governing regions and the Capital City of Prague, which also has the status of region in the Czech Republic. Municipalities may also be the recipient of the grant, however, only when the self-governing regions have redistributed the funds within grant programmes organised by the self-governing regions. According to the recommendation of the Ministry of the Interior of the Czech Republic, these regions should take the following criteria into account, when redistributing subsides to municipalities [30]:

(1) compliance with the methodological materials of the Ministry of the Interior of the Czech Republic;
(2) symbolic significance of the soft target;
(3) international character of the soft target;
(4) significance of the soft target within the self-governing region;
(5) number and concentration of people in the soft target.

As can be seen in Table 2, the funding support that may be provided to municipalities from the Czech Republic's state budget under the grant programme of the Ministry of the Interior of the Czech Republic reached 52 million CZK in 2020. However, the total amount of funding support for the protection of soft targets is also given by the budgets earmarked for this purpose by the municipalities themselves from their own budgets. Although exact figures are not available, recent data indicate that the share of expenditures on crisis prevention in general within municipal budgets is decreasing in the Czech Republic [35, 36].

4 Conclusion

The state of soft targets protection in the Czech Republic was examined based on the analysis of the Czech Framework of Soft Targets Protection and related methodological materials. The degree of compliance with EU Good Practices to Support the Protection of Public Spaces was assessed. The absence of requirement for the assessment of security and physical protection issues in the design process of a new facility in the Czech Framework and related methodological materials was identified as being the most significant discrepancy. In addition, the current state of the establishment of contact points between the Police of the Czech Republic and owners or operators of soft targets, as well as clarification of respective roles and responsibilities in public–private cooperation, was assessed as insufficient. It would be appropriate to resolve these shortcomings by issuing new methodological materials related to these issues.

The planned grant support for the protection of soft targets amounted to 139 million CZK in 2019 and 121 million CZK in 2020, which was earmarked by the Ministry of the Interior of the Czech Republic, the Ministry of Culture of the Czech Republic and the Ministry of Health of the Czech Republic. Based on the given criteria for recipient selection, it can be concluded that to date, the grant funding announced in the Framework has been almost completely restricted (with significant exemption of a grant programme of the Ministry of Culture of the Czech Republic) to institutions established by the state, regions or municipalities. It can be predicted that the impact of such grant support on the security of privately-owned soft targets will be negligible. Moreover, as can be seen in Table 3, the activities that may be supported from these grant projects are limited. It is particularly advisable to modify the criterions for recipient selection so that the private owners and operators of soft targets could benefit more from grant support. This would increase their motivation to invest in securing soft targets owned or operated by them.

Therefore, on the basis of the obtained findings, it can be concluded that increased efforts should be devoted to the unification and standardization of the system of soft targets protection within the EU. As demonstrated, even in the case of the Czech Republic, the compatibility of Czech and EU methodological materials is not complete.

References

1. Act No. 240/2000 Coll., on Crisis Management and amending certain acts (Crisis Act)
2. Australia-New Zealand Counter-Terrorism Committee (ANZCTC) (2017) Australia's strategy for protecting crowded places from terrorism. ISBN 978-1-925593-95-2
3. Břeň J, Zeman T (2017) Fault tree analysis of terrorist attacks against places of worship. In: 2nd international conference on system reliability and safety (ICSRS), pp 531–535. IEEE, New York. ISBN 978-1-5386-3322-9
4. Centre Against Terrorism and Hybrid Threats (2017) 10 principles of soft target hardening. https://www.mvcr.cz/cthh/soubor/brochure-10-principles-of-soft-target-hardening-pdf.aspx
5. Centre Against Terrorism and Hybrid Threats (2018) Evaluation of the vulnerability of soft targets or what, when, where and from whom you are threatened, in Czech: Vyhodnocení ohroženosti měkkého cíle aneb co, kdy, kde a od koho vám hrozí. https://www.mvcr.cz/cthh/ soubor/vyhodnoceni-ohrozenosti-mekkeho-cile.aspx
6. Centre Against Terrorism and Hybrid Threats (2019a) The security plan of the soft target or what should not be neglected during its processing, in Czech: Bezpečnostní plán měkkého cíle aneb co by nemělo být opomenuto při jeho zpracování. https://www.mvcr.cz/cthh/soubor/bez pecnostni-plan-mekkeho-cile-nove-2-b2-samostatne-strany-pdf.aspx
7. Centre Against Terrorism and Hybrid Threats (2019b) Security standards for organizers, in Czech: Bezpečnostní standardy pro pořadatele. https://www.mvcr.cz/cthh/soubor/brozura-bez pecnostni-standardy-pro-poradatele-sportovnich-kulturnich-a-spolecenskych-akci.aspx
8. Czech Office for Standards Metrology and Testing (2016) ČSN 73 4400: Crime Prevention— security management in planning, implementing and using schools and educational institutions. Prague
9. European Commission (2013) Regulation (EU) No 1290/2013 of the European Parliament and of the Council of 11 December 2013 laying down the rules for participation and dissemination in "Horizon 2020 - the Framework Programme for Research and Innovation (2014–2020)" and repealing Regulation (EC) No 1906/2006. Official J Eur Union L 347:81–103
10. European Commission (2017a) COM(2017) 41 Communication from the Commission to the European Parliament, the European Council and the Council: Fourth progress report towards an effective and genuine Security Union. Brussels
11. European Commission (2017b) COM(2017) 612 Communication from the Commission to the European Parliament, the European Economic and Social Committee and the Committee of the Regions: Action Plan to support the protection of public spaces
12. European Commission (2017c) Protection ID: ISFP-2017-AG-PROTECT. https://ec.europa. eu/info/fundingtenders/opportunities/portal/screen/opportunities/topic-details/isfp-2017-ag- protect
13. European Commission (2017d) Security for smart and safe cities, including for public spaces ID: SU-INFRA02–2019. https://ec.europa.eu/info/funding-tenders/opportunities/por tal/screen/opportunities/topic-details/su-infra02-2019
14. European Commission (2018a) COM(2018) 470 Communication from the commission to the European Parliament, the European Council and the Council: fifteenth progress report towards an effective and genuine Security Union
15. European Commission (2018b) Commission decision of 29 June 2018 setting up the EU rail passenger security platform (2018/C 232/03). Official J Eur Union C 232:10–13
16. European Commission (2018c) Protection ID: ISFP-2018-AG-CT-PROTECT. https://ec. europa.eu/info/funding-tenders/opportunities/portal/screen/opportunities/topic-details/isfp- 2018-ag-ct-protect
17. European Commission (2018d) SWD(2018) 400 commission staff working on document synopsis report: summary of the consultation on improving security of rail passengers
18. European Commission (2019a) Internal Security Fund—Police. https://ec.europa.eu/homeaf fairs/financing/fundings/security-and-safeguarding-liberties/internal-security-fund-police_en
19. European Commission (2019b) SWD(2019) 140 Commission staff working document: Good practices to support the protection of public spaces

20. Framework of education in population protection and crisis management. Praha (2017).
21. Jakubcová L, Kovařík Z, Šesták B (2017) Exact estimation of factor composition of security threats for the Czech Republic. Secur Theory Pract 4(2017):24–30
22. Karlos V, Larcher M, Solomos G (2018) Review on soft target/public space protection guidance. Publications Office of the European Union, Luxembourg. https://doi.org/10.2760/553545
23. Soft Targets Protection Institute (2016) Basics of soft targets protection—guidelines (2nd version). Prague. https://www.mvcr.cz/cthh/soubor/basics-of-soft-target-protection-gui delines.aspx
24. The Ministry of Culture of the Czech Republic (2018) Grant selection procedure for 2019— announcement, in Czech: Výběrové dotační řízení na rok 2019 – vyhlášení. https://www.mkcr. cz/vyberove-dotacni-rizeni-na-rok-2019-vyhlaseni-1948.html
25. The Ministry of Finance of the Czech Republic (2018) Report on the state budget of the Czech Republic for 2019, in Czech: Zpráva ke státnímu rozpočtu České republiky na rok 2019
26. The Ministry of Finance of the Czech Republic (2019) Report on the state budget of the Czech Republic for 2020, in Czech: Zpráva ke státnímu rozpočtu České republiky na rok 2020
27. The Ministry of Health of the Czech Republic (2018) Grant program to enhance the protection of soft targets in the health sector, in Czech: Dotační program na zvýšení ochrany měkkých cílů v resortu zdravotnictví. https://www.mzcr.cz/dokumenty/dotacni-program-na-zvyseni-och rany-mekkych-cilu-v-resortu-zdravotnictvi-_16036_3913_1.html
28. The Ministry of the Interior of the Czech Republic (2017) Framework of soft targets protection for 2017–2020, in Czech: Koncepce ochrany měkkých cílů pro roky 2017–2020. Prague
29. The Ministry of the Interior of the Czech Republic (2018a) Announcement of the Grant program to enhance the protection of public spaces and objects (actions) of public administration, schools and school facilities as soft targets—2019, in Czech: Vyhlášení Dotačního programu pro zvýšení ochrany veřejných prostranství a objektů (akcí) veřejné správy, škol a školských zařízení jako měkkých cílů – 2019. https://www.mvcr.cz/cthh/clanek/dotacni-program-pro-och ranu-mekkych-cilu.aspx
30. The Ministry of the Interior of the Czech Republic (2018b) Principles for the provision of grants from the state budget for expenditures implemented under the Grant program to enhance the protection of public spaces and objects (actions) of public administration, schools and school facilities as soft targets—2019, in Czech: Zásady pro poskytování dotací ze státního rozpočtu na výdaje realizované v rámci Dotačního programu pro zvýšení ochrany veřejných prostranství a objektů (akcí) veřejné správy, škol a školských zařízení jako měkkých cílů - 2019. https:// www.mvcr.cz/cthh/clanek/dotacni-program-pro-ochranu-mekkych-cilu.aspx
31. Tušer I (2020) The development of education in emergency management. In: Flaut D, Hošková-Mayerová Š, Ispas C, Maturo F, Flaut C (eds) Decision making in social sciences: between traditions and innovations. Stud Syst, Dec Control 247:169–175. Springer, Cham. https://doi. org/10.1007/978-3-030-30659-5_10
32. U.S. Department of Homeland Security (DHS) (2018) Soft targets and crowded places security plan overview
33. Urban Innovative Actions (2018) UIA - Guidance (Version 4). https://www.uia-initiative.eu/ sites/default/files/2018-10/UIAguidance_V4.pdf
34. Urban Innovative Actions (2019) What is urban innovative actions? https://www.uia-initiative. eu/en
35. Urban R, Kudlák A (2017) Allocation of resources as a management-security risk in the level of municipalities. Krízový Manažment 1(2017):24–30
36. Urban R, Kudlák A (2019) Security and protection of soft targets, in Czech: Bezpečnost a ochrana měkkých cílů. In: Riešenie krízových situácií v špecifickom prostredí 2019. Žilinská univerzita, Žilina
37. Resolution of the Government of the Czech Republic of 24 July 2017 No. 527 on the proposal to establish grant programs to enhance the protection of soft targets, in Czech: Usnesení vlády České republiky ze dne 24. července 2017 č. 527 k návrhu na zřízení dotačních programů za účelem zvýšení ochrany měkkých cílů

Proposal of a Group-Specific Risk Assessment Procedure for Soft Targets: A Data-Based Approach

Tomáš Zeman⬤, Jan Břeň⬤, Pavel Foltin⬤, and Rudolf Urban⬤

Abstract According to available definitions, a soft target is a location that combines high vulnerability with a low level of protection. Such a general definition, however, cannot be used to determine whether an object is a soft target or not. The aim of this chapter is to propose a procedure for group-specific risk assessment of soft targets. Risk of terrorist attack against each type of soft target was calculated based on the relative frequency of incidents and relative number of casualties per attack. The suggested procedure was tested using data on 390 cases of terrorist attacks aimed against soft targets in the European Union from the years 2000 to 2019. The chapter is a revision and extension of the previously published article Soft Targets: Definition and Identification (Zeman in Acad Appl Res Mil Public Manag Sci 19(1):109–119, 2020).

Keywords Soft targets · Terrorist attacks · Crisis management · Risk assessment

T. Zeman (✉)
Department of Military Science Theory, Faculty of Military Leadership, University of Defence, Kounicova 65, 662 10 Brno, Czech Republic
e-mail: tomas.zeman2@unob.cz

J. Břeň
Centre for Security and Military Strategic Studies, University of Defence, Kounicova 65, 662 10 Brno, Czech Republic
e-mail: jan.bren@unob.cz

P. Foltin
Department of Logistics, Faculty of Military Leadership, University of Defence, Kounicova 65, 662 10 Brno, Czech Republic
e-mail: pavel.foltin@unob.cz

R. Urban
Department of Security and Law, AMBIS College, Lindnerova 575/1, 180 00 Praha 8, Czech Republic
e-mail: ubn17@centrum.cz

1   Introduction

A great deal of attention has been paid recently to the issue of soft targets and measures for the increase of their security. In the United States (US) and Europe, there is an increasing number of violent attacks on soft targets (e.g., [2, 10]). Due to their attractiveness, ease of access and accessibility, terrorist groups increasingly seek them. Libicki et al. [8] presumed that this trend is caused by the hardening of prominent targets such as the Pentagon or White House after September 11, 2001. The difficulty of attacking these prominent targets leads terrorist groups to focus their attacks against soft targets, which are far more vulnerable.

Unfortunately, there is no universally recognised definition of soft targets to date. According to [5], a soft target is generally "any person or thing that is vulnerable to attack but not protected". Recently, in its Fourth progress report towards an effective and genuine Security Union, the European Commission [4] defined soft targets as locations that "are vulnerable and difficult to protect and are also characterised by the high likelihood of mass casualties in the event of an attack". Nevertheless, both definitions are very common, which does not allow their use in the process of soft targets identification. The creation of more specific definition of soft targets would significantly facilitate the process of soft targets identification and contribute to a better understanding of terrorist aims and targets selection.

In terms of soft targets, the greatest attention nowadays is paid to objects or events that involve a large number of people in relatively small areas such as temples, schools, universities, hospitals, sport events, concerts, restaurants, hotels, bus/train stations etc. Although all these facilities or events are similar in many aspects, "not all such targets are equally vulnerable", as noted by [1]. In addition, it can be assumed that individual soft targets differ also in their popularity among terrorists. In fact, several complex methods for risk assessment of soft targets were recently created (e.g. [3, 6, 16]. However, implementation of these methods is often quite demanding, not only due to their complicated calculation, but also because of the need to provide expert judgement.

The aim of this chapter is to propose a simple data-based method for group-specific risk assessment of soft targets based on the relative frequency of terrorist attacks and relative number of casualties caused by these attacks. In order to achieve this goal, a statistical analysis of data about terrorist attacks committed in the European Union (EU) between 2000 and 2019 was performed.

2   Methods

The Global Terrorism Database [12] was utilised as a data source. All terrorist acts committed in the EU between 2000 and 2019 were selected from the Global Terrorism Database (GTD) in the first step. Subsequently, only terrorist acts targeted against targets listed in the Table 1 were selected using the variable "targsubtype1" according

Table 1 Coding of selected soft targets categories according to the GTD Codebook [11], with the number of incidents and the number of casualties between 2000 and 2019 in the European Union

Coding	Category of soft targets	Number of incidents (N)	Number of wounded	Number of dead	Number of casualties (NC)
2	Restaurant/bar/café	41	110	40	150
8	Hotel/resort	19	57	5	62
11	Entertainment/cultural/stadium/casino	37	352	91	443
44	Airport	9	138	16	154
49	School/university/educational building	22	56	12	68
57	Civilian maritime	1	0	0	0
60	Port	0	0	0	0
74	Marketplace/plaza/square	13	97	20	117
78	Procession/gathering	6	457	87	544
79	Public areas	20	22	2	24
81	Museum/cultural centre/cultural house	12	0	4	4
86	Place of worship	119	27	3	30
96	Tour bus/van/vehicle	1	30	6	36
99	Bus (excluding tour bus)	13	2	0	2
100	Train/train tracks/trolley	65	1821	197	2018
101	Bus station/stop	2	0	0	0
102	Subway	4	135	16	151
103	Bridge/car tunnel	1	0	0	0
104	Highway/road/toll/traffic signal	5	0	0	0
	Total	390	3304	499	3803

Note Based on the data from the GTD [12]

to the GTD Codebook [11]. The result of this selection was 390 cases of terrorist attacks against soft targets from 19 target categories according to the GTD. The number of casualties (NC) were calculated as the sum of persons killed or wounded during the attacks based on the variables "nkill" and "nwound" from the GTD. For each terrorist attack, information of the weapon used in the attack was based on GTD variables "weaptype1". However, information on evidence of terrorist group engagement during preparation of the attack was retrieved from various publicly available sources, particularly from news media websites, such as BBC News, The New York Times etc., to ensure that the data was up to date.

The mean number of casualties per one attack (MNC) was computed for each category of soft targets separately, using the equation

$$\text{MNC}_i = \frac{\text{NC}_i}{\text{N}_i}, \tag{1}$$

where MNC_i is the mean number of casualties for ith category of soft targets (see Table 1 for an overview of categories), NC_i the total number of casualties for i-th category of soft targets and N_i the number of terrorist attacks against i-th category of soft targets. Based on MNC values, the relative numbers of casualties per incident (RNC) were subsequently calculated using the equation

$$RNC_i = \frac{MNC_i}{\sum MNC_i}, \tag{2}$$

where RNC_i is the relative number of casualties per incident for ith category of soft targets.

Risk of terrorist attack (RTA) was calculated as

$$RTA_i = RN_i \cdot RNC_i, \tag{3}$$

where RTA_i is the risk of terrorist attack against i-th category of soft targets and RN_i the relative number of terrorist attacks against i-th category of soft targets calculated using the equation

$$RN_i = \frac{N_i}{\sum N_i}. \tag{4}$$

A terrorist attack was considered to had been organised by a terrorist group in two cases: (a) A terrorist attack was claimed by the group and this claim was not questioned by any relevant source, e.g. conclusions from a police investigation; (b) Involvement of a terrorist group was proved during a police investigation. In cases when two or more terrorist groups claimed one terrorist attack, but it was not clear which claim was true, the terrorist attack was considered to had been organised by a terrorist group. Any other terrorist act not corresponding to any of the aforementioned criteria was not considered to had been organised by a terrorist group. This procedure led to the division of all terrorist attacks into two groups: (a) terrorist attacks demonstrably organised by a terrorist group; (b) terrorist attacks committed by an individual, i.e. lone wolves or lone actors, and terrorist attacks organised by a terrorist group, but with a lack of evidence of the terrorist group's engagement.

Detailed statistical analysis of the type of weapon used, as well as the involvement of a terrorist organization was performed for soft target categories with at least 15 incidents documented in the GTD [12] between 2000 and 2019. All calculations were performed in statistical software R [15]. The relationship between variables was assessed using Spearman's rank correlation coefficient (R_S).

3   Results

First, the risk of terrorist attack was calculated for each soft target category with non-zero values of both incidence and number of casualties (Table 2, Figs. 1 and 2). Afterwards, the relative frequency of the types of weapons used for the attacks (Fig. 3), as well as the relative frequency of attacks with the involvement of a terrorist group (Fig. 4) were computed for all GTD categories with at least 15 documented terrorist attacks between 2000 and 2019.

As can be seen in Fig. 2, the highest risk of terrorist attack was calculated for the GTD category Train/train tracks/trolley. The number of attacks against targets from this category was, however, also very high. There were 65 attacks documented by the GTD [12] between 2000 and 2019. The absolute majority of them was bombing (52%) or arson (35%) attacks as seen in Fig. 3.

The category of Processions/gatherings had the second highest risk of terrorist attack. In contrast to the Train/train tracks/trolley category, terrorist attacks against Processions/gatherings were very rare; in fact, only six such terrorist attacks were documented in the selected period. However, the highest mean number of casualties among all soft targets categories has been reported for this category (91 dead or wounded people per attack). This high number of victims is due to a vehicle attack

Table 2 Group-specific risk calculated for the categories of soft targets with non-zero incidence and number of fatalities

Coding	Category of soft targets	Relative number of incidents (RN)	Relative number of casualties (RNC)	Risk (RTA)
2	Restaurant/bar/café	0.1051	0.0149	0.0016
8	Hotel/resort	0.0487	0.0133	0.0006
11	Entertainment/cultural/stadium/casino	0.0949	0.0488	0.0046
44	Airport	0.0231	0.0697	0.0016
49	School/university/educational building	0.0564	0.0126	0.0007
74	Marketplace/plaza/square	0.0333	0.0367	0.0012
78	Procession/gathering	0.0154	0.3693	0.0057
79	Public areas	0.0513	0.0049	0.0003
81	Museum/cultural centre/cultural house	0.0308	0.0014	<0.0001
86	Place of worship	0.3051	0.001	0.0003
96	Tour bus/van/vehicle	0.0026	0.1466	0.0004
99	Bus (excluding tour bus)	0.0333	0.0006	<0.0001
100	Train/train tracks/trolley	0.1667	0.1265	0.0211
102	Subway	0.0103	0.1538	0.0016

Note Based on the data from the GTD [12]

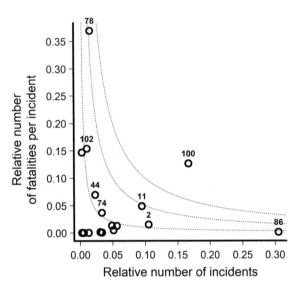

carried out by Mohamed Lahouaiej-Bouhlel in Nice, on 15 July 2016. According to
the GTD [12], 86 people were killed and 433 wounded during the attack.

The third highest risk of terrorist attack was calculated for the Entertain-
ment/cultural/stadium/casino category. The attacks were relatively frequent (n = 37).
In most cases, however, there was evidently no intention to kill: Bomb devices were
usually detonated at night outside opening hours and is often preceded by a telephone
call of the upcoming bomb attack. The reason why these soft targets have the second
highest mean number of casualties caused by terrorist attacks can be found in the
Paris attacks on 13 November 2015, specifically the Bataclan concert hall massacre.
The Bataclan attack has shown the vulnerability of this kind of soft target. In this
case, three perpetrators armed with firearms were able to penetrate the building with
six security agents on duty that night being unable to stop them. The massacre led
to 90 people killed and 217 wounded according to GTD [12]. On the other hand,
security measures proved to be effective at another terrorist attack performed that
day in Paris, the suicide bombing at Stade de France, when three suicide bombers
attempted to get inside the stadium where 79,000 people were watching a friendly
football game between France and Germany. This plan failed after a security guard

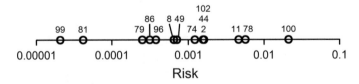

Fig. 2 Calculated risk of terrorist attacks for categories of soft targets. Each type of soft target is
marked using coding of variable "targsubtype1" from GTD ([11]; for details see Table 1)

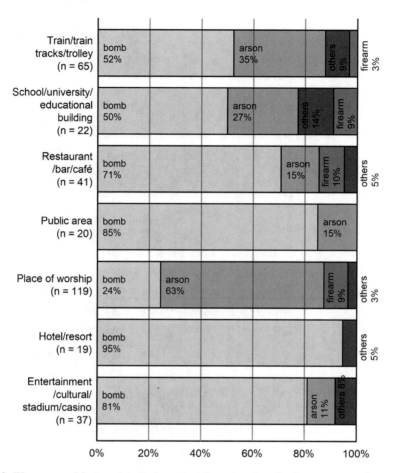

Fig. 3 Weapons used in terrorist attacks against the categories of soft targets most frequently exposed to terrorist attacks. Based on data from GTD [12]

discovered the suicide vest of the first bomber and prevented him from entering the stadium. As a result, instead of hundreds of dead, only one person was killed, when all three perpetrators detonated themselves near the entrance gates to the stadium.

The GTD category with the fourth highest risk of terrorist attack was Restaurant/bar/café. These are relatively common targets, yet the lethality of these attacks is relatively low. In these places, it is unlikely that there would be security guards, cameras, etc. that could prevent a terrorist attack. According to the GTD [12], there were 41 terrorist attacks against these types of soft targets. However, the actual reasons why these targets have the fourth highest risk are due to the Paris attacks on 13 November 2015; specifically on restaurants in the area of the 10th arrondissement. The remaining terrorist attacks against the targets from this category were far from being so devastating. The modus operandi of these attacks was quite similar to the attacks against targets from the Entertainment/cultural/stadium/casino category: In

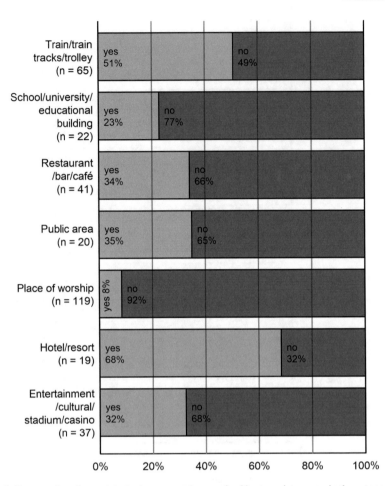

Fig. 4 Frequencies of terrorist attacks apparently organised by terrorist groups in the categories of soft targets most frequently exposed to terrorist attacks

most cases, some kind of explosive device was used (71%). The bomb attacks were often carried out outside opening hours, indicating that the primary goal of these attacks was not to kill civilians.

The fifth and sixth highest risks were calculated for the Subway and Airport categories. Like the Processions/gatherings category, there were very few attacks against these targets; however, some of them were highly deadly. This was the result of the suicide bomber attacks carried out on 22 March 2016 at the Maalbeek Metro Station in Brussels, as well as at Brussels Airport in Zaventem, Belgium. According to the GTD [12, 32 people died and 270 were wounded during these attacks.

For the Marketplace/plaza/square category, a relatively high risk is also associated primarily with one terrorist attack with high lethality. In this case, it was a vehicle attack at a Christmas market in Berlin on 19 December 2016. According to the GTD [12], this attack led to 12 deaths and 48 wounded persons. Terrorist attacks against soft targets from the GTD category of School/university/educational buildings were moderately frequent and lethal. Half of these cases were bomb attacks (50%). However, there is also a relatively high percentage of direct assaults carried out by assailants armed with firearms or knives (18%). Similar risk value was calculated for the GTD category Hotels/resorts. Interestingly enough, the most frequent targets were Spanish or French hotels and resorts. This corresponds with the fact that in most cases, the attacks were carried out either by Basque or Corsican separatist groups, e.g. the ETA (Euskadi ta Askatasuna). Detonated explosives were used as the primary method of attack in all cases (95%).

Terrorist attacks against places of worship as a soft target were frequent in all over the European Union (n = 119), they are in fact the most frequently attacked soft targets [12]. In spite of this, their lethality was very low (0.025 dead and 0.227 wounded people per attack). Regarding these soft targets, an interesting geographical distribution can be observed: attacks against places of worship in France and Germany give 49% of all the terrorist attacks against these types of soft targets. The most targeted places are synagogues and mosques in all of Europe. In most cases, no terrorist organisation claimed responsibility for these attacks and there was no convincing evidence indicating that the attack was committed by any terrorist organisation (92%) as seen in Fig. 4. The most widespread techniques of attack were explosives and arson attacks (setting fires or throwing Molotov cocktails).

4 Discussion

One of two basic features of soft targets according to the common definitions is vulnerability. In suggested procedure, vulnerability was measured by lethality, i.e. the relative number of casualties caused by terrorist attacks against each soft target category. Vulnerability is given by many factors, such as the concentration of people, efficiency of security measures as well as professionalism of the terrorist attack itself [17]. A good argument for evaluating not only the frequency of attacks but also the relative number of victims is an example of soft targets from a places of worship category. During the selected period of twenty years, the GTD [12] registered 119 terrorist attacks against soft targets from this category in the EU. However, all these attacks ended with only 3 persons died and 27 people wounded [18]. This is probably caused by the low level of professionalism of these terrorist attacks. If we only evaluated the frequency of attacks, places of worship would be given the highest priority in most countries (i.e. most investments in security). However, this would lead to a neglect of security of soft targets with a low frequency of terrorist attacks, but a greater severity of these attacks. For these reasons, both the frequency and lethality of terrorist attacks against soft targets were considered in suggested method for risk assessment. Together, they reflect all the important aspects of terrorist attacks against soft targets, i.e. the concentration of people on site, the efficiency of security measures and the target preferences of terrorists.

The method of risk calculation described in the Methods section may be effectively used to assess the group-specific risk of terrorist attack against soft targets from a specific group of targets (regardless of how this group is defined) in a given country or region. On the other hand, this method is not suitable for the calculation of risk for individual soft targets, e.g. for a particular school. This is due to the fact that the procedure does not take into account those significant factors which influence the risk of attacking an individual soft target, e.g. the position of the soft target. For example, the marketplace in the capital or large city has a significantly higher probability of a terrorist attack than a marketplace in a village [14].

Another possible confounding factor is the rate of involvement of terrorist groups in attacks. The significant correlation between the terrorist group involvement rate and the relative number of casualties per attack, which was observed in the analysed sample ($R_S = 0.46$), indicates that terrorist attacks organised by terrorist groups are deadlier than terrorist attacks committed by unaffiliated individuals. This corresponds with a higher rate of bomb attacks in terrorist attacks committed by terrorist organisations (70% of all terrorist attacks organised by terrorist organisations) compared to terrorist attacks committed by individuals or by an unknown perpetrator (45% of all terrorist attacks performed by individuals or an unknown perpetrator). The willingness of terrorist organisations to target civilians primarily varies significantly between groups, depending on, for example, whether an organisation relies on local or international support, as reported by [13]. The proposed procedure also does not consider direct damage to property or indirect economic damage caused by a terrorist attack, e.g. in the tourism sector (for review, see [20]).

To perform a risk assessment for individual soft targets, more complex methods are required. For this purpose, it is possible to use, for example, the method suggested by [3]. This method is elaborated in great detail, since it combines quantitative data with expert judgements, however, its implementation is also significantly more demanding than the procedure presented in this chapter. One promising possibility for the future is the use of computer simulations for risk assessments of soft targets, e.g. [9].

However, even the group-specific assessment of soft targets can be effectively used in safety management in many ways, e.g. by deciding on investments in the security of groups of soft targets or the allocation of subsidies (for review of possible counter-terrorism measures, see [14] As can be seen in Table 3, there are significant differences in group-specific risks of terrorist attack against soft targets between countries. For example, while soft targets from the Marketplace/plaza/square category are most endangered in Germany, the risk of terrorist attacks against these soft targets is much less significant in Spain and France.

Calculated group-specific risk of terrorist attack may be, however, used for comparison between countries or regions as well. As can be seen in Table 3, risk score for the Train/train tracks/trolley category in Spain is an order of magnitude higher than other scores (even the highest risk scores observed in Germany or France). This indicates that the risk of terrorist attack is much more concentrated in one category in Spain than in Germany and France.

In addition to the impossibility of determining risks for individual soft targets, this approach has another significant weakness: It is based solely on historical data. The

Table 3 Group-specific risk calculated for the categories of soft targets in Germany, Spain and France

Coding	Category of soft targets	Germany	Spain	France
2	Restaurant/bar/café	0.0007	0.0007	0.0002
8	Hotel/resort	0	0.0099	0
11	Entertainment/cultural/stadium/casino	0.011	0.0038	0.0006
44	Airport	0	0	0
49	School/university/educational building	0	0.0067	0
74	Marketplace/plaza/square	0.0147	0.0008	0.0002
78	Procession/gathering	0	0	0.0123
79	Public areas	0	0.0023	0
81	Museum/cultural centre/cultural house	0	0	0
86	Place of worship	0.0001	0	0
96	Tour bus/van/vehicle	0	0	0
99	Bus (excluding tour bus)	0	0.0002	0
100	Train/train tracks/trolley	0.0004	0.1583	0
102	Subway	0	0	0

Note Based on the data from the GTD [12]

modus operandi of terrorist attacks, as well as target preferences of terrorists, changes quickly. For this reason, soft target identification based solely on historical data is necessarily not entirely accurate. For example, vehicle attacks carried out repeatedly after 2016 which caused a high number of victims led to a significant increase in the risk score for Marketplace/plaza/square and Procession/gathering categories.

5 Conclusions

In this chapter, a procedure for group-specific risk assessment was proposed. This procedure may be used to assess the group-specific risk of terrorist attack against soft targets for a specific group of targets in selected country or region. In security management, the risk score may be used, for example, as a key to reallocate investments or subsidies in order to increase the security of selected soft targets. The proposed procedure is, however, unsuitable for assessing the risks of an individual soft target. Moreover, another significant weakness of this method is the reliance on historical data on terrorist attacks only, thereby taking into account only the frequency of incidents and number of victims caused by the attack and ignoring other important variables, such as the position of the soft target or adopted security measures. More sophisticated methods should be used for the risk assessment of individual soft targets. Despite this, the proposed method could become a useful tool for security management at the state level due to its simplicity and data availability, e.g. GTD [12].

References

1. Asal VH, Rethemeyer RK, Anderson I, Stein A, Rizzo J, Rozea M (2009) The Softest of targets: a study on terrorist target selection. J Appl Security Res 4(3):258–278
2. Beňová P, Hošková-Mayerová Š, Navrátil J (2019) Terrorist attacks on selected soft targets. J Security Sustain Issues 8(3):453–471. https://doi.org/10.9770/jssi.2019.8.3(13)
3. Cuesta A, Abreu O, Balboa A, Alvear D (2019) A new approach to protect soft-targets from terrorist attacks. Saf Sci 120:877–885
4. European Commission (2017) COM (2017) 41 Communication from the Commission to the European Parliament, the European Council and the Council: Fourth progress report towards an effective and genuine Security Union. Brussels
5. Fagel MJ, Hesterman J (2016) Soft targets and crisis management: what emergency planners and security professionals need to know. Routledge, New York
6. Grant MJ, Stewart MG (2017) Modelling improvised explosive device attacks in the West—Assessing the hazard. Reliab Eng Syst Saf 165:345–354
7. Hošková-Mayerová Š, Bekesiene S, Beňová P (2021) Securing Schools against Terrorist Attacks. Safety 7(1):13. https://doi.org/10.3390/safety7010013
8. Libicki MC, Chalk P, Sission M (2007) Exploring terrorist targeting preferences. RAND Corporation, Santa Monica
9. Liu Q (2020) A social force approach for the defensive strategy of security guards in a terrorist attack. Int J Disaster Risk Reduct 46:101605
10. Martin RH (2016) Soft targets are easy terror targets: increased frequency of attacks, practical preparation, and prevention. Forensic Res Criminol Int J 3(2):273–278
11. National Consortium for the Study of Terrorism and Responses to Terrorism (2019) Global Terrorism Database – Codebook: inclusion criteria and variables. University of Maryland, Maryland [online 10–03–2021]. Available online at: https://www.start.umd.edu/gtd/
12. National Consortium for the Study of Terrorism and Responses to Terrorism (2021) Global Terrorism Database [online 10–03–2021]. Available online at: https://www.start.umd.edu/gtd/
13. Polo SMT (2020) The quality of terrorist violence: Explaining the logic of terrorist target choice. J Peace Res 57(2):235–250
14. Quagliarini E, Fatiguso F, Lucesoli M, Bernardini G, Cantatore E (2021) Risk reduction strategies against terrorist acts in urban built environments: towards sustainable and human-centred challenges. Sustainability 13(2):901
15. R Core Team (2019) R: a language and environment for statistical computing. R Foundation for Statistical Computing, Vienna
16. Tušer I, Bekešienė S, Navrátil J (2020) Emergency management and internal audit of emergency preparedness of pre-hospital emergency care. Qual Quant. https://doi.org/10.1007/s11135-020-01039-w
17. Tušer I (2020) The development of education in emergency management. In: Flaut D., Hošková-Mayerová Š., Ispas C., Maturo F., Flaut C. (eds) Decision making in social sciences: between traditions and innovations. Studies in systems, decision and control, vol 247. Springer, Cham. https://doi.org/10.1007/978-3-030-30659-5_10
18. Tušer I, Jánský J (2021) Security management in the emergency medical services of the Czech Republic—Pre-case study. In: Soitu D., Hošková-Mayerová Š., Maturo F. (eds) Decisions and

trends in social systems. Lecture notes in networks and systems, vol 189. Springer, Cham. https://doi.org/10.1007/978-3-030-69094-6_32

19. Zeman T (2020) Soft targets: definition and identification. Acad Appl Res Mil Public Manag Sci 19(1):109–119. https://doi.org/10.32565/aarms.2020.1.10

20. Zeman T, Urban R (2019) The negative impact of terrorism on tourism: not just a problem for developing countries? Deturope 11(2):75–91

Preparation for an Extraordinary Event in the Railway Area

Jiří Barta and Michaela Henzlova

Abstract Current time and modern technology bring new security risks. Although the situation is currently stable the Czech Republic, the risks and threats are not disappearing, and it is necessary to be prepared for them. The chapter deals with the issue of soft targets in relation to contemporary terrorism. In order to eliminate the threats of terrorist attacks and their impact on human activities, it is necessary to increase the crisis preparedness of the public administration for extraordinary events (hereinafter also "emergency"). For the elaboration of emergency plans, it is necessary to perform analysis of all subject's possibilities for the needs of dealing with an emergency, verification of possibilities of emergency communication and the ability to respond to an emergency. Verification of the ability to react is carried out through risk analysis, processing of planning documentation and subsequent practical exercises of the crisis management, who is involved.

Keywords Emergency · Crisis management · Terrorism · Terrorist attack · Soft target · Preparedness · Training

1 Introduction

In recent years, there is a significantly increasing number of attacks on soft targets. Soft targets are generally understood as places with a high concentration of people and at the same time with a low degree of security [7]. The reasons why soft targets are target of attacks are based on its general characteristics. The attractiveness of soft targets for attackers is enhanced by the almost constant movement of large

J. Barta (✉)
Department of Military Science Theory, University of Defence, Kounicova 65, 61200 Brno, Czech Republic
e-mail: jiri.barta@unob.cz

M. Henzlova
Department of Security and Law, AMBIS College, Lindnerova 575/1, 180 00 Prague, Czech Republic
e-mail: michaela.henzlova@ambis.cz

© The Author(s), under exclusive license to Springer Nature Switzerland AG 2022
I. Tušer and Š. Hošková-Mayerová (eds.), *Trends and Future Directions in Security and Emergency Management*, Lecture Notes in Networks and Systems 257,
https://doi.org/10.1007/978-3-030-88907-4_9

numbers of people. In addition, there is another characteristic of the soft targets, it is easy availability and a low level of protection. In general, the weaker protection and higher concentration of people, the greater likelihood of attack. Due to characteristic of the soft targets, they are often selected as a target of terrorist attacks [8].

The level of security of the environment differs from soft targets to so-called hard targets. The hard targets are well-protected and guarded objects. This group of hard targets includes for example state administration objects (especially the objects of ministries), military objects or objects essential to meet the needs of the population [8].

The Czech Republic is trying to approach the topic of protection of soft targets comprehensively. It addresses this topic within the framework of the Strategy of the Czech Republic for the Fight against Terrorism, the Methodology Fundamentals of Soft Target Protection or the Czech Technical Standard 73 4400 Crime Prevention—Security Management in the Planning, Implementation and Use of Schools and School Facilities. This standard is applicable to public, private or state institutions. This standard can be used only for risks associated with the protection of individuals and property. The design and implementation of security measures is based on security risk management systems. Within the preventive management of security risks, it is necessary to consider the changing social and cultural specifics of the given areas in which the buildings are located [9].

These security documents are one of many examples of the Czech Republic's active approach to the protection of soft targets. However, some attacks show that soft targets become targets not only for terrorist attackers, but also for ordinary perpetrators of violent crime, such as the attack in Uherský Brod or the shooting incident in a shopping centre in Munich [8].

2 Threat to Soft Targets

Threats that can affect soft targets can be divided into natural and anthropogenic. Examples of natural origin emergencies are large-scale floods, prolonged droughts or extreme weather events. Examples of emergencies of anthropogenic origin (i.e. emergencies caused by human activity) are serious accidents, leakage of biological agents or terrorism. Influences of anthropogenic threats are relatively very dynamic. On the other hand, natural threats are gradual and barely manageable. Their occurrence can be relatively reliably predicted and preventive measures adapted accordingly.

The scope and size of the impact of emergencies is influenced by various factors such as population density, infrastructure cohesion or the risk of a domino effect. The issue of securing soft infrastructure, especially from the point of view of terrorism, is very topical, timeless and hides a few security consequences. Soft targets can cover a very wide range of public and socially important sites in view of the increased risk of a terrorist attack [6]. These can be, for example:

- night clubs, restaurants,

- parks and squares, tourist monuments,
- shopping centres,
- parades, demonstrations,
- sports halls,
- theatres, concert halls,
- major transport hubs, train and bus stations, airport terminals,
- hospital,
- school facilities.

Based on Mr. Zeman analysis [24], which drew data from the Global Terrorism Database (National Consortium for the Study of Terrorism and Responses to Terrorism), it was found that 90.2% of terrorist attacks carried out in 2000–2015 were carried out through bombing attack. The data also showed that the largest number of people were killed and an average of most injured in terrorist attacks on rail transport infrastructure (44 attacks), as shown in Fig. 1. In the context of the attacks on the area of railway infrastructure, there was 36% of the attacks carried out on the area of the railway station (36% of attacks on railway lines and 18% of attacks on trains), so this is an area that we need to pay attention [18]. There are few

Coding	Target	Number of incidents	Mean number of wounded	Mean number of dead
2	Restaurant/bar/café	29	3.759	1.345
8	Hotel/resort	19	3.353	0.278
11	Entertainment/cultural/stadium/casino	36	9.629	2.6
44	Airport	8	0.375	0
49	School/university/educational building	15	2.333	0.733
57	Civilian maritime	0	0	0
60	Port	0	0	0
74	Marketplace/plaza/square	7	5	0.143
78	Procession/gathering	4	5.75	0.25
79	Public areas	23	0.957	0.087
81	Museum/cultural centre/cultural house	6	0	0.667
86	Place of worship	63	0.27	0.016
96	Tour bus/van/vehicle	1	30	6
99	Bus (excluding tour bus)	9	0.111	0
100	Train/train tracks/trolley	44	40.977	4.341
101	Bus station/stop	2	0	0
102	Subway	3	0	0
103	Bridge/car tunnel	1	0	0
104	Highway/road/toll/traffic signal	5	0	0
	Total	275	8.989	1.28

Note: Mean numbers were calculated as the sum of wounded or dead people divided by the number of incidents.

Fig. 1 Selected soft targets with the number of incidents and the average number of victims between 2000 and 2015 [24]

Security	Group 1	Group 2
Camera system	X	X
Safety frame		
Sensors that signal the violation of external parts		
Security of publicly inaccessible areas of the hall (bars, security doors, authenticated entrance, ...)	X	X
Security Service	X	
Safety foils and safety glasses	X	X
Electrical security systems	X	X
Electronic fire alarm	X	X

Fig. 2 Comparison of security of railway station halls

security features at the railway station, as opposed to entertainment venues, sports stadiums or casinos, and for this reason they can be an attractive target for a terrorist attack [4, 24].

For the purpose of this study was chosen the railway station hall as an area of research in railway infrastructure. The railway stations halls are highly exposed sites and most of them are places with a low level of security. Four railway stations in the Czech Republic were visited as part of the survey. Two of these stations are situated in larger cities (hereinafter also "group 1"), another 2 in cities with a population of up to 15,000 (hereinafter also "group 2"). Due to the non-approval of the facility manager, it is not possible to specify the railway station in more detail. The differences in the security of these two groups are summarized in Fig. 2.

As can be seen from Fig. 2, the railway stations in both groups are secured at the similar level. The difference of the group 2 is mainly in the provision of security services, or security guards. This is one of the most effective tools for conducting inspections within the building, given that there is free movement of persons.

Security guard carries out inspections (a tour of):

- at fixed locations (e.g. at the entrance),
- at mobile stations (control activity of e.g. train sets),
- at patrol stations (regular patrols in the protected area).

Due to the fact that station halls fall into the category of soft targets, there are almost no preventive or other measures stipulated by Act No. 240/2000 Coll. (Crisis Act) or by other legal regulations of the Czech Republic dealing with security.

It is possible to take an example from objects of critical infrastructure (hereinafter also "critical infrastructure") and apply the security measures used there to soft targets, at least partially. Crisis preparedness requirements for critical infrastructure have been defined in [6], as most critical infrastructure assets in OECD countries are privately owned. It is estimated that 80% of critical infrastructure is owned or operated by the private sector [2, 17]. When using preventive measures that are in the Czech Republic commonly used, the study deals with the possibility of preparing professional security personnel and workers moving on the premises of the railway station, as people who would have to react to an emergency. For the study it was chosen as a model of emergency a terrorist attack.

There are several possibilities for the preparation of both mentioned groups of workers. One of the possibilities is regular safety training. The great disadvantage of it is the need for high specialization of the trainer, high costs for this training and also the need to transport employees to this training. Due to the high cost of these training, it is almost impossible for employers to let all employees train.

Another option that can save a considerable amount of time is e-learning, which is being used still more and more. E-learning is often associated with a practical part of training. This connection is probably one of the best options to choose. Employees can't gain only theoretical knowledge through e-learning, which they can check for example with a test, but thanks to the practical part of the training, they can also test the newly acquired knowledge in practice, using a simulated emergency [4, 11].

Despite little experience with this emergency in the Czech Republic, it is necessary to approach the preparedness process responsibly and take into account all aspects of life, such as the readiness of the population itself, rescue services, social and medical facilities in relation to communication skills [5].

3   Used Methods

The security audit method was used as the main one to identify possible threats to the railway station hall, as one of the possible methods of risk analysis. This method allowed us to identify threats to the object which were studied.

For the implementation of the exercise was further elaborated the threat of a terrorist attack using explosives (i.e. booby-traps). In conjunction with other methods (attack tree method), an emergency scenario was created, including the course of activation of a selected threat. On this basis a simulation was created for practical training (e-training).

Another method used for the purpose of the study was the method of a controlled interview with some employees of the railway station.

For the preparation of practical exercises and its implementation was used the method of constructive simulation, where the simulated entities are controlled by a simulated operator. Constructive simulation is a type of simulation where the model contains everything needed to replace the original during the simulation, even a human. In constructive simulation, is a man expressed by a sub model. In the events of constructive simulation, the decisions of these simulated subjects are then applied. The control of this type of simulation is realized by means of the user interface.

The display of the synthetic environment is similar to a topographic map. Constructive simulation is used in different resolution levels for different types of operations in dealing with emergencies [10, 11].

4 Possibilities of Practical Exercises of Interested Workers

The following part of the text focuses on the proposal to provide practical exercises for crisis management bodies, especially for employees of the concerned organizations.

On the basis of type plan of the Czech Republic "Violation of large-scale legality" to which this extraordinary event belongs, the crisis management bodies are obliged to incorporate the methods of resolving the situation into the emergency documentation of regions and municipalities with extended powers. Each stakeholder should check and ensure preparedness for all types of emergencies that were evaluated during the analysis and that may occur in its facilities [20]. Within the prepared scenario, each interested party can practice the tasks for which it is responsible, and at the same time it can test the coordination with other entities with which it cooperates in dealing with the emergency [19].

Each practical exercise can be divided into four consecutive stages forming a cycle. The first stage is preparatory, in which a scenario of the course of an emergency with a timeline of the exercise is created. In this stage, the goals of the practical exercise are also set, together with the procedures of individual actors for their successful fulfilment. A logistic support is defined to ensure the implementation of the scenario and complete documentation is created for the implementation of practical exercises [3]. An integral part of the preparatory documentation is also a plan for the merger of individual stakeholders and the possibility of crisis communication [15]. The next stage of the practical exercise is the execution of the exercise itself. According to their knowledge and skills, the instructor tries to achieve the successful fulfilment of the goals of the practical exercise within the defined scenario. After performing the practical exercise, the evaluation stage follows. Within this stage, the achieved results are compared with the planned goals according to the set scenario. The strengths and weaknesses of the trainers and their abilities and skills in performing the assigned tasks within the solution of the emergency are identified. The final stage is to create a plan of adequate measures. It is developed on basis of a previous evaluation and the identified weaknesses. Applied measures and their implementation into emergency planning processes should lead to improved crisis management procedures [14].

The basic documents of the preparatory stage of the practical exercise include determining the probable extent of the emergency. To do this, it is advisable to use a suitable modelling program, which is designed to calculate the explosion range of booby-trapped explosive systems. Based on previous analyzes [16] and [21], the TerEx simulation program was selected, which serves for immediate evaluation of the extent of leakage of hazardous chemical and toxic substances or detonation of a booby-trapped explosive system.

4.1 Explosion Range Model of a Booby-Trapped Explosive System

The EXPLOSIVE module can be used in the TerEx program to simulate the extent of the explosion of a booby-trapped explosive system. It simulates possible impacts of explosive detonation systems based on condensed phase used to threats around the detonation.

Within the calculation, it is possible to enter a precisely defined amount of specific explosive, or to misuse the possibility of an unknown substance with a professionally estimated amount of explosive according to the size of the luggage in which the unknown explosive is transported. Details of entering the simulation parameters can be seen in Fig. 3.

Based on the performed analyzes [22] and [23] it turned out that the most frequently used booby-trapped explosive system is brought in a backpack or plastic bag. This corresponds to the equivalent of 5 kg of explosive. Due to the variety of

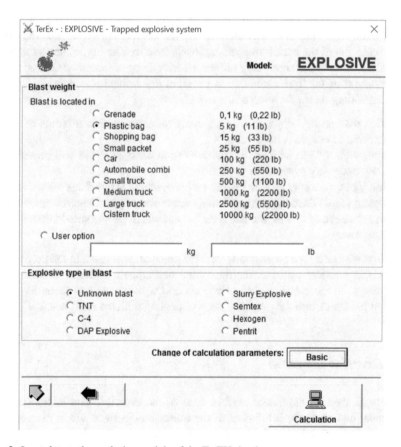

Fig. 3 Input data to the explosive module of the TerEX simulator

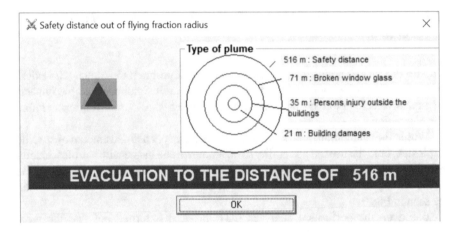

Fig. 4 Output data from modeling in TerEx

explosives used in attacks on soft targets, it is not possible to determine the exact type of explosive. Therefore, an unknown explosive was used to simulate the extent of the explosion of the booby-trapped explosive system. The Fig. 4 shows the result of the simulation in the simulation program TerEx (5 kg of unknown explosive).

The output of the TerEx simulation program determined the individual hazard zones depending on the location of the explosion:

- solid walls and buildings will be seriously damaged within a radius of 21 m; evacuation of people is necessary,
- persons within 35 m who are not hidden inside fixed buildings will be seriously injured; necessary evacuation of persons,
- within 71 m, windows will be broken and people inside buildings will be injured by flying shards from broken windows, evacuation of people is recommended,
- from a distance of 516 m is a safe zone behind which people should no longer be endangered.

By that was given the basic parameters for practical exercises. In the preparatory stage of the practical exercise, the individual participants of the exercise are also acquainted with the scope of the emergency and with the created scenario, which preceded the attack and the potential steps of dealing with the emergency.

4.2 Exercise Scenario

The second stage is the actual performance of the exercise. The instructor tried to achieve the successful fulfillment of the exercise objectives within the prepared emergency scenario. The individual steps of the scenario were defined on the basis of the emergency documentation of the affected object. The basis of the scenario brought the bodies exercising to the point of emergency. The location of the explosion was defined and the previous simulation in the TerEx program results in individual affected areas—danger zones. To create the exercise, it was best to start from an emergency that has happened in the past [14].

A possible emergency scenario could start as follows: On December 29, 2019, at 4:45 p.m., an unspecified substance exploded in the entrance hall of the railway station. The extent of environmental damage is shown in Fig. 4. The subsequent development of the scenario already depends on the trainees, how they will proceed in dealing with this emergency. Due to cost savings and simplification of processes, it is appropriate to use a simulation program for practical exercises [10, 12].

4.3 Practical Exercises Using a Simulator

It is always necessary to choose a suitable and available simulation program for the implementation of the exercise. Within the research project, this issue was addressed and the basic criteria that the simulation program for performing practical exercises should meet [3, 11, 12] were determined. The basic criteria included user friendliness and variability of the simulator:

- scene editor;
- implementation of external data;
- simulation level (teams or individuals);
- communication possibilities;
- continuity in relation to the surrounding environment.

Based on the set criteria, the SIMEX simulator was selected. This simulator made it possible to use a constructive simulation to simulate the activities of individual components of the integrated rescue system and other interested entities. The simulator must contain a closed configurable communication system that can fully replace the communication lines of the trainees. All communication must be recorded synchronously with the simulation, so that the recording can be used for subsequent evaluation of the exercise [3].

In terms of controlling the simulation environment in the simulation workplace, there are two possible approaches to performing practical exercises:

A. Forces and resources in the simulator instructors are controlled directly during exercise.
B. The trainers only issue instructions within their scope and the forces and resources in the simulator are controlled by a trained operator based on the trainers' instructions.

Due to the fact that the practical exercises take place individually and in different compositions of the trainees, departments and institutions, it is inefficient to use the "A" approach, where it would be necessary to teach the trainees to operate the simulator. This would require considerable financial costs in the form of time for trainers and lecturers and would disproportionately increase the time allocated to practical exercises.

Solution "B" is more suitable, where the forces and means in the simulator are controlled by a trained operator on the basis of the instructors' instructions. If the scenario in the simulator is well prepared, the individual entities in the simulator can be programmed in advance and the actual control of the simulator during the practical exercise can be done by one trained worker [12].

Advantages of solution "B":

- shorter training time—it is not necessary to train the trainees in the control of the simulator;
- the trainee focuses only on his tasks—decision-making processes;
- lower frequency of user errors in simulation control.

Disadvantages of solution "B":

- higher time required to create exercise scenarios in the simulator;
- higher staffing requirements for simulator operation;
- higher personnel costs for operating the simulator.

Although sufficient funding for crisis management must be allocated in the budget of each organization [13], its size plays an important role incurred at work and the training of personnel for crisis management. By using the simulator, costs are effectively spent and the simulator provides materials for the management and decision-making functions of the trainees. An important role in the simulation is played by the communication system, which is essential for practical exercises. Based on the variability of the closed communication system of the SIMEX simulator, communication means (telephone, e-mail, walkie-talkie) can be assigned to each instructor or each group of trainees (based on the supplied data on the real equipment of individual workplaces). In a similar way, information technology (computers, internet connection, access to databases,...) can be assigned to trainers, depending on what these employees actually have at their workplace.

This determined the parameters of the exercise in which the trainers can move. The basis is to realize that most trainees have never been in a situation where the railway station appears to an attacker who is prepared to commit a terrorist attack.

5   Evaluation of the Exercise Possibilities

The third stage of the practical exercise is its evaluation. Within this stage, it is necessary to compare the achieved results with the set goals of the planned scenario of dealing with an emergency. An important role in the evaluation of practical exercises is played by immediate initial evaluation, in which all available communication is synchronously recorded with the course of the simulation. The evaluation can be supplemented with selected sections of the simulation, or display a visual 3D projection of the simulation with the communication that was taking place at that moment.

As part of the evaluation, it is necessary to identify the strengths and weaknesses of the trainers and their dispositions and abilities to perform tasks during the solution of the emergency. This evaluation is the basic basis for the development of a plan of adequate measures focusing on the identified weaknesses and shortcomings.

6 Conclusion

To meet the requirements of the Concept of Protection of the Population of the Czech Republic, it is necessary to focus on the prevention of emergencies. The basis of prevention is primarily preparedness, which is a part of crisis management. The best way to train employees is an implementation of practical exercises. Thanks to that it is possible to test knowledge, skills and abilities of employees in dealing with an emergency. The acquisition of new skills and deepening of existing skills of employees will enable a timely response that minimizes losses in the event of a terrorist attack. Based on the evaluation of the exercises, it is possible to design a plan of measures that will improve decision-making processes in dealing with real emergencies.

The chapter also describes the preparatory phase of the practical exercise, including parts of the proposed scenario and the set criteria. The preparatory phase of the practical exercise is focused on the specification of the schedule and defined scenario. The prepared scenario was aimed at simulating the basic possibilities of responding to an emergency and allowing the trainers to solve the situation according to their abilities. The scenario focused mainly on the communication skills of individual practitioners and their professional terminology.

Within the used simulator, communication interfaces for emergency communication were applied according to the prepared connection plan for defined emergency. Also the movements of forces and means were simulated according to the decision-making processes of the trainees. Based on the performed simulations, it is confirmed that the use of the simulator for the needs of practical exercises supports the decision-making processes of the trainees. It brings the simulated scenarios closer to reality. The practical exercises themselves and the skills acquired contribute to increasing the ability of all concerned to respond quickly and more effectively to possible emergencies of this nature.

References

1. Act No. 240/2000 Sb., on crisis management and amending certain Laws (Crisis Law). Collection of law No. 118/2011, No. 44, pp. 1114–1135. ISSN 1211–1244
2. Arroyo J, Alguacil N, Carrio X (2010) A risk-based approach for transmission network expansion planning under deliberate outages. IEEE Trans Power Syst 25(3): 1759–1766
3. Barta J (2017) Comparison of simulators used for education and practical training of the critical infrastructure staff. E-learning 9:279–293. ISSN 2451–3644
4. Barta J, Vašková M, Urbánek J (2016) Evaluation of simulation programs applicable to the support of decision-making processes in crisis management of critical infrastructure. Int J Educ Learn Syst 2016(1):74–80. ISSN 2367–8933
5. Brehovská L, Freitinger Skalická Z, Šimák-Líbalová K, Líbal L (2015) Safety research of population according to population differentiation in Czech Republic. Int J Educ Inform Technol 9:12–20. ISSN 2074–1316. Available at: http://www.naun.org/cms.action?id=10200
6. Council Directive 2008/114/EC of 8 December 2008 on the identification and designation of European critical infrastructures and the assessment of the need to improve their protection. Offi J Eur Union L 345/75–82
7. Czech Republic (2016) Terminological dictionary of terms from the field of crisis management, population protection, environmental security and national defense planning. (In Czech: Terminologický slovník pojmů z oblasti krizového řízení, ochrany obyvatelstva, environmentální bezpečnosti a plánování obrany státu.) [online] [cit. 2020–08–20]. Ministry of Interior. Available at: http://www.mvcr.cz/
8. Czech Republic (2020) Protection of soft targets. (In Czech: Ochrana měkkých cílů.) [online] [cit. 2020–08–20] Ministry of Interior. Available at: https://www.mvcr.cz/clanek/ochrana-mekkych-cilu.aspx
9. ČSN 73 4400 (2016) Prevence kriminality – řízení bezpečnosti při plánování, realizaci a užívání škol a školských zařízení. Praha: Úřad pro technickou normalizaci, metrologii a státní zkušebnictví, 44 s
10. Hubáček M, Vráb V (2012) The use of constructive simulation for policemen training. The Science for Population Protection [online]. 3/2012 [cit. 2020–08–16]. ISSN 1803–635X. Available at: http://www.population-protection.eu/prilohy/casopis/19/130.pdf
11. Hubáček M, Řezáč D (2013) Simulation technology and training of rescue services. The Science for Population Protection [online]. 3/2013 [cit. 2020–08–16]. ISSN 1803–635X. Available at: http://www.population-protection.eu/prilohy/casopis/16/118.pdf
12. Kincl P, Oulehlová A (2018) Comparison of methods of student education and implementation of exercises of rescue services and crisis management bodies in preparation for dealing with emergencies and crisis situations in a constructive simulation environment. In: 13. doctoral conference: New approaches to ensuring state security. (In Czech: Komparace metod edukace studentů a provádění cvičení záchranných složek a orgánů krizového řízení při přípravě na řešení mimořádných událostí a krizových situací v prostředí konstruktivní simulace. In: 13. doktorandská konference: Nové přístupy k zajištění bezpečnosti státu). Brno: University of Defence in Brno, s. 78–85. ISBN 978–80–7582–037–2
13. Kudlák A, Urban R, Hošková-Mayerová Š (2020) Determination of the financial minimum in a municipal budget to deal with crisis situations. Soft Comput 24:8607–8616. https://doi.org/10.1007/s00500-019-04527-w
14. Oulehlová A, Malachová H, Kincl P, Navrátil J (2016) Simulated Exercise—"Gale" Crisis Scenario. In: Vision 2020: innovation management, development sustainability and competitive economix growth. VOLS I–VII. International Business Information Management Association (IBIMA), Seville, pp 3867–3876. ISBN 978–0–9860419–8–3
15. Oulehlová A, Kavan Š (2017) Preparation for providing crisis communication during blackout occurrence. In: Proceedings of the 30th international business information management association conference, vision 2020: sustainable economic development, innovation management, and global growth. International Business Information Management Association (IBIMA), Madrid, pp 1416–1425. ISBN 978–0–9860419–9–0

16. Ptáček M, Ščurek R, Holubová V (2018) The improvised explosive device (IED) as a threat to infectious hospital wards. Sci Popul Protect 1(10):41–50. ISSN 1803–568X
17. Rehak D, Novotny P (2016) Bases for modelling the impacts of the critical infrastructure failure. Chem Eng Trans 53:91–96. ISBN 978–88–95608–44–0; ISSN 2283–9216. https://doi. org/10.3303/CET1653016
18. Rehak D, Slivkova S, Pittner R, Dvorak Z (2020) Integral approach to assessing the criticality of railway infrastructure elements. Int J Crit Infrastruct 16(2):107–129. https://doi.org/10.1504/ IJCIS.2020.107256
19. Tušer I, Bekešienė S, Navrátil J (2020) Emergency management and internal audit of emergency preparedness of pre-hospital emergency care. Qual Quant. https://doi.org/10.1007/s11135-020-01039-w
20. Urbánek JF, Johanidesová J, Urbánek JJ, Barta J (2017) Operational improvement of systems and processes at critical infrastructure detection check sites. In: Risk, reliability and safety: innovating theory and practice. Taylor & Francis Group, London, pp 1831–1835. ISBN 978–1–138–02997–2
21. Vašková M, Náplavová M, Barta J (2018) Awareness and preparation of the population for emergencies. In: Safety and reliability—safe societies in a changing world. Taylor & Francis Group, London, pp 45–52. ISBN 978–0–8153–8682–7
22. Zeman T, Břeň J, Urban R (2018) Profile of a lone wolf terrorist: a crisis management perspective. J Secur Sustain Issues 8(1):5–18. ISSN 2029–7017
23. Zeman T, Urban R (2019) The negative impact of terrorism on tourism: not just a problem for developing countries? Deturope 11(2):75–91. ISSN 1821–2506
24. Zeman T (2020) Soft targets: definition and identification. Acad Appl Res Military Public Manage Sci 18(3). ISSN 2498–5392

Assessment of Security Aspects of Immigration

Olga Hararova and Alena Oulehlova🄳

Abstract Despite the positive aspects of migration, it is necessary to pay attention to the negative effects of migration on countries and regions, as well as on individuals. In the 1990s, migration began to be associated with the threat of crime and the infiltration of Western countries by Islamist terrorist groups. The chapter deals with a specific part of the protection of the population in connection with immigration and the subsequent assessment of criminal risks resulting from the security aspects of immigration. Time series of data from important crime groups were used to assess criminal risks and trends in the development of the security environment in the Czech Republic, including impacts on security, were taken into account. Based on the results, the chapter presents a proposal for the methodology of criminal risk assessment in the environment of the Czech Republic and in relation to the Slovak Republic and the Republic of Poland based on a set of risk factors.

Keywords Crime · Immigration · Methodology · Migration · Risk factors · Security aspects · Time series

1 Introduction

Migration represents one of social processes that has accompanied humans since time immemorial. It can be defined as movement of people between two territorial units. Its causes are varied and depend on the motivation of migrants. These can be, for example, political, economic, religious or environmental motives. Throughout history, human migration has changed aspects of all countries and continents, the racial and ethnic structure of the population, together with the linguistic structure [14].

O. Hararova · A. Oulehlova (✉)
Department of Military Science Theory, University of Defence, Kounicova 65, 662 10 Brno, Czech Republic
e-mail: alena.oulehlova@unob.cz

O. Hararova
e-mail: olga.hararova@seznam.cz

© The Author(s), under exclusive license to Springer Nature Switzerland AG 2022
I. Tušer and Š. Hošková-Mayerová (eds.), *Trends and Future Directions in Security and Emergency Management*, Lecture Notes in Networks and Systems 257,
https://doi.org/10.1007/978-3-030-88907-4_10

Humans migrated for food and resources from the very beginning of their exis-tence. The first rests of human ancestors, from whom Homo sapiens later evolved, were discovered in Africa. People gradually spread from this continent to the whole world. Significant historical migrations were the Great Greek colonization in the 8th–sixth centuries, migration of nations in the 4th–seventh centuries AD, and the Arab expansion into the Middle East and Europe in the 7th–eighth centuries AD [5]. The goal was always similar: to acquire new territories, land, raw materials and space for the sale of goods [14].

The largest population movement occurred after the discovery of America. The period of the first wave was in the 1830s and the second wave began in the 1880s, later migration related to the First World War. After the First World War, the old states disintegrated and new ones emerged. In the interwar period, the demand for labour almost ceased and states began to introduce measures in the form of migration laws, passports and regulation of free movement of people. Forced migration was typical for the period of the Second World War; in the second half of the twentieth century, migration was mainly influenced by the development of the policies of individual states [14].

Rapid development of globalization, especially communication and information technologies and transport in the first decade of the twenty-first century, contributed to the increase in the free movement of people. Service markets expanded, new investment opportunities emerged and labour markets opened up. As a result of this, all forms of migration were on the rise [14].

The issue of migration came to the forefront of global attention, especially in the second decade of the twenty-first century. In 2015, Europe faced the greatest migra-tion challenge since World War II. The crisis did not weaken until the end of 2018. Massive numbers of refugees and illegal migrants who came to Europe highlighted the shortcomings and gaps in the migration policies of individual countries. The European Union appealed on individual states to reform their asylum and migration policies [7, 14].

This experience has shown that migration policy needs to be addressed inten-sively and both immediate and long-term measures need to be developed and imple-mented. Due to national and international disputes and conflicts, global poverty, environmental changes and other factors, migration crisis can be expected to recur. Migration in the Czech Republic, as in all countries of the world, has economic, social and security impacts. Arrival, stay and integration of foreigners in the Czech Republic represent processes that bring possible positive and negative consequences for the Czech society. Active and flexible approach of the Czech Republic can signif-icantly influence these processes. In order to minimize negative risks associated with migration, which were also pointed out by the National Security Audit of the Ministry of the Interior of the Czech Republic (MOI CR), the Czech Republic takes measures in the form of normative legal acts and strategic documents. The key principles of the migration policy of the Czech Republic are defined in the Strategy on Migration Policy of the Czech Republic [14, 21].

The aim of the chapter is to present a proposal for a methodological procedure for the assessment of crime risks resulting from migration and to verify the procedure on statistical data on the crime committed by both foreigners and citizens in the Czech Republic. Statistical data was also obtained for citizens and foreigners in the Republic of Poland and the Slovak Republic and it was processed using the same methodological procedure. Statistically processed results of crime risks and their trends were compared between individual states. Statistical analysis of crime data of citizens and foreigners on the territory of the above-mentioned states was performed for the 2005–2015 period. This period provides continuous and mutually comparable data, as the methodology of data collection and its publication was subsequently modified and the input data after 2016 does not provide its consistency and reliability with the previous period. For the period after 2015, a prediction of crime risks resulting from migration was carried out.

2 Current State

The chapter on current state defines migration typology and explains the basis of migration policy applied by the European Union. Based on it, the approach of the Czech Republic to dealing with the issue of migration is presented and explained, including the identified security aspects resulting from migration.

2.1 Migration Typology

Migration can be defined and analysed from different points of view. Migration has not yet had a uniform legal definition established by an international agreement [10]. At the international level, solely the term migrants has been defined. According to the UN definition, migrants are persons who stay outside their country of origin for three to twelve months [16]. The definition does not take into account the reasons for which the migrant left the country of origin [14].

Migration is a dynamic process of moving individuals or groups of inhabitants in a territory that could be short-term, long-term, permanent or repeated in terms of time [27]. Furthermore, migration can be described as individual, collective and mass according to the number of migrants [31]. The basic division of migration is set according to the location into internal and international migration. Internal migration represents the movement of people from rural areas to cities within the home country, international migration means movement of people from one country to another [18]. Another division takes into account the direction and differentiate emigration and immigration. Immigration means that people from abroad settle on the territory of a particular state. Emigration represents the opposite phenomenon, people leave or flee from the state [14, 31].

Migration is influenced by motivational effects referred to as push and pull factors or a combination of them. For example, economic instability in the home country, changes in living standards, war, national and religious clashes or rapid demographic

growth can be considered as push factors. Pull factors motivate migrants to seek destination countries with the assumption of economic prosperity, higher life quality,
political stability and freedom [4]. Based on these factors, migration can be divided
into voluntary and forced. Voluntary migration arises from person's own initiative,
such as job opportunities, studies, family reunification or other personal reasons.
Forced (involuntary) migration forces the people to leave their home country, for
example due to political power and strength or unequal power relations between
people [31]. According to the cause, migration can be divided into legal and illegal
migration. Legal migration is a process of organized, state-controlled immigration
regulated through residence permits which is generally related to labour migration,
which can meet the needs of the domestic labour market through the regulated admission of foreigners. The purpose of legal migration is, for example, study or family
reunification. Illegal migration takes place without control and organization by the
target countries. Foreigners enter or stay in destination countries without a proper
residence permit. Illegal migration is related to the refugee category and is usually
organized by networks of human smugglers and may be attributed the aspect of
organized crime [14, 22].

Protection of both internal and external borders plays an important role in the
migration process. The EU internal border means the common land borders of the
Member States, airports located on the territory of the Member States and seaports
designated for regular ferry services, all exclusively from and to the territory of
Member States, within the Schengen[1] area. Internal borders form common land
borders with neighbouring states, including river and lake borders and airports located
on the territory of the corresponding state. The EU external border, in the context of
the Schengen area, represent the land, air and sea border of Member States, unless
they are internal borders. These can be, for example, international airports [14, 27].

Migration flows currently include not only refugees but also migrants, especially
work migrants. There are fundamental differences between these concepts. A migrant
is a person who travels voluntarily within their own state or from one country to
another. If a foreigner stays in the destination country for the purpose of a longer-
term stay, he or she is called an immigrant [23]. The United Nations (UN) defines an
immigrant as someone who comes to the destination country for more than twelve
months. Economic migrants are foreigners who left their country voluntarily in order
to improve their economic situation. Depending on the residence permit, an economic
migrant can occur in both the legal and illegal status in the destination country [28].
An illegal migrant is a foreigner who enters the territory of the destination country
or resides on the territory of the destination country without a valid authorization
[27]. A refugee is a foreigner who had to leave his or her country and who meets the
defining characteristics of a refugee within the meaning of the Geneva Convention
[3, 14].

Refugees or some migrants may be granted international protection in the form
of asylum or subsidiary protection upon arrival in the destination country. Asylum

---

[1] Schengen is a term used to describe Schengen cooperation, which created the missing instrument
to fulfil the free movement of persons, transport and goods [27].

is a protection status that the state, under specific legal conditions, grants to a third-country[2] national or stateless person in connection with the risk of persecution in the country of origin. An asylum seeker is a foreigner who has been granted asylum in the destination country [14].

Migration has economic, political, legal, social, population, psychological, cultural, religious and security impacts on both home and transit countries, however especially on destination countries [17]. Integration of immigrants in the target countries is crucial for maintaining social cohesion of the society and for economic, social and cultural development. Integration of immigrants is supported by the state in cooperation with helping organizations that provide integration services such as social and legal counselling and also monitor the situation of foreigners [14, 23].

Increasing migration dynamics can also lead to migration crises. The European migration crisis, which was an international political crisis in the European Union (EU), reached its peak in 2015 and was caused by the arrival of large numbers of migrants. They came to Europe in various migration waves, in mass, mixed groups including both refugees and economic migrants, or illegal migrants heading for the EU. Migrants reached the central parts of Europe using the Western Mediterranean route, the Central Mediterranean route and the Eastern Mediterranean route [2]. In the second quarter of 2020, migration activity on the Eastern Mediterranean route to Greece was at a very low level, which was related to the measures introduced in connection with the COVID-19 pandemic. Arrivals of migrants by sea and by land decreased. The arrivals of migrants by the central Mediterranean route to Italy gradually increased again, compared to 2019, which means 2.5 times more migrants. However, these arrivals of migrants are lower than in the 2016–2018 period. On the western Mediterranean route to Spain, a decrease of 50% of incoming migrants was recorded in comparison with the first quarter of 2020 [14, 26].

Migration crisis was closely connected to migration quotas, which were part of the proposal of the European Commission in 2015 and aimed at distributing migrant asylum seekers, who entered the EU during the migration crisis, among individual EU Member States. In 2016, the proposal was extended by a corrective allocation mechanism, which set permanent quotas for the redistribution of asylum seekers among all EU Member States [8]. Until 2019, quotas did not work as planned by the European Commission, however, a debate about the possibility of introducing a mandatory redistribution mechanism across the EU was going on. In 2020, the European Commission presented a proposal for a new asylum and migration system involving all EU countries. The European Commission wants to lay down rules to speed up and make the return of migrants to their place of origin more effective. In crisis situations, such as during a migration crisis, the European Commission may issue a regulation on the admission of part of migrants from another state [7, 14].

---

[2] A third-country national is a national of a non-EU country, which is not a citizen of Iceland, Liechtenstein, Norway or Switzerland at the same time.

2.2 Security Aspects of Migration in the Czech Republic

The situation in the field of migration in the Czech Republic (CR) has been stable for a long time. The number of foreigners in the Czech Republic almost doubled in the 2005–2019 period as can be seen from Fig. 1. 2008 was followed by four years of continuous decline [6]. This was probably the result of the global economic crisis, which manifested itself in limited employment opportunities on the Czech labour market [20]. Since 2014, the increase has been more significant, due to the increasing number of EU citizens who registered their residence in the Czech Republic. A significant increase in the number of foreigners was evident in the 2015–2019 period, with 40,000 more foreigners living in CR in 2018 than in 2017. Foreigners in the Czech Republic accounted for 5.5% of the population in 2019 and their number has been growing slowly and continuously [6]. The most numerous groups of foreigners in the Czech Republic in the given period were citizens of Ukraine, followed by the citizens of the Slovak Republic (SR), Vietnam, Russia and the Republic of Poland (PL). Labour migration and migration for the purpose of family reunification, study and business predominated [14, 23].

EU law has a fundamental influence on the migration policy of the Czech Republic. The Treaty on the Functioning of the European Union established common policies in the areas of migration, international protection (asylum) and borders. Common policies were implemented through EU legal instruments in the form of specific regulations and directives [27]. Practical cooperation at EU level is carried out, inter alia, through EU agencies such as the European Asylum Support Office (EASO), the European Border and Coast Guard (FRONTEX) and Europol. In the field of

Year	CR citizens	Foreigners in CR
2005	9.973.079	278.000
2006	9.966.189	321.000
2007	9.989.130	392.000
2008	10.029.542	438.000
2009	10.073.813	433.000
2010	10.108.770	424.000
2011	10.071.445	434.000
2012	10.080.125	436.000
2013	10.073.419	439.000
2014	10.088.275	450.000
2015	10.088.843	465.000
2016	10.579.000	493.000
2017	10.610.000	524.000
2018	10.650.000	564.000
2019	10.694.640	593.000

Fig. 1 Development in the number of foreigners living in the Czech Republic in the 2005–2019 period. *Source* [6, 14]

international law, it is necessary to mention the UN Convention related to the Status of Refugees. According to the Act No. 2/1969 Coll. on the Competences migration falls under the responsibility of the MOI CR, which has a comprehensive legislative framework within migration policy, which covers extensive migration issues and, if necessary, is subject to partial modifications. In addition to general standards, important laws in the field of migration are, for example, Act No. 326/1999 Coll., on the Residence of Foreigners in the Czech Republic, Act No. 216/2002 Coll., on the Protection of State Borders of the Czech Republic, Act No. 325/1999 Coll., on Asylum and Act No. 221/2003 Coll., on Temporary Protection of Aliens [13, 14].

The strategy of migration policy of the Czech Republic issued in 2015 represent the basic strategic framework of migration policy. It presents principles that are prioritized with regard to the security aspects of migration. These are the areas of integration of immigrants, illegal and return policy, international protection— asylum, external dimension of migration policy, free movement of persons in the EU and Schengen area, legal migration and international and European obligations of the Czech Republic in the field of migration [14].

The integration of immigrants is a key security aspect of migration according to this strategy. It is a two-way process which is entered by both immigrants and the majority society. Integration is a complex issue that affects many areas, especially internal security. Insufficient integration of immigrants can raise concerns among the domestic population, which can result in the threat of extremism [19]. Cultural practices of immigrants which are incompatible with the legal order of the corresponding country or a reduced willingness to integrate can cause threats arising from the uncontrolled growth of Islam [24]. The integration of a larger number of immigrants can be challenging and there is a high probability that certain closed communities or excluded localities associated with increased crime risk may emerge [19]. Security of the state may be threatened by mass uncontrolled immigration, which could result in social unrest or radicalism, both on the part of the minority and the majority [24].

In terms of integration, the Czech Republic, has an Updated Concept of Integration of Foreigners issued in 2016, which proposed needs within the integration process involving the integration of all foreigners in the Czech Republic [12]. A new State Integration Programme was approved in 2015 to support integration, which focuses on the process of integration of persons with granted international protection [30]. Successful integration of immigrants into the society of the host country is essential for maximizing the opportunities offered by legal migration [7, 14].

The Security Strategy of the Czech Republic (SS CR) issued in 2015 defines, as one of the security threats, an increased rate of uncontrolled migration and insufficient integration of foreigners. In the points 25 and 26 of the SS CR, it is more specifically stated that "The security impacts of demographic change will continue to increase, especially the risks arising from the aging of the population in developed countries and uncontrollable migration. Problems related to poverty, long-term social exclusion and lack of basic needs and services can significantly increase the probability of extremism, crime, local armed conflict and mass uncontrollable migration. The same applies to the social and economic lagging behind of the large areas and regions, especially in the developing world" ([21], p. 9). The 2015 risk analysis for the Czech

Republic identified "Large-scale migration wave" as one of 22 unacceptable threats [14, 29].

In 2016, the Government of the Czech Republic approved the National Security Audit. It is a complex material that redefines the security policy of the state in current security areas. This material assesses the resilience of the Czech Republic to current security threats. In accordance with the SS CR of 2015, the research team identified the threat of illegal migration "as a consequence of the increased number of local armed conflicts and also the threat of insufficient integration of legal migrants, which may represent a source of social tension and especially the threat of uncontrolled migration. under certain circumstances one of the elements of the hybrid threat" ([13], p. 62). In view of the significant interconnectedness of the security aspects of migration with other areas and the difficult separability of the identified threats, the research team did not use a specific method for their identification. The team research was based on the available conceptual and strategic materials at its disposal. Two main threats of illegal migration have been defined due to the increased number of local armed conflicts in the world and insufficient integration of foreigners, which can become a source of social tension [13]. An Action Plan of the National Security Audit [14] was created to assess the fulfilment of individual measures set by the audit.

3 The Methods

The What-if method, which is based on structured brainstorming, was chosen to identify the risks arising from immigration in the Czech Republic. A qualified research team examines unexpected events that may occur in the area by the form of questions and answers. The formulated questions start with the characteristic "What will happen if …?" and the consequences of the situation or situation itself are estimated, measures and recommendations are proposed. The What-if method is popular in practical research since it does not place high demands on time. However, it must be taken into account that lower time demandingness of the study has its roots in an intuitive, unsystematic process. This method can be very effective and efficient if the research team has both experience and can apply this method practically. Otherwise the result of the study may be controversial [1, 15].

Preparation for the application of the What-if method first consisted in collecting all available data. These were mainly conceptual materials, laws, studies, methodologies and reports in the field of migration and crime prevention. An implementation team was set up to implement the What-if method. Both basic and specific criteria were set for the selection of members of the implementation team. The implementation team had 6 members, 5 of which had to meet both the basic and the specific criteria, solely the representative of lay public did not have to meet these criteria. Participation of a representative of lay public was recommended by brainstorming. The most experienced member of the implementation team was chosen as a facilitator. The course of the What-if method research was recorded by the secretary. The implementation team set a rating of 0, 1, 2, 3 to determine the value of the risk level for identified threats resulting from immigration.

In order to carry out the assessment of individual risks of security aspects of immigration, time series were calculated using linear regression with a regression equation according to a linear trend and a logarithmic trend. Time series of data from important crime groups were used to assess crime risks and trends in the development of the security environment in the Czech Republic, including impacts on security which were taken into account. The chapter presents a proposal for the methodology for crime risk assessment in the environment of the Czech Republic resulting from the research results and set of risk factors, in relation to the Slovak Republic and the Republic of Poland.

Furthermore, statistical methods were used to calculate individual risk. The time series calculation was performed using linear regression with a regression equation according to the linear trend and logarithmic trend, the regression t-test, determination coefficient, calculation of 95% confidence interval and correlation analysis: Pearson correlation coefficient, Spearman correlation coefficient and Shapiro–Wilk test [32]. The statistical apparatus used is described in Sect. 4.2.

4   The Practical Part

Practical part of the research described in the chapter is divided into 4 parts. The first part deals with the results of identifying the risks resulting from immigration. Based on them, for the highest estimated risk, crime and its forms, a methodological procedure was proposed for predicting their development and verified on the basis of data from the Czech Republic, the Slovak Republic and the Republic of Poland. Correlation of mutual results was summarized in the last section of the practical part.

4.1   What-If Method Results

The What-if method implementation team identified 6 threats resulting from immigration and divided the impacts into 5 possible categories. The results of the risk assessment resulting from migration are shown in Fig. 2.

The highest level of risk was achieved by the threat of crime. Therefore, another round of the What-if method research was undertaken, where the types of crime were identified and a risk assessment was performed using the same impact categories. The results showed that the most affected group of assets are people's lives and health, their safety and property.

Threats	Target areas					
	Lives and health of people	Safety of people	Property	Public welfare	State administration and self-government	Summary risk
Migration waves	2	2	2	2	2	10
Diseases	3	2	1	1	1	8
Unemployment	3	3	2	2	1	11
Crime rate	3	3	3	3	2	14
Islamisation	2	2	1	2	1	8
Threat to traditions	1	3	1	2	1	8

Fig. 2 Results of risk assessment resulting from immigration using the What-if method. *Source* [14]

4.2 Proposal of a Methodological Procedure for the Assessment of Crime Risks Resulting from Immigration

It was necessary to assess the development of the situation and estimate trends resulting from immigration in the Czech Republic in order to support a decision-making process and set up quality risk management processes. The highest crime risk was estimated based on the results of the risk assessment using the What-if method performed by the implementation team. For this reason, a key section of the practical part of the chapter is devoted to the procedure of assessing crime risks resulting from immigration. The procedure established in Fig. 3 was verified on the assessment of the total crime prosecuted in the Czech Republic.

Prior to start with the methodological procedure for assessing crime resulting from immigration, it was necessary to collect data from publicly available databases of the Czech Statistical Office (CZSO), the Police of the Czech Republic (PP CR) and the MOI CR.

Based on them, it was possible to calculate individual risk according to formula (1):

$$\text{Individual risk in year } t = \frac{\text{Number of prosecuted persons in year } t}{\text{Number of inhabitants in year } t} \quad (1)$$

Time series were created based on the results of individual risk in individual years. Time series of individual risk of crime types were calculated for the 2005–2015 period. Time series trend was assessed and predictions for the years 2020–2023 were calculated based on it.

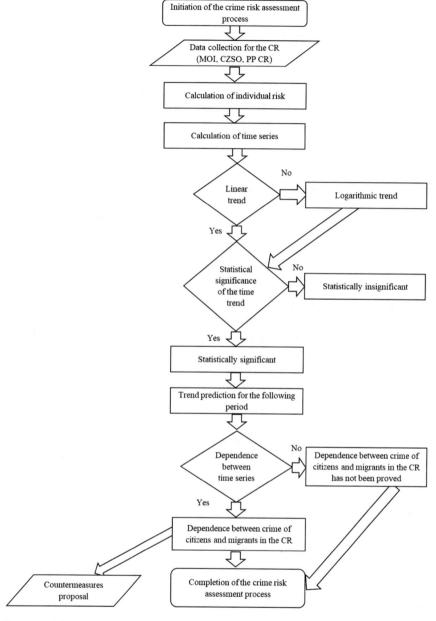

Fig. 3 Decision-making diagram of the methodological assessment procedure for crime resulting from immigration in the Czech Republic. *Source* own

The trend of time series was estimated using linear regression, while the estimated regression equations were for the following types of trends as follows:

Linear trend according to formula (2):

$$y_t = b_o + b_1 t \tag{2}$$

Logarithmic trend according to formula (3):

$$y_t = b_o + b_1 \log t \tag{3}$$

where y_t is the value of the time series in time $t = 1, \ldots 11$ (corresponding to the years 2005–2015) and b_o, b_1 are the values of the estimated regression coefficients.

The statistical significance of the selected time trend was assessed using a t-test for regression according to formula (4):

$$t = \frac{b_1}{s(b_1)} \tag{4}$$

where b_1 is the value of the estimated regression coefficient, $s(b_1)$ is standard error of the regression coefficient estimation and t is the value of the test criterion. For a given test criterion, a p-value was calculated using software. It was subsequently compared with a significance level of 0.05, and in case it was lower, the time trend was marked as statistically significant at a significance level of 0.05.

The quality of the time series interpolation by the trend was assessed using the coefficient of determination (5):

$$R^2 = 1 - \frac{ESS}{TSS} \tag{5}$$

where ESS is the residual sum of squares and TSS is the total sum of squares.

The predictions of the development of the time series for the years 2020–2023 were calculated by substituting into the estimated regression equation $t = 12, 13, \ldots 19$ (it corresponds gradually to the years 2020–2023). A 95% confidence interval was calculated for the prediction according to formula (6):

$$(d, h) = (\hat{y} - t_{krit} SE_{HAC}; \hat{y} + t_{krit} SE_{HAC}) \tag{6}$$

where d is the lower limit of the confidence interval, h is the upper limit of the confidence interval, $\hat{y}$ is the prediction, t_{krit} is the quantile value of the Student distribution and SE_{HAC} are the standard errors robust to heteroskedasticity and autocorrelation.

Dependence between the time series of individual risks was assessed using correlation coefficients. Prior to the analysis, the time series were stripped of the trend in order to avoid the problem of so-called apparent dependence. Apparent dependence ("spurious regression", or "spurious relationship") between two time series occurs when both have a monotonous trend. For example, when both time series grow, there is a strong positive correlation between them just because they have a trend, not because they are really related. Therefore, when assessing the dependence of two time series, it is more appropriate to analyse the fluctuations around their trend. It is necessary to observe whether: when one time series is above the trend, the second time series is usually above the trend (direct dependence; positive correlation) or when one time series is above the trend, the second time series is usually below the trend (indirect dependence, negative correlation) or whether the fluctuation of two time series above and below the trend shows no systematic dependence (independence, statistically insignificant correlation). Thus, the time series residues obtained by subtracting the trend from the original values were used for the correlation analysis (7):

$$e_t = y_t - \widehat{y}_t \tag{7}$$

where y_t is the value of individual risk at time t, $\widehat{y}_t$ is the value of the balanced time series according to the trend function at time t a e_t is the residual of the time series at time t.

Positive residue values at time t mean that the time series value was above its trend at time t, and on the other hand negative residues at time t mean that the time series value was below its trend at time t.

Pearson and Spearman correlation coefficients were used to test the dependence of time series pairs of individual risk residues. Pearson correlation coefficient was calculated according to formula (8):

$$r = \frac{c(x, y)}{s(x)s(y)} \tag{8}$$

where $c(x, y)$ is the selection covariance of the time series x_t and y_t, $s(x)$ and $s(y)$ are the standard deviations of the time series x_t and y_t and r is the value of the Pearson correlation coefficient.

The Spearman correlation coefficient was calculated according to formula (9):

$$r_S = 1 - \frac{6 \sum d_t^2}{n(n^2 - 1)} \tag{9}$$

where n is the number of observations in the time series, d_t is the difference in the order of the residual values between two time series and r_S is the value of the Spearman correlation coefficient.

Pearson correlation coefficient was used when a normal distribution could be assumed for the data. Assessment of the assumption of a normal distribution was performed using the Shapiro–Wilk test. If a normal distribution could not be assumed, Spearman correlation coefficient was used.

When proving the dependence between the crime of the citizens and immigrants in the Czech Republic, proposals for the minimization of risks were implemented and the process of crime risk assessment was completed.

4.3 Verification of the Proposed Methodology for the Total Crime Prosecuted in the Czech Republic

In accordance with the methodological procedure presented in Sect. 4.2, its functionality was verified. For verification, 4 types of crime were investigated (total, property, violent and murder). Solely the outputs for the total crime prosecuted in the Czech Republic are presented due to the limited extension of the chapter. The individual risk for the total crime prosecuted for the citizens and foreigners in the Czech Republic was estimated and a time series was created in accordance with the procedure. Figure 4 presents the dependences of types of individual risk on time.

Figure 4 shows that the individual risk of total crime prosecuted of the citizens in the Czech Republic—red colour, in the years 2005–2015 was situated in lower values

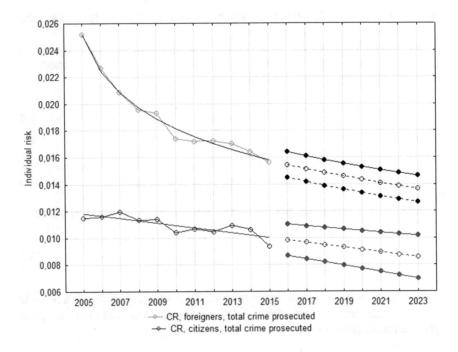

Fig. 4 Total crime prosecuted in the Czech Republic. *Source* [14]

Parameter	Estimation	Standard error	t	p
Constant (b_0)	0.0120137	0.000223103	53.85	<0.0001
Time trend (b_1)	−0.000180465	3.93633e-05	−4.585	0.0013
F	21.01853	p(F)	0.001319	R^2	0.683112

Fig. 5 Results for the total crime prosecuted in the Czech Republic (citizens). *Source* [14]

than that of foreigners in the Czech Republic—green colour. For the citizens of the Czech Republic, the individual risk had an evenly decreasing tendency according to the linear trend, while for foreigners the rate of decline was higher at the beginning of the observed period and gradually slowed down according to the logarithmic trend. The prediction of the individual risk of the total crime prosecuted was marked in red for the citizens of the Czech Republic and in black for foreigners in the Czech Republic [14].

Figure 5 contains results of the estimated regression coefficients, t-tests, F-test and coefficient of determination according to the calculations performed using the Gretl program.

The trend curve of individual risk for the total crime prosecuted for the citizens in the Czech Republic was estimated using linear regression according to formula (2). In the observed time period, the individual risk decreased, as the value of b_1 was negative and the direction of the line was slightly decreasing. According to the t-test, the linear time trend was statistically significant ($p < 0.05$) and according to the index of determination, the model explained 68% of the variability of the time series.

The trend curve of individual risk for the total crime prosecuted for the citizens in the Czech Republic was estimated using linear regression according to formula (2). In the observed time period, the individual risk decreased, as the value of b1 was negative and the direction of the line was slightly decreasing. According to the t-test, the linear time trend was statistically significant ($p < 0.05$) and according to the index of determination, the model explained 68% of the variability of the time series.

A prediction was made with a confidence interval of 95% of the lower limit of the confidence interval and the upper limit of the confidence interval based on these calculations. Predictions of the development of total crime prosecuted in the Czech Republic fell with 95% confidence between those outermost bands. The results can be seen from Fig. 6.

Figure 7 contains the results of estimated regression coefficients, t-test, F-test and coefficient of determination, according to calculations performed using the Gretl program, for the total crime prosecuted for foreigners in the Czech Republic.

The trend curve of individual risk for the total crime prosecuted for foreigners in the Czech Republic was estimated using linear regression according to formula (3). In the observed time period, the individual risk decreased, as the value of b_1 was negative and the direction of the line was decreasing. The logarithmic time

Year	Lower limit for 95% CI	Prediction	Upper limit for 95% CI
2020	0.0077	**0.0091**	0.0105
2021	0.0075	**0.0089**	0.0104
2022	0.0072	**0.0088**	0.0103
2023	0.0070	**0.0086**	0.0102

Fig. 6 Prediction of total crime prosecuted in the Czech Republic (citizens). *Source* [14]

Parameter	Estimation	Standard error	t	p
Constant (b_0)	0.0251817	9.53040e-05	264.2	<0.0001
Time trend (b_1)	−0.00391398	9.04927e-05	−43.25	<0.0001
F	1870.727	p(F)	9.43e-12	R^2	0.984996

Fig. 7 Prediction of total crime prosecuted in the Czech Republic (foreigners). *Source* [14]

Year	Lower limit for 95% CI	Prediction	Upper limit for 95% CI
2020	0.013	**0.014**	0.015
2021	0.013	**0.014**	0.015
2022	0.013	**0.014**	0.015
2023	0.013	**0.014**	0.015

Fig. 8 Prediction of total crime prosecuted in the Czech Republic (foreigners). *Source* [14]

trend was statistically significant (p < 0.05) according to the t-test. According to the determination index, the model explained 98% of the time series variability.

A prediction was made with a confidence interval of 95% of the lower limit of the confidence interval and the upper limit of the confidence interval based on these calculations. Predictions of the development of total crime prosecuted of foreigners in the Czech Republic fell with 95% confidence between those extreme bands. The results can be seen from Fig. 8.

The individual risk both for the Czech citizens and foreigners in the Czech Republic for the total crime prosecuted can be expected to decrease in the forth-coming years, assuming estimated trends, with the individual risk of foreigners declining faster and slowly approaching the values of the Czech citizens.

The same methodological approach was chosen for the assessment of crime risks in the Republic of Poland (PL) and in the Slovak Republic (SR) as for the assessment of crime risks resulting from immigration in the Czech Republic. In the Slovak Republic, individual risks for general crime, property crime and violent crime were assessed. In PL, individual risks for general crime, property crime and murder were assessed.

4.4 Correlation

Trends in time series of individual types of crime correlated strongly and positively. After detrending the time series (subtracting the global trend), the interdependencies of the residues were tested using an independence test based on the Pearson correlation coefficient. For most residual time series, a normal distribution could be assumed according to the Shapiro–Wilk test. For one time series (total crime of foreigners prosecuted and murders by foreigners detected), this assumption was rejected by the Shapiro–Wilk test and the dependence was tested on the basis of the nonparametric Spearman coefficient of rank correlation. A significance level of 0.01 was chosen for assessing the correlations. Due to the fact that several hundred correlation coefficients were calculated, the risk of obtaining a statistically significant result increased only based on chance. More precisely, at a significance level of 0.05, only 5 statistically significant results caused by chance can be expected when calculating 100 correlations. To minimize this risk, a significance level of 0.01 was chosen for assessing the correlations. By assessing the dependence of individual types of crime in the Czech Republic, it was found that 11 pairs of individual risk had positive values of the Pearson correlation coefficient, and thus meant a positive linear correlation. The obtained values of correlation coefficients R showed a strong dependence in all cases of crime [14].

The main relationship found was that property crime in the Czech Republic has a significant effect on reducing or increasing total crime in the Czech Republic. Demonstrating this strong correlation between property and total crime was important for creating specific countermeasures. Another significant result resulting from the correlations was the fact that reducing violent crime, murders and property crime in the Czech Republic will contribute to reducing the total crime of foreigners in the Czech Republic. Demonstrating the dependence between the crime of Czech citizens and immigrants made it possible to formulate procedures and recommendations for minimizing the risks [14].

Assessment of crime in the Republic of Poland and in the Slovak Republic was proceeded according to the methodological procedure created for the assessment of crime in the Czech Republic. The results were compared between countries. Facts and correlations found in the Czech Republic were also proven in the Republic of Poland and in the Slovak Republic. These countries face the same problems, it has been especially proven that the level of property crime has a significant impact on reducing or increasing total crime in these countries [14].

The chapter presents results of the research carried out as dissertation research on the topic of assessment of security aspects of immigration by Olga Harárová, under the leadership of the dissertation supervisor Alena Oulehlová.

5 Conclusion

Globalization process deepens all types of migration. The danger of this phenomenon has been and will continue mainly in uncontrollable migration, insufficient integration of foreigners and large-scale migration waves.

Methodology for assessing crime risk resulting from immigration in the Czech Republic was proposed as one of the security aspects in connection with increased migration to Europe after 2015 and growing concerns about the negative effects of migration. Crime belongs among the most important negative security aspects of immigration. The assessed groups of crime were total, property, violent, murders, robbery murders, motivated murders and other murders. The crimes were divided into crimes detected and prosecuted. The proposal of methodology was tested for the Czech Republic, the Slovak Republic and the Republic of Poland due to the availability of data.

The proposed methodology is based on the use of regression analysis and correlation analysis. Calculations to predict the development of crime for citizens of a given state and foreigners of a given state until 2023 were performed using a linear regression analysis. In all cases the prediction followed a decreasing trend.

A key contribution to the assessment of security aspects of immigration is the proposal of a methodological procedure for risk assessment and the potential of its use for assessment of other types of crime, for assessment of crime in selected regions of the Czech Republic and for the prevention of consequences that may arise from immigration. Dependences of some types of crime in the Czech Republic for citizens and foreigners have been proven. Currently, a positive trend has been set, which, however, can change instantly, and therefore it is necessary for the security system of the Czech Republic to be able to react to it well in advance. The results of the research can be further used to carry out analyses, national security audits and action plans at the national level.

References

1. Babinec F (2005) Risk management. In Czech: Management rizika, [online]. Slezská univerzita [cit. 12/3/2016]. Available online at: https://www.slu.cz/math/cz/knihovna/ucebni-texty/Ana lyza-rizik/Analyza-rizik-1.pdf/
2. Břicháček T (2016) Union in the wake of the migration crisis. In Czech: Unie ve víru migrační krize. Institut Václava Klause, Praha, 264 s. ISBN 978–80–7542–023–7
3. *Convention relating to the Status of Refugees*, (1951). In Czech: Úmluva o právním postavení uprchlíků ze dne 28. července 1951, ve znění Newyorského protokolu ze dne 31. ledna 1967, publikována pod č. 208/1993 Sb.
4. Čerňanská B (2018) Theory of push-pull effects on migration. In: Czech: Teorie push-pull vlivů na migraci, [online]. Encyklopedie migrace [cit: 30/7/2020]. Available online at: https://www.encyclopediaofmigration.org/teorie-pushpull/
5. Čapek V, Pátek J (1992) World History I. In: Czech: Světové dějiny I. Fortuna, Praha, 176 s. ISBN 80–7168–147–4

6. ČSÚ (2020) Development of the number of foreigners in the Czech Republic. In: Czech: Vývoj počtu cizinců v ČR [online]. Český statistický úřad. [cit. 27/7/2020]. Available online at: http://www.statistikaamy.cz/2016/02/vyvoj-poctu-cizincu-v-cr/

7. EC (2020) Migration and home affairs. [online]. European Commission. [cit. 6/10/2020]. Available online at: https://ec.europa.eu/home-affairs/what-we-do/policies/legal-migration_en

8. EC (2016) Towards a sustainable and fair common European Asylum System. [online]. European Commission. [cit. 28/7/2020]. Available online at: https://ec.europa.eu/commission/pre sscorner/detail/en/IP_16_1620

9. EC (2020) Press statement by President von der Leyen on the New Pact on Migration and Asylum [online]. European Commission. [cit. 5/10/2020]. Available online at: https://ec.eur opa.eu/commission/presscorner/detail/cs/statement_20_1727

10. Euractiv (2017) Refugees and migration: Impressions and concepts. In: Czech: Uprchlíci a migrace: Dojmy a pojmy, [online]. EA. [cit. 30/7/2020]. Available online at: https://euractiv. cz/section/aktualne-v-eu/opinion/uprchlici-a-migracedojmy-a-pojmy/

11. Fiala P (2017) Reason and courage: how to face Europe's current challenges and crises. In: Czech: Rozum a odvaha: jak čelit současným výzvám a krizím Evropy. Barrister & Principal, Brno. 132 s. ISBN 978–80–7485–131–5

12. Government of the Czech Republic (2015) Updated concept of integration of foreigners 2016— In mutual respect. In: Czech: Aktualizovaná koncepce integrace cizinců 2016 - Ve vzájemném respektu. Praha: Ministerstvo vnitra ČR

13. Government of the Czech Republic (2016) National security audit. In: Czech: Audit národní bezpečnosti. Ministerstvo vnitra ČR, Praha

14. Harárová O (2020) Assessment of security aspects of immigration. In: Czech: Hodnocení bezpečnostních aspektů imigrace. Disertační práce. Univerzita obrany, Brno, 156 s

15. ISO 31010 (2019) Risk management—risk assessment techniques. 2. International Organiza- tion for Standardization, Geneva

16. IOM (2020) Key Migration Terms [online]. [cit. 3/10/2020]. Available online at: https://www. iom.int/key-migration-terms#Migrant

17. Keller J (2017) European contradictions in the light of migration. In Czech: Evropské rozpory ve světle migrace. SLON, Praha. 227 s. ISBN 978–80–7419–249–4

18. Klufová R, Poláková Z (2010) Demographic methods and analyzes: demography of the Czech and Slovak population. In: Czech: Demografické metody a analýzy: demografie české a slovenské populace. Wolters Kluwer Česká republika, Praha. 306 s. ISBN 978–80–7357–546–5

19. MV ČR (2009) Strategies to combat extremism. In: Czech: Strategie boje proti extremismu. Ministerstvo vnitra, Praha. 67 s

20. MV ČR (2013) Report on the situation in the area of migration and integration of foreigners in the Czech Republic in 2012. In: Czech Zpráva o situaci v oblasti migrace a integrace cizinců na území České republiky v roce 2012. Ministerstvo vnitra, Praha,182 s

21. MV ČR (2015) Strategy on migration policy of the Czech Republic. In: Czech: Strategie migrační politiky České republiky. Ministerstvo vnitra, Praha. 24 s. ISBN 978–80–86466–83–5

22. MV ČR (2018) Report on the situation in the area of migration and integration of foreigners in the Czech Republic in 2017. In: Czech: Zpráva o situaci v oblasti migrace a integrace cizinců na území České republiky v roce 2017. Ministerstvo vnitra, Praha 217 s

23. MV ČR (2019) Report on the situation in the area of migration and integration of foreigners in the Czech Republic in 2018. In: Czech: Zpráva o situaci v oblasti migrace a integrace cizinců na území České republiky v roce 2018. Ministerstvo vnitra, Praha. 204 s

24. MV ČR (2020) Security aspects of migration. In: Czech: Bezpečnostní aspekty migrace. [online]. [cit. 3/10/2020]. Available online at: https://www.mvcr.cz/cthh/clanek/bezpecnostni- aspekty-migrace.aspx

25. MZV (2015) Security strategy of the Czech Republic 2015. In: Czech: Bezpečnostní strategie České republiky 2015. Ministerstvo zahraničních věcí České republiky, Praha 24 s. ISBN 978–80–7441–005–5

26. OAMP MV ČR (2020) Quarterly Migration Report for the fourth quarter of 2019. In: Czech: Čtvrtletní zpráva o migraci za čtvrté čtvrtletí roku 2019. [online]. [cit. 30/9/2020]. Available online at: https://www.mvcr.cz/migrace/clanek/ctvrtletni-zprava-omigraci-za-ctvrte-ctvrtl eti-roku2019.aspx2020
27. OAMP (2020) Terminology. In: Czech: Slovníček pojmů. [online]. [cit. 28/7/2020]. Available online at: https://www.mvcr.cz/migrace/clanek/slovnicek-pojmu.aspx
28. Palát M (2013) Economic aspects of international migration: theory and practice in the European Union. In: Czech: Ekonomické aspekty mezinárodní migrace: teorie a praxe v Evropské unii. Key Publishing, Ostrava. 92 s. ISBN 978–80–7418–161–0
29. Paulus F, Kromer A, Petr J, Černý J (2015) Risk analysis for the Czech Republic. In: Czech: Analýza hrozeb pro Českou republiku. HZS, Praha
30. SIP (2020) State integration program. In: Czech: Státní integrační program. [online]. Státní integrační program. [cit: 27/7/2020]. Available at: http://www.integracniprogram.cz/2020
31. Uherek Z, Honusková V, Ošťádalová Š, Günter V (2016) Migration: history and present. In: Czech: Migrace: historie a současnost. PANT, Ostrava, 148 s. ISBN 978–80–905942–9–6
32. Vose D (2009) Risk analysis: a quantitative guide. 3. Wiley, Chichester. ISBN 978–0–470–51284–5

Risks and Threats

Assessment Criteria for Municipality Territory Resilience to Anthropogenic Threats

Pavel Kincl and Alena Oulehlová

Abstract This chapter deals with the requirements for assessment criteria for municipality territory resilience to anthropogenic threats. The theoretical part of the chapter defines the role and importance of territory resilience, presents the context of a resilience assessment, and describes the United Nations Office for Disaster Risk Reduction (UNISDR) Disaster Resilience Scorecard for Cities tool, which can also be used as a methodological basis for assessing the municipality resilience in the Czech Republic. The practical part of the chapter sets out the authors' approach to creating resilience assessment criteria as one of ten key factors in increasing and building municipality resilience (Ten Essentials) designed for conditions in the Czech Republic. Essential 9, presented here, examines how to "Ensure Effective Disaster Response". It is submitted in the chapter as a practical example of the Integrated Rescue System's incorporation of the functioning in the emergency response field under Czech conditions into the criteria for resilience assessment in accordance with the principles of UNISDR's Disaster Resilience Scorecard for Cities.

Keywords Resilience · Anthropogenic threats · Municipality · Ten Essentials for Making Cities Resilient · Disaster Resilience Scorecard for Cities · Emergency response

1 Introduction

In response to the increasing occurrence of the natural and anthropogenic impacts of emergencies on the lives and health of the population, on property and the environment, both individual countries and international organizations have become aware of the importance of preventing and mitigating disaster risks. During the last twenty six years, international obligations have been confirmed by three frameworks of

P. Kincl (✉) · A. Oulehlová
University of Defence, Kounicova 65, 662 10 Brno, Czech Republic

A. Oulehlová
e-mail: alena.oulehlova@unob.cz

© The Author(s), under exclusive license to Springer Nature Switzerland AG 2022
I. Tušer and Š. Hošková-Mayerová (eds.), *Trends and Future Directions in Security and Emergency Management*, Lecture Notes in Networks and Systems 257,
https://doi.org/10.1007/978-3-030-88907-4_11

disaster risk mitigation: the Yokohama Strategy [23], Hyogo Framework for Action [24] and the Sendai Framework [25], which were adapted to take on board developing knowledge about disaster risk mitigation. Their objective is to achieve appropriate crisis readiness and the ability to react to emergencies, including area redevelopment according to the principle Build Back Better. The Czech Republic has also accepted the obligations of the Sendai Framework and pledged to fulfill them.

To achieve country resilience to emergencies, it is necessary for smaller regional administrative units (municipalities) to achieve resilience. To support disaster risk reduction, in 2010 the United Nations Office for Disaster Risk Reduction (UNDRR) founded the initiative Making Cities Resilient—My City is Getting Ready! [19]. The campaign is designated for municipalities, irrespective of size or number of residents. The initiative to build area resilience makes use of the Ten Essentials for Making Cities Resilient, which were formulated to accelerate implementation of the Sendai Framework.

The chapter is based on the assumption that to incorporate the Czech Republic into a disaster risk reduction policy, it is necessary to deal with disaster risk reduction at the place where risks occur (municipalities). Currently, there is no universal method for assessing territory resilience in the Czech Republic. For assessing territory resilience, it is necessary to define the criteria and indicators that will be in accordance with the rules and principles for international development of city resilience.

The objective of this chapter is to apply qualitative research and identify a useful tool for assessing territory resilience at the regional level and adapt it to the specific conditions of the Czech Republic. Based on the defined criteria, the tool Disaster Resilience Scorecard for Cities [20, 21] was selected; this tool uses the Ten Essentials for Making Cities Resilient that cover all phases of crisis management for comprehensive assessment of territory resilience. Due to the extent and complexity of the topic, the chapter deals only with Essential 9, which is dedicated to the phase of ensuring effective disaster response. Selected semiquantitative criteria for resilience assessment have been designed and justified for Essential 9.

2 Theoretical Part

Resilience is the ability of a system or society to resist, mitigate, accept and redevelop the impacts of dangers in a timely and effective manner, including the preservation and renovation of its essential basic infrastructure and functions [11].

During research on the resilience of municipalities in the Czech Republic to anthropogenic threats [7], which is analyzed in this chapter, territory resilience was investigated as a concept that closely relates to risk management, crisis management, environmental security, and theory of systems. In relation to the theory of systems, the municipal territory was investigated as a "system of systems", i.e. a complex system that consists of other mutually connected systems. The theory and application of the "system of systems" is used e.g. in publications by the authors [8], and [18]. Also, [26] mentions this term in connection with resilience assessment. In the theory of systems, municipality assets and their interconnection in relation to the functioning of the municipality can be considered as individual system elements (or its subsystems). In connection to the municipality assets defined above, their vulnerability in relation to risk management [4] was investigated, as was subsequently the inversive relationship between resilience and the vulnerability of municipality assets. In the context of crisis management, achieving resilience was investigated in relation to the timing and progress of the onset and course of the emergency. Crisis management can be understood as a cycle of prevention, readiness, response and redevelopment phases. Achieving resilience should cover all these phases of crisis management, where the most cost effective are considered ex-ante measures that ensure resilience during the prevention and readiness phases, together with applying the principle "Build Back Better" when renovating the impacted area [15, 20]. In relation to risk management and crisis management, assessing and building area resilience is considered to be a tool for strengthening the environmental security. From the environmental point of view, security is defined as the ability of the system to resist security threats that may potentially have an undesirable impact on the system, where the interaction and mutual relationships between two elements (subsystems) of the whole system— human society and ecological system—are key to the ability to resist the natural and anthropogenic threats [11]. In the context of defining environmental security and the importance of crisis management, territory resilience can be linked to the issue of sustainable development. Assessing and improving territory resilience, the principal part of the Sendai Framework for Disaster Risk Reduction 2015–2030 [25], contributes to increased comprehensive territorial security and stability. By this sustainable development can be achieved.

On the basis of acquired knowledge, territory resilience has been defined as a systemic, multidisciplinary, holistic and cost-effective approach to crisis preparedness. The given definition extends the general definition and characterizes the resilience concept and its assessment for the purpose of the presented research.

It was necessary to find a useful tool for investigating the territory resilience of municipalities in the Czech Republic against anthropogenic threats, one that would be used as a methodical base for designing resilience assessments oriented towards municipalities in the Czech Republic. By summarizing the approaches and tools for resilience assessment available in the technical literature, i.e. [5, 26], and the [12], knowledge was acquired that made it possible to define the following criteria for tool selection:

1. Multidisciplinary, complex and systematic assessment.
2. Semiquantitative resilience assessment (i.e. an assessment using a numeric scale based on an interval or qualitative categorization).
3. Assessment oriented towards the local level (municipality).
4. Oriented towards anthropogenic threats, or with potential to assess anthropogenic threats.

A multidisciplinary, complex and systematic assessment is required to achieve the above-defined territory resilience. A semiquantitative assessment is preferred due to the greater relevance of its results in comparison with a qualitative assessment, and the possibility to take the interval quantitative indicators into account. In comparison with a solely quantitative resilience assessment, the semiquantitative method is preferred due to the possibility it affords of adjusting or excluding evaluation criteria or parts of the assessment process—this would be hard or impossible to achieve if a complex mathematical model were necessary for the quantitative assessment. Another advantage of the semiquantitative assessment is the possibility to assess resilience no matter which units of the assessed variables or types of criteria categorization are used (qualitative or interval categorization) [4]—thus there are significantly wider possibilities for assessing area resilience across a much wider spectrum. The reasons to focus on resilience assessment at a local level (municipality) are as follows:

1. Municipalities are characterized by a high population density in one place, which increases vulnerability [10].
2. Multiple sources of threats in one place, potentially high exposure level when the source of threat is activated (see point 1), and potential escalation of events as part of synergic and domino effects.
3. People in municipalities need the local infrastructure whose resilience/vulnerability is crucial for the assessment [10].
4. Application of the subsidiarity principle.
5. Taking into account local factors and conditions in relation to assessing and developing resilience [25].
6. The municipality is the basic territorial unit of public administration in the Czech Republic engaged in tasks concerning crisis management and protecting the population.

It is absolutely necessary to perform a territory resilience assessment, regardless of the sources of threat. In the Czech Republic, the analysis of threats for country was elaborated in 2015 [16]. The analysis of threats was performed as a two-phase analysis. During the first phase, 72 threats were identified, where 54% of them were of anthropogenic origin. Selected threats whose level exceeded the defined reference level were subsequently assessed in more detail. From the 49 threats selected for a detailed risk assessment, 61% were anthropogenic threats, of which 51% were technogenic threats and 10% were sociogenic threats. From the risk estimation in the detailed assessment, 22 risks were assessed as unacceptable, of which 13 risks were of anthropogenic origin. The specified risk analysis procedure [16] was subsequently applied also at the region level. Due to the significant number of anthropogenic threats, the authors decided to focus on assessing territory resilience to threats of anthropogenic origin. It is assumed that the proposed criteria will be utilizable (without adjustment or with minor changes) for assessing territory resilience to natural threats, and thus be used as a universal tool for assessing territory resilience.

Based on the specified criteria, the tool UNISDR's Disaster Resilience Scorecard for Cities [20, 21] was selected as the methodic base for further research.

2.1 UNISDR's Disaster Resilience Scorecard for Cities

The Disaster Resilience Scorecard for Cities [20, 21] is a tool for assessing territory resilience at a local level that provides the evaluator with a set of evaluation criteria based on UNISDR's Ten Essentials for Making Cities Resilient (hereinafter referred to as the "Essentials"). The tool also helps with monitoring and assessing progress in fulfilling the requirements of the Sendai Framework for Disaster Risk Reduction 2015–2030 ([20], p. 3).

The Disaster Resilience Scorecard for Cities [20, 21] exists in two variants, with different levels of detail for performing the resilience assessment. The first (preliminary) assessment variant is more general [21] and enables an assessment of territory resilience during one- or two-day workshops. The second variant is the detailed assessment methodology [20]. As the title suggests, this document is used for more detailed resilience assessments [22] that take 1 to 4 months.

The document specifying the Disaster Resilience Scorecard for Cities [20, 21] is divided into three parts. The first part contains a brief introduction to the issue of resilience assessment, as well as specifying the role of the Disaster Resilience Scorecard for Cities in this context and sets out information and instructions for using the tool and defining the Essentials. The second part contains the Essentials and the related evaluation criteria, which are the core of the document for the actual resilience assessment. The last part of the document contains annexes with a list of terms, the history of the development of the Disaster Resilience Scorecard for Cities, and an overview of the interlinked objectives and indicators of the Sendai Framework in relation to the Essentials and other international documents concerning the issue [20, 21].

2.2 Ten Essentials for Making Cities Resilient and Their Evaluation Criteria

In the Disaster Resilience Scorecard, the criteria for resilience evaluation are systematized in relation to the Essentials, using their present definition to implement the Sendai Framework for Disaster Risk Reduction 2015–2030 [25]. The multidisciplinary character of the assessment of territory resilience affects the characteristics as well as the scope of tasks and actions of the individual Essentials that subjects have to take into account when assessing and increasing resilience at local level ([20], p. 4). According to UNISDR ([20], p. 4) and the terminology of risk management in the Czech Republic, the specified tasks can be divided into the following categories: governance and financial capacity (Essentials 1–3), planning and disaster preparation (Essentials 4–8), disaster response and post-event recovery (Essentials 9–10). The list of Essentials is as follows ([20], p. 4):

1. Organize for disaster resilience.
2. Identify, understand and use current and future risk scenarios.

3.   Strengthen financial capacity for resilience.
4.   Pursue resilient urban development.
5.   Safeguard natural buffers to enhance the protective functions offered by natural ecosystems.
6.   Strengthen institutional capacity for resilience.
7.   Strengthen and understand societal capacity for resilience.
8.   Increase infrastructure resilience.
9.   Ensure effective disaster response.
10. Expedite recovery and build back better.

For each Essential there is a set of maximizing evaluation criteria divided into subcategories with respect to achieving resilience. For each evaluation criterion there is a semiquantitative evaluation scale. Each point of the scale contains a point value and a written description of the level of fulfilment of the requirements for achieving resilience that have to be reached to achieve the assigned point value [20, 21].

3   Methods

The solution to the problem of how to evaluate territory resilience to emergencies is based on the research process. The performed literature search reveals that there is no suitable universal tool for assessing the resilience of municipal territories in all countries worldwide. It is necessary to adapt the tool Disaster Resilience Scorecard for Cities [20, 21] to national conditions with respect to the defined population protection system and crisis management. This knowledge was, utilizing all publicly available information, followed by an analysis of the population protection and crisis management in the Czech Republic, with an emphasis on the emergency response phase. The analysis also examined the implementation scope of the emergency response at the level of regions and municipalities in the Czech Republic. Individual semiquantitative, maximizing evaluation criteria for municipality areas were set out by drawing upon an analysis of the internal national environment and benchmarking.

4   Practical Part

The practical part of the chapter introduces the approach of the authors when setting up criteria for assessing the resilience of municipal territories in the Czech Republic to anthropogenic threats. First of all, the authors specify the principles for setting up the criteria on the basis of the UNISDR's Disaster Resilience Scorecard for Cities [20, 21] with respect to the conditions of municipalities in the Czech Republic. The requirements of individual Essentials on the territory resilience assessment are extensive, so the chapter contains the practical application of creating evaluation criteria for Essential 9, which focuses on responses to emergencies and crises.

The authors of the chapter had to utilize their knowledge and define principles for creating the criteria on the basis of the UNISDR's Disaster Resilience Scorecard for Cities [20, 21] for conditions in the Czech Republic. The principles are as follows:

- When setting up the evaluation criteria, the specific aspects of the public administration, local self-administration, and the system of crisis management and population protection in the Czech Republic must be respected. During this process, the original criteria specified in the Disaster Resilience Scorecard for Cities [20, 21] may be selected, merged or omitted due to the specific conditions in the Czech Republic.
- The objective of the criterial evaluation is not to assess whether the obligations defined by normative legal acts and legally enforced in the Czech Republic are met. These criteria above all deal with the possibilities of municipalities to improve resilience on their own initiative, above and beyond statutory obligations—this corresponds to the proactive character of the assessment, building and increasing the resilience of municipalities.
- The thematic range and focus of individual Essentials must be observed as much as possible.
- The number, range and focus of criteria are defined according to data available at the municipality level in the Czech Republic.
- Criteria for which the municipality does not achieve the maximum point evaluation are used as a motivation for improvements in that area (benchmarking). Taking additional measures to achieve a higher point evaluation should be first analyzed in a feasibility study and cost-effectiveness analysis for specific territories—different municipalities may have different practical applicability for measures derived from the evaluation criteria.
- It is not desirable to introduce a universal reference level for the point evaluation of resilience for individual Essentials or for the evaluation process as a whole, because such level is based on the practical applicability of measures derived from evaluation criteria for specific municipal territories.
- The criteria are evaluated in relation to specific threats, i.e. for each threat (or each threat impact scenario) separately. Some criteria may be more general and are practically applicable to all sorts of threats. Nevertheless, reviewing the evaluation for each threat separately is desirable, and taking into account the threat impact scenario in relation to the criterion.
- When designing the set of criteria and assessment procedure, a combination of both versions of the tool Disaster Resilience Scorecard for Cities will be used—the version for detailed assessment [20] and the version for more general (preliminary) assessment [21].
- The term "disaster" used by the tool Disaster Resilience Scorecard for Cities was generalized, so it can be used for evaluation; so were the terms "emergency" and "crisis" that are used in the Czech Republic for emergencies with potential impacts (and the appropriate character of response).

In addition to the tool Disaster Resilience Scorecard for Cities [20, 21], the territory resilience indicators created by the author Oulehlová [14] as a part of assessing resilience to natural threats are also used for defining some evaluation criteria. These indicators of territory resilience to natural threats [14] are in accordance with the research orientation and applicable to the municipality level in the Czech Republic. Some of them are also useful for assessing resilience to anthropogenic threats, or they can be adapted to become applicable to anthropogenic threats. In addition to the mentioned tools [14, 20, 21], other knowledge, publications and legal or other documents are also used to create the criteria, provided they make it possible to assess resilience in accordance with the research orientation and create the possibility for the proposed criteria to be used with municipalities in the Czech Republic. When setting up the criteria, the authors have also taken advantage of knowledge based on the ongoing COVID-19 pandemic.

4.1 Essential 9: Ensure Effective Disaster Response

The essential specific aspects that are important when transposing Essential 9 into conditions in the Czech Republic reflect the readiness and response of the Integrated Rescue System (IRS) to emergencies, the structure of the crisis management authorities, and the function of municipalities in crisis management, as well as in population protection as defined by the responsibilities, tasks and competences of their authorities. For Essential 9, a set of 9 evaluation criteria for assessing resilience was defined, as listed in Table 1.

As the IRS is the considered the co-ordinated proceedings of its bodies during preparations for emergencies, and during rescue and clean-up operations [2]. The IRS is divided into basic IRS bodies and other IRS bodies. The basic bodies of the IRS are the Fire Rescue Service of the Czech Republic (FRS) and fire units, based on fire cover, the Medical Rescue Service and the Police of the Czech Republic. The other bodies of the IRS are the specified forces and means of armed bodies, other armed security services, other rescue services, public health protection authorities, emergency, stand-by, specialised and other services, civil protection establishments, nongovernmental, non-profit organizations and civil associations, which can be used for rescue and clean-up operations [2] (Fig. 1).

The activities of the IRS when responding to emergencies and crises in the Czech Republic create specific aspects for defining resilience criteria in comparison with the general definitions of the tool Disaster Resilience Scorecard for Cities [20, 21]. The first three criteria (9.1, 9.2 a 9.3) deal with the basic bodies of the IRS (except for the municipal police and FPU not designated for covering territory which are also part of the criteria), and for each unit they focus on specific areas that concern ensuring municipality resilience.

ID	Criterion	Scale
9.1	Police of the Czech Republic and the municipal police	3 – The local division of the Police of the Czech Republic with forces and resources for operations is located at a distance < 2 km or directly in the municipality being assessed. Or possibly the municipality establishes municipal police or a specifically designated patrol service of the Police of the Czech Republic. 2 – The local division of the Police of the Czech Republic with forces and resources for operations is located at a distance (12;2) km from the municipality being assessed. 1 – The local division of the Police of the Czech Republic with forces and resources for operations is located at a distance (22;12) km from the municipality being assessed. 0 – The local division of the Police of the Czech Republic with forces and resources for operations is located at a distance > 22 km from the municipality being assessed.
9.2	Forces of the Fire Rescue Services (FRS) and Fire Protection Units (FPU)	3 – The municipality has an FPU I located in its area, or also other levels of FPU. 2 – The municipality: • has more than 1,000 residents and an FPU II or FPU III designated for covering territory is located there. • has fewer than 1,000 residents and an FPU II, III or V designated for covering territory is located there. 1 – The municipality: • has more than 1,000 residents and an FPU V designated for covering territory or an FPU not designated for covering territory is located there. • has fewer than 1,000 residents and an FPU not designated for

ID	Criterion	Scale
		covering territory is located there. 0 – The municipality does not have its own FPU.
9.3	Medical Rescue Service (MRS), possible obstacles when providing urgent pre-hospital care	2 – No obstacles when providing urgent pre-hospital care. 1 – Slight obstacles when providing urgent pre-hospital care. 0 – Significant obstacles when providing urgent pre-hospital care.
9.4	Danger level of municipality	3 – Danger level IV of municipality. 2 – Danger level III of municipality. 1 – Danger level II of municipality. 0 – Danger level I of municipality.

Fig. 1 Evaluation criteria—Essential 9. *Source:* [7, 14] with the help of the [1, 3, 9, 13, 17] and the [6]

9.5	Voluntary activities	2 – The municipality has its own extensive voluntary services that provide significant support during responses to emergencies and crises. 1 – The municipality has its own voluntary services that provide partial support during responses to emergencies and crises. 0 – The municipality does not have any relevant voluntary services.
9.6	Adequacy of forces and resources, complexity of mission	4 – Rescue and liquidation services are provided by basic bodies, there is no need to constantly coordinate them during joint missions. 3 – Rescue and liquidation services are provided by basic and other bodies at the local level where the emergency occurs, or bodies must be constantly coordinated by a mission commander. 2 – Rescue and liquidation services are provided by basic and other bodies at the regional level where the emergency occurs, or during their joint mission the bodies must be constantly coordinated by a mission commander with help of the commander's staff, convening a municipal crisis staff. 1 – Rescue and liquidation services are provided by basic and other bodies including forces and resources from other regions, coordinated by mayors of multiple municipalities with extended competences (convening a crisis staff of the municipality with extended competences) or by the region's governor (convening a regional crisis staff). 0 – Convening a Central Crisis Staff – declaration of an emergency (Oulehlová 2017a).
9.7	Training IRS bodies and crisis management authorities	3 – The existing authorities of municipality crisis management, or the forces and resources of IRS bodies designated for operations in the municipality, have already participated in some form of tactical training, or the training of crisis management authorities focused on responses to emergencies related to the assessed threat. 2 – The existing authorities of municipality crisis management, or the forces and resources of IRS bodies designated for operations in the municipality, participate not only in tactical training or the training of crisis management authorities in the field, but also in other possible innovative forms of response training – e.g. training in a constructive simulation environment. 1 – The existing authorities of municipality crisis management, or the forces and resources of IRS bodies designated for operations in the municipality, have already participated in tactical training or the training of crisis management authorities. 0 – The existing authorities of municipality crisis management, or the forces and resources of IRS bodies designated for operations in the municipality, have not yet participated in tactical training or the training of crisis management authorities.

Fig. 1 (continued)

ID	Criterion	Scale
9.8	Predictability and impact rapidity of dangers	2 – There are extensive prediction possibilities and the impact of dangers may be reduced to minimum (e.g. legal violations and rowdiness during transfers of soccer fans, or simultaneously announced demonstrations of antagonistic factions). 1 – Dangers can be partially predicted, or their impact rapidity is distributed over a longer time period (e.g. leakage of dangerous substances with slow spreading into the surroundings). 0 – Dangers are difficult to predict or cannot be predicted at all, impacts are very rapid or instant (e.g. a traffic accident followed by an explosion).
9.9	Population protection – warnings and announcements	4 – The municipality fulfills the levels defined below and, in addition, it is also developing in line with trends concerning population warnings and announcements as defined by the General Directorate of the Fire Rescue Service (2015) as follows: • measuring terminal elements (e.g. concentrations of dangerous substances). • information terminals used to notify dangers relating to important buildings (e.g. education, social and medical facilities). • application of a selective two-way system for controlling and monitoring the status of terminal elements, to measure and warn, for system control and diagnostics. 3 – There are no places not covered by a unified system of warnings and notifications in the municipality area. The municipality uses electronic sirens that are able to transmit verbal information and a local information system into which the electronic sirens are integrated. 2 – There are no places not covered by a unified system of warnings and notifications in the municipality area. The municipality uses electronic sirens that are able to transmit verbal information. 1 – There are no places not covered by a unified system of warnings and notifications in the municipality area. The municipality uses only rotary sirens. 0 – There are some places not covered by a unified system of warnings and notifications in the municipality area, where a substitutive system of warnings must be provided. The municipality uses only rotary sirens.

Fig. 1 (continued)

Criterion 9.1 evaluates municipality resilience in relation to the distance between the relevant local division of the Police of the Czech Republic and the assessed municipality. The criterion is based on the organizational structure of the Police of the Czech Republic. At the regional level, the Police of the Czech Republic is divided into territorial divisions. Subsequently, these territorial divisions are divided into local divisions with a designated territorial scope for a specific number of municipalities [17]. Each local division of the Police of the Czech Republic covers a large area and (unlike other basic IRS bodies) police divisions also have to patrol their areas. Municipalities are not permitted to establish their own divisions of the Police of the Czech Republic. Unlike the MRS, the Police of the Czech Republic does not "guarantee" arrival times, nor does it have a multilevel system (combination of the regional FRS units and Voluntary Fire Units (VFU) established by municipalities) of territorial coverage as in the case of FPU units. Due to these reasons, geographical distance is crucial for the availability of the forces and resources of the Police of the Czech Republic in specific municipalities. The intervals used in criterion 9.1 were defined by analyzing the arrival times among local divisions to municipalities within their territorial scope for the territorial division Brno-venkov. The territorial division Brno-venkov covers 8 local divisions with 223 municipalities within their territorial scope (the municipalities in which the local divisions are located were not counted). To measure the distance, the server Mapy.cz was used [9]—the shortest arrival time for a car route from the appropriate local division to the municipality within the territorial scope was measured. The shortest measured distance was 1.6 km, while the longest was 21.6 km—the rounded values (2 and 22 km) were used to set up the criterion intervals. The number of intervals was set up according to the scale range 0 to 3. A municipality can also achieve a higher point evaluation if it establishes its own municipal police or a specifically designated patrol service of the Police of the Czech Republic. The criterion does not deal with situations, when patrols of the Police of the Czech Republic drive to the place of emergency from somewhere in the field—it is not possible to define a scale for such situations because the arrival time varies. Criterion 9.1 does not deal with specialized forces of the Police of the Czech Republic.

Criterion 9.2 deals with the FRS and FPU. Criterion 9.2 uses different scales than criterion 9.1. The scale of criterion 9.2 focuses not only on strengthening municipality resilience thanks to territorial coverage by the FPU, but also on establishing and developing VFU managed by municipalities. In addition to the IRS, supporting VFU in municipalities also helps develop voluntary firefighters (VF), i.e. interest groups dealing with fire prevention without getting involved in missions—operations of VF are usually coordinated with the operations of VFU, and VF significantly contribute to the municipality's voluntary capacities. To make the scale more useful, municipalities are also divided into categories according to their number of inhabitants. The categories of FPU and territorial coverage are defined with respect to the [1] and the [13]. The arrival times of FPU units are based on these categories of territorial coverage and the FPU according to the norms listed above.

The criterion 9.3 focuses on MRS activities. The arrival time of the MRS (up to 20 min) is defined by the [3], thus the distribution of mission bases should respect the requirements defined by that act. Instead of arrival times and distances, criterion 9.3 focuses on potential obstacles when providing urgent pre-hospital care, which may include e.g. requirements for decontamination, rescue services or violence against rescuers. MRS operations are also assessed in Essential 8 (in relation to the number of injured persons and their necessary triage).

Criterion 9.4 is based on the [13], which defines criteria for calculating the municipality danger level. The danger level determines the claims to territorial coverage by the FPU and the appropriate arrival times of individual FPU types. The municipality danger level is determined with respect to the number of inhabitants, geographical aspects of the territory, and the number of missions within the area [13].

Criterion 9.5 focuses on voluntary activities in the municipality. It is evaluated empirically—on the basis of previous experience. Both organized (e.g. nongovernmental, non-profit organizations) and unorganized (initiatives of individual persons and legal entities) voluntary activities are taken into account. Despite not being an anthropogenic threat, this criterion can be evaluated on the basis of the extent of support and solidarity during the current COVID-19 pandemic, because voluntary help has turned out to be a decisive factor in managing the crisis. In addition, this domain can be significantly developed and supported by municipal authorities.

Criterion 9.6 focuses on the adequacy of forces and resources, and the complexity of missions. The criterion has already been used before during research on resilience to natural threats [14]; with minor alterations, it can be also used for assessing resilience to anthropogenic threats.

Criterion 9.7 focuses on the tactical training of IRS bodies and crisis management authorities. In the Czech Republic, tactical training is used as a standard tool to verify and strengthen readiness, and to practice the response of IRS bodies or crisis management authorities in relation to specific threats and their impact scenarios.

Criterion 9.8 assesses possible responses with respect to the predictability and impact rapidity of dangers. The criterion was derived from the natural threat indicator which was created to assess territory resilience to natural threats [14]. The criterion's character and scale had to be altered to ensure its suitability for the impact mechanism and possible prediction of anthropogenic threats.

Criterion 9.9 assesses the municipality's development level with respect to warning and notifying residents when any emergency or crises occurs. The criterion is based on information contained in sources of the [6].

5   Conclusion

This chapter has focused on evaluation criteria that can be used when assessing the resilience of municipalities in the Czech Republic to anthropogenic threats. The theoretical part of the chapter defined the term "territory resilience". In addition, resilience was put into the context of risk management, crisis management, environmental security and theory of systems as a systematic, multidisciplinary, holistic and cost-effective approach to crisis preparedness.

There was a need to find a useful tool for investigating the territory resilience of municipalities in the Czech Republic to anthropogenic threats, which would be used as the methodological basis for designing resilience assessment of municipalities in the Czech Republic. To find an appropriate tool, criteria were created, setting out the requirements the tool should meet. These criteria for selecting the appropriate tool were created with the help of knowledge summarizing the approaches and tools for assessing resilience available in the technical literature: [5, 26], and the [12]. Following the criteria, the tool UNISDR's Disaster Resilience Scorecard for Cities [20, 21] was selected, since it meets all the defined criteria. The tool UNISDR's Disaster Resilience Scorecard for Cities [20, 21] is described at the end of the theoretic part of the chapter.

The practical part of the chapter deals with the approach of the authors when creating the criteria with respect to the tool UNISDR's Disaster Resilience Scorecard for Cities [20, 21]. The crucial principles for creating the criteria for the ultimate purpose—assessing the resilience of municipal territories in the Czech Republic to anthropogenic threats—were defined with respect to the specific territorial aspects of municipalities in the Czech Republic and the character of the threat to these areas. As a result of the practical application of these principles was presented Essential 9, which deals with the response to emergencies and crises.

The essential specifics that were important when transposing Essential 9 into conditions in the Czech Republic reflect the readiness and response of the Integrated Rescue System, as well as of the crisis management system, to emergencies, and the function of municipalities in crisis management, as well as in population protection as defined by the responsibilities, tasks and competences of their authorities. For Essential 9, a set of 9 evaluation criteria for assessing resilience was created, as described in the chapter.

The proposal for the specifics of the Czech Republic follows the structure of the tool Disaster Resilience Scorecard for Cities [20, 21], i.e. it is divided into ten Essentials. Essential 9 is one of ten partial sections that, together with the approach created during the research [7], will be applied to increase the level of municipality resilience in the Czech Republic to anthropogenic threats. By increasing the resilience level, it is possible to improve the crisis readiness of the municipality area in a cost-effective manner. In addition, increasing the resilience level of the territory makes it possible to fulfill some of the requirements defined in the Sendai Framework [25] and to contribute to sustainable development, the highest value of society.

References

1. Act on Fire Prevention (1985). In: Czech: Zákon o požární ochraně. In: Sbírka zákonů České republiky. Praha, ročník 1985, číslo 133, částka 34/1985. Available at: https://www.zakonypro lidi.cz/cs/1985-133
2. Act on the Integrated Rescue System (2000). In: Czech: Zákon o integrovaném záchranném systému. In: Sbírka zákonů České republiky. Praha, ročník 2000, číslo 239, částka 73/2000. Available at: https://www.zakonyprolidi.cz/cs/2000-239

3. Act on the Medical Rescue Service (2011). In: Czech: Zákon o zdravotnické záchranné službě. In: Sbírka zákonů České republiky. Praha, ročník 2011, číslo 374, částka 131/2011. Available at: https://www.zakonyprolidi.cz/cs/2011-374

4. Božek F, Urban R (2008) Risk management: general part. In: Czech: Management rizika: obecná část. Univerzita obrany, Brno. ISBN 978–80–7231–259–7

5. Cutter S (2015) The landscape of disaster resilience indicators in the USA. Nat Hazards. 80(2):741–758. ISSN (online): 1573–0840. Available at: https://link.springer.com/article/10. 1007%2Fs11069-015-1993-2

6. General Directorate of the Fire Rescue Service (2015) Ochrana obyvatelstva a krizové řízení: skripta. Tiskárna Ministerstva vnitra ČR, Praha, p. o. ISBN 978–80–86466–62–0

7. Kincl P (2020) Resilience assessment of the territory to the selected anthropogenic threats. In: Czech: Hodnocení odolnosti území na vybraná antropogenní nebezpečí. Univerzita obrany, Brno. Disertační práce. Školitelka práce Alena Oulehlová

8. Lane JA (2013) What is a system of systems and why should i care? Center for systems and software engineering [online, cit. 14. 03. 2019]. University of Southern California. Available at: http://csse.usc.edu/TECHRPTS/2013/reports/usc-csse-2013-500.pdf

9. Mapy.cz (2020) Mapy.cz [online, cit. 19. 04. 2020]. Available at: https://mapy.cz/zakladni?x= 16.5317000&y=49.2989000&z=11

10. Martin A (2015) A framework to understand the relationship between social factors that reduce resilience in cities: Application to the City of Boston. Int J Disaster Risk Reduct 2015(12):53–80. ISSN 2212–4209

11. Ministry of the Interior of the Czech Republic (2016) Terminological dictionary of terms from the field of crisis management, population protection, environmental security and national defense planning. In: Czech: Terminologický slovník pojmů z oblasti krizového řízení, ochrany obyvatelstva, environmentální bezpečnosti a plánování obrany státu. Ministerstvo vnitra České republiky [online, cit. 05. 03. 2019]. Available at: https://www.mvcr.cz/clanek/terminologicky-slovnik-krizove-rizeni-a-planovani-obrany-statu.aspx

12. National Research Council (2015) Developing a framework for measuring community resilience: summary of a workshop. National Academies Press, Washington, D. C

13. Ordinance of the Ministry of the Interior of the Czech Republic on the Organization and Activities of Fire Protection Units (2001). In: Czech: Vyhláška Ministerstva vnitra o organizaci a činnosti jednotek požární ochrany. In: Sbírka zákonů České republiky. Praha, ročník 2001, číslo 247, částka 95/2001. Available at: https://www.zakonyprolidi.cz/cs/2001-247

14. Oulehlová A (2017) Resilience of the territory to the occurrence of natural disasters. Resilience území na výskyt přírodních katastrof. Brno. Habilitační práce. Univerzita obrany, In Czech

15. Oulehlová A (2017b) The role of risk management in ensuring security: habilitation lecture. In: Czech: Úloha managementu rizik v zajištění bezpečnosti: habilitační přednáška. Univerzita obrany, Brno

16. Paulus F, Krömer A, Petr J, Černý J (2015) Threat analysis for the Czech Republic: final report. In: Czech: Analýza hrozeb pro Českou republiku: závěrečná zpráva. Hasičský záchranný sbor České republiky, Praha. Available at: https://www.hzscr.cz/clanek/strategicke-a-koncepcni-mat erialy.aspx

17. Police of the Czech Republic (2020) Policie ČR [online, cit. 20. 04. 2020]. Available at: https:// www.policie.cz/policie-cr.aspx

18. Rainey L, Tolk A (2015) Modeling and simulation support for system of systems engineering applications. Wiley, New Jersey. ISBN 978–1–118–46031–3

19. UNDRR (2019) The making cities resilient—my city is getting ready! https://www.unisdr.org [online]. Geneva: UNISDR, 2019 [cit. 2020–02–01]. Available at: https://www.unisdr.org/cam paign/resilientcities/

20. UNISDR (2017a) Disaster resilience scorecard for cities: detailed level assessment. UNISDR [online, cit. 05. 03. 2019]. Available at: https://www.unisdr.org/campaign/resili entcities/assets/documents/guidelines/04%20Detailed%20Assessment_Disaster%20resilie nce%20scorecard%20for%20cities_UNISDR.pdf

21. UNISDR (2017b) Disaster resilience scorecard for cities: preliminary level assessment. UNISDR [online, cit. 08. 02. 2020]. Available at: https://www.unisdr.org/campaign/resilient cities/assets/toolkit/Scorecard/UNDRR_Disaster%20resilience%20%20scorecard%20for% 20cities_Preliminary_English.pdf

22. UNISDR (2017c) Disaster resilience scorecard for cities. UNISDR [online, cit. 28. 09. 2018]. Available at: https://www.unisdr.org/we/inform/publications/53349

23. United Nations (1994) E/1994/85 international decade for natural disaster reduction: Yokohama strategy for a safer world. Switzerland, United Nations, Geneva

24. United Nations (2005) Hyogo framework for action 2005–2015: building the resilience of nations and communities to disasters. Switzerland, United Nations, Geneva

25. United Nations (2015) A/RES/69/283 Sendai framework for disaster risk reduction 2015–2030. Switzerland, United Nations, Geneva

26. Winderl T (2014) Disaster resilience measurements: stocktaking of ongoing efforts in developing systems for measuring ressilience. United Nations Development Programme. Available at: https://www.preventionweb.net/files/37916_disasterresiliencemeasurementsundpt.pdf

Security Management—Quantitative Determination of Crime Risk in a Selected Region: Case Study

Jiří Jánský⊙ and Irena Tušer⊙

Abstract The chapter deals with the possibilities of analysis of security risks at the regional level, including their identification, classification and possibilities of prevention of their occurrence. Based on a research survey in a selected region, the authors address the issue of strengthening security with respect to a selected type of threat—crime. The results of the assessment of security risks resulting from crime can be used in favor of ensuring greater efficiency in promoting security in the region. The chapter deals with the application of the Saaty's method to determine the severity of impacts in more than twenty of the most common types of crime. The aim of the authors was not only to determine the risks and cumulative risks that arise from individual types of crime, but especially to identify and separate the most risky types of crime from the less risky ones, and to propose measures to reduce the risk of crime at regional level.

Keywords Crime · Crisis management · Regional security · Risk · Risk analysis · Risk management · Security assessment · Security management · Threat

1 Introduction

Due to its geographical and geopolitical location, the Czech Republic (CR) has always been influenced by the development of the security situation in the world and will probably continue to do so. Recent events, such as the migration crisis or the Russia-Ukraine conflict, foreshadow the deteriorating security situation in the world, and thus in Europe and the Czech Republic. The security system of the state and regions must be prepared to face these new threats and the resulting risks [7]. At

J. Jánský (✉)
Faculty of Military Technology, University of Defence, Kounicova 65, Brno 66210, Czech Republic
e-mail: jiri.jansky@unob.cz

I. Tušer
Department of Security and Law, AMBIS College, Lindnerova 1, Prague 180 00, Czech Republic
e-mail: irena.tuser@ambis.cz

I. Tušer and Š. Hošková-Mayerová (eds.), *Trends and Future Directions in Security and Emergency Management*, Lecture Notes in Networks and Systems 257, https://doi.org/10.1007/978-3-030-88907-4_12

the same time, it must be prepared to face potential internal threats, such as crime or anthropogenic and naturogenic emergency situations [9]. The first step in ensuring the security of a given region or territory should be the ability to identify, analyze, characterize threats and then take proactive action against them.

One of the long-term and recurring threats to the internal security of the territory is crime. Criminality has always been a security threat to human society as a whole. The Czech Republic approaches the issue like most of the Western world, using modern, democratic, and legal methods. In order to understand crime as much as possible, it is necessary to analyze it in detail, examine it, and then apply functional preventive measures. A system based on a functional strategy to fight and prevent crime is the basis for building a safe environment and a determinant of the further development of society. An integral part of a functional strategy for the fight against crime is the ability to assess the current state of risks arising from crime in the examined area/region [6].

The aim of this study was set based on the above-mentioned facts and to strengthen the security of the selected region (South Moravian Region). The objective was to assess the risks arising from crime and their effects at the regional level. Based on the assessment of these risks, possible measures for their mitigation were subsequently proposed. The authors also asked a research question:

What kind of crime is most risky for the South Moravian region?

1.1 Current State of Crime in the South Moravian Region

The selected region for the research survey was the South Moravian Region (SMR). It is one of the largest and most populous regions in the Czech Republic (CR). The population in 2019 was 11,91,989, which is 11% of the total population of the CR.

In 2019, a total of 1,99,221 criminal offenses were registered throughout the Czech Republic and only 46.8% were solved. 86,209 people were prosecuted and the amount of damages reached CZK 24.26 billion. The South Moravian Region does not differ statistically in any way from the overall state of crime in the CR. In 2019, 19,757 crimes were registered and the clear-up rate was 42.5% [2]. Figure 1 shows the development trend of registered crime in the South Moravian Region over the last ten years.

The curve of development trends of total crime in the South Moravian Region shows a declining trend in the number of registered crime. The main reason for the declining trend is the gradual year-on-year decline in property and economic crime. On the contrary, moral crime, as the only major form of crime, remains in slight year-on-year growth.

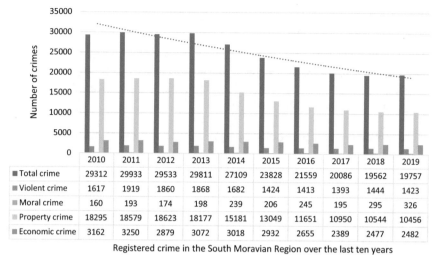

	2010	2011	2012	2013	2014	2015	2016	2017	2018	2019
■ Total crime	29312	29933	29533	29811	27109	23828	21559	20086	19562	19757
■ Violent crime	1617	1919	1860	1868	1682	1424	1413	1393	1444	1423
■ Moral crime	160	193	174	198	239	206	245	195	295	326
■ Property crime	18295	18579	18623	18177	15181	13049	11651	10950	10544	10456
■ Economic crime	3162	3250	2879	3072	3018	2932	2655	2389	2477	2482

Registered crime in the South Moravian Region over the last ten years

Fig. 1 Development trends of crime in the South Moravian Region. *Source* Pavlík, 2021

2 Data and Methods

The research survey dealt with the determination of risks in 26 types of crime in the SMR. Input data on the number of individual crimes and offences were drawn from official statistics for 2019 [4], presented in Supplement 1 (Table 7). The aim of this chapter is to present a procedure enabling a multi-criteria comparison of a larger number of items (it is effective for > 15 items). This procedure will be illustrated in determining the risks of individual types of crime. Their assessment, comparison, and evaluation will then be the subject of Chap. 3. The analysis carried out in these two chapters is based on the paper [3]. The source data that were analyzed are listed in Table 7

2.1 Methodology Used

The risk of R(t) is defined by the relation

$$R(t) = p(t) \cdot d, \tag{1}$$

where t denotes the investigated time period, $p(t)$ is the probability of event activation and d is the impact of the given event [1]. In this work, however, only crime in the year 2019 is studied, and it is assumed that the probability of occurrence of the given type of crime is constant throughout this year. Thus, relation (1) is in the form

$$R = p \cdot d, \tag{2}$$

where p is the probability of occurrence of the crime, which is obtained from police statistics, and d is the relevant impact (or severity) of the crime. When calculating the risk of a group of crimes k, the risk of the whole group of crime R must be the sum of all risks of individual crimes R_i, $i = 1, 2, \ldots, k$, i.e.
$R = \sum_{i=1}^{k} R_i$.

The probability of victimization by a given crime was calculated by the relation

$$p = \frac{n}{m}, \tag{3}$$

where n is the number of crimes of a given type of crime and m is the number of inhabitants. The total probability p that any of the given k crimes occurs is the sum of the probabilities p_i, $i = 1, 2, \ldots, k$ that the individual crimes will occur:

$$p = \sum_{i=1}^{k} p_i$$

The impact of a given type of crime was determined by the Saaty's method. In this method, respondents pairwise compare the impacts $d_1, \ldots, d_k$ of individual types of crime. More precisely, they compare the impact d_i with the impact d_j for each $i = 1, \ldots, k$, $j = 1, \ldots, k$. In this comparison, they fill in the elements of the Saaty's matrix. These elements are called s_{ij} and are an estimate of the proportions of the sought impacts d_i and d_j, so the following holds:

$$s_{ij} \approx \frac{d_i}{d_j}, i = 1, \ldots, k, j = 1, \ldots, k$$

After finding the unknown values of d_i, the geometric diameters of the rows of the Saaty's matrix are calculated first

$$G_i = \sqrt[k]{s_{i1} \cdot s_{i2} \cdots \cdots s_{ik}}, i = 1, \ldots, k$$

and then the rows are normalized. This means that the geometric means G_i are divided by their sum. This is how we obtain the required values of impacts.

$$d_i = \frac{G_i}{\sum_j G_j}, i = 1, \ldots, k$$

In the direct application of this method, it would be necessary to compare all 26 investigated types of crime in pairs. Even if the respondent filled in only the upper half of the Saaty's table (because the elements below the main diagonal are the inverse values of the elements above it), each respondent would have to make a total of $\frac{26^2-26}{2} = 325$ comparisons. This number of comparisons would be unacceptable for the following reasons:

- Respondents would not be willing to make so many comparisons.
- Respondents would not be able to keep attention for responsible evaluation of so many comparisons.
- It is better to compare only crimes within subgroups of a similar nature. For example, the effects of individual types of theft separately, and the effects of individual types of violent crime also separately.

For these reasons, it was necessary to divide the problem into several parts, which were analyzed separately [5]. This significantly reduced the number of pairwise comparisons. The method of distribution will be explained in Chaps. 2.2–2.4 and the exact number of comparisons that respondents had to make will be calculated in Chap. 2.5.

2.2 Identifying the Impacts and Risks of the Main Types of Crime

All offenses against the law are divided into 9 main types of crime, which are marked $C_1, .., C_9$ and described in Table 1. The types of crime are sorted here according to the number of cases in descending order. The four most common types of crime $C_1, .., C_4$ will be further divided into other subgroups, which will be analyzed in more detail in other subchapters. The presented division is adaptem from the distribution of types of crimes from the official statistics of the Police of the Czech Republic [4].

Table 1 Probability of victimization by the main group of investigated crime

Crime		probability	impact	risk
i	C_i	$p_i \cdot 10^{-5}$	$d_i \cdot 10^{-5}$	$R_i \cdot 10^{-5}$
1	Property—other	528.8	6800	36
2	Property—simple thefts	476.6	07,700	36.7
3	Economic crime	208.2	12,200	25.4
4	Violent crime	119.4	38,300	45.7
5	Addiction	101.4	04,100	4.19
6	Obstruction to authority's decision	78.4	03,500	2.77
7	Road accidents	45	03,200	1.43
8	Moral crime	27.3	20,900	5.7
9	Others in C	71.5	03,400	2.43
Total		$p = 0.1656$	$d = 0.209$	$R = 0.0016$

Source Pavlík 2021

The p_i probability of victimization by a given crime was calculated using relation (3), which is now in the form of

$$p_i = \frac{n_i}{m}, i = 1, 2, \ldots, 9,$$

where n_i are the numbers a given type of crime $C_i, i \in \{1, 2, \ldots, 9\}$ and $m = 1191989$ is the number of inhabitants of the SMR. The resulting values are shown in Table 1.

Saaty's pairwise comparison method was used to calculate the impacts (or severity) of individual types of crime $C_i, i \in \{1, 2, \ldots, 9\}$. During its application, ten experts from the armed security forces were approached. Each respondent anonymously filled in Saaty's table, in which he compared the impacts of the types $C_i, i \in \{1, 2, \ldots, 9\}$ with each other and determined the ratio of their severity. A table with the decisions of each respondent $j \in \{1, 2, \ldots, 10\}$ was subsequently created and the impacts that the jth respondent attributes to the given events were determined [8]. The calculated values of impacts from individual respondents are given in Table 2.

The resulting impacts $d_1, \ldots, d_9$ of the main types of crime were calculated as the arithmetic average of the impact values determined by individual respondents and are given in the last column of Table 2.

The risks R_i of the main types of crime $C_1, \ldots, C_9$ were calculated according to relation (2), which is now in the form

$$R_i = p_i d_i, i = 1, 2, \ldots, 9.$$

The resulting values are given in the last column of Table 1. Since relation (2) also applies to the total risk, we obtain the value of the total impact of all types of crime as follows:

Table 2 Calculated values of impacts from individual respondents

	resp. 1	resp. 2	resp. 3	resp. 4	resp. 5	resp. 6	resp. 7	resp. 8	resp. 9	resp. 10	d_i
C_1	0.065	0.081	0.040	0.053	0.070	0.080	0.088	0.052	0.076	0.070	0.068
C_2	0.082	0.075	0.041	0.079	0.075	0.067	0.093	0.085	0.070	0.101	0.077
C_3	0.121	0.073	0.043	0.137	0.131	0.120	0.158	0.134	0.128	0.177	0.122
C_4	0.377	0.403	0.376	0.394	0.393	0.402	0.351	0.397	0.384	0.354	0.383
C_5	0.039	0.033	0.112	0.027	0.037	0.037	0.036	0.026	0.024	0.034	0.041
C_6	0.016	0.061	0.043	0.031	0.029	0.028	0.029	0.039	0.046	0.029	0.035
C_7	0.024	0.051	0.094	0.012	0.016	0.018	0.023	0.014	0.048	0.017	0.032
C_8	0.256	0.147	0.208	0.247	0.228	0.221	0.202	0.230	0.159	0.193	0.209
C_9	0.020	0.077	0.043	0.019	0.020	0.028	0.020	0.023	0.065	0.025	0.034
$\sum$	1	1	1	1	1	1	1	1	1	1	1

Source Pavlík 2021

$$d = \frac{R}{p} = 0,209$$

Thus, the total impact is not just the sum of the impacts $d_1, \ldots, d_9$.

The aim of this article is to describe and illustrate the method for the analysis of a total of 26 types of crime. Its principle is the division of the most common types of crime $C_1, \ldots, C_4$ into other subgroups. Each subgroup will first be analyzed separately. The intermediate results thus obtained will then be recalculated to correspond to the values given in Table 1.

2.3 Analysis of the Most Common Type of Crime

Most crimes fall into the category of "Property—Other" C_1). The probability of its occurrence is $p_1 = 528, 8 \cdot 10^{-5}$, the magnitude of its impact $d_1 = 0.068$, and the risk that this type of crime poses $R_1 = p_1 d_1 = 36 \cdot 10^{-5}$. This type of crime will now be divided into 6 subcategories, designated $C_1^1, C_1^2, \ldots, C_1^6$ and described in Table 7. The aim of this chapter is to determine the probabilities of the occurence of $p_1^1, p_1^2, \ldots, p_1^6$, the magnitudes of the impact $d_1^1, d_1^2, \ldots, d_1^6$, and the risks $R_1^1, R_1^2, \ldots, R_1^6$ they pose.

First, for each subgroup $C_1^1, C_1^2, \ldots, C_1^6$, the numbers of crimes (denoted $n_1^1, n_1^2, \ldots, n_1^6$) were determined from the statistics. Then, using the relation (3), which is now in shape.

Nejprve byly u každého poddruhu $C_1^1, C_1^2, \ldots, C_1^6$ ze statistik zjištěny počty skutků (značené $n_1^1, n_1^2, \ldots, n_1^6$). Dále pak byla pomocí vztahu (3), který je nyní ve tvaru

$$p_1^j = \frac{n_1^j}{m}, j = 1, 2, \ldots, 6$$

the probability $p_1^j, j \in \{1, 2, \ldots, 6\}$ of victimization by a given crime was calculated for each act, $m = 1191989$ is the population of the South Moravian Region. The calculated values of these probabilities are given in the penultimate column of Table 3.

Now it is necessary to determine the severity of the impacts $\overline{d}_1^j, j \in \{1, 2, \ldots, 6\}$ of the crimes $C_1^1, C_1^2, \ldots, C_1^6$. The impacts were determined using the Saaty's method and the resulting values are given in the last column of Table 3. However, in the Saaty's method, the respondents compared the impacts $\overline{d}_1^j, j \in \{1, 2, \ldots, 6\}$ only with each other. Saaty's method determines the correct ratios of the values $\overline{d}_1^j, j \in \{1, 2, \ldots, 6\}$, but it says nothing about their absolute size. Therefore, it is necessary to multiply these values by the coefficient a_1 so that they correspond to the meaning that the type of crime C_1 has among the other types $C_2, \ldots, C_6$. When calculating the risks arising from the subgroups of crime $C_{1,\ldots,}^1 C_1^6$, it is necessary to keep in mind

Table 3 Number of crimes and probability of victimization of other property crime

C_1: Property—other

j	C_1^j	$p_1^j \cdot 10^{-5}$	$\overline{d}_1^j$
1	Burglary	229.75	0.349
2	Fraud	109.11	0.336
3	Graffiti tagging	51.85	0.048
4	Evading alimony	7.63	0.089
5	Vandalism	91.28	0.103
6	Others in C_1	30.45	0.075
Total:	C_1	$p_1 = 528.8 \cdot 10^{-5}$	1

Source Pavlík 2021

that they are examined in the context of crime C_1. It is therefore necessary to find such a constant a_1 that the relations hold

$$d_1^j = a_1 \overline{d}_1^j, \ R_1^j = p_1^j d_1^j \ j = 1, 2, \ldots, 6$$

and also relations

$$R_1 = \sum_{j=1}^{6} R_1^j, \ p_1 = \sum_{j=1}^{6} p_1^j.$$

From this we get the constant a_1 in the form

$$a_1 = \frac{R_1}{\sum_{j=1}^{6} p_1^j \overline{d}_1^j,} = \frac{36 \cdot 10^{-5}}{131,71 \cdot 10^{-5}} = 0,273.$$

Thus, the effects of crimes $C_{1,\ldots}^1, C_1^6$ in the context of crimes $C_1, \ldots, C_6$ are

$$d_1^j = 0,273 \overline{d}_1^j, \ j = 1, 2, \ldots, 6.$$

The interpretation of the constant $a_1 = 0,273$, which must be multiplied by the impacts $\overline{d}_1^j$ from Table 3, is such that it is a measure of the representation of "Property—other" crime in total crime. The resulting probabilities of occurrence, magnitude of impact and risks of crimes $C_1^1, C_1^2, \ldots, C_1^6$ are given in Table 4.

In the following chapter, the analysis of the remaining important types of crimes C_2, C_3, C_4 will be performed in an analogous way.

Table 4 The resulting probabilities of occurrence, magnitude of impact and risks of crimes

j		1	2	3	4	5	6	Celkem
C_1	$p_1^j \cdot 10^{-5}$	229.75	109.11	51.85	5185	91.28	30.45	$528.8 \cdot 10^{-5}$
	d_1^j		0.1070	0.1031	0.0147	0.0273	0.0316	0.02301	0.068
	$R_1^j \cdot 10^{-5}$	21.393	8.137	0.755	2.0973	2.936	0.671	$36 \cdot 10^{-5}$

Source Pavlík 2021

2.4 Analysis of Other Important Types of Crime

In this chapter, each of the other three major types of crime C_2, C_3, C_4 is divided into 5 subgroups, and for each of them the probability of occurrence, impact and risk it poses to society will be determined. The calculation procedure is completely analogous to the previous subchapter. Again, the numbers of relevant crimes were first determined. These are listed in Table 5.

Table 5 The number of crimes concerned

j	C_2^j: Property—simple thefts	n_2^j	$\overline{d}_2^j$
1	Pickpocketing	965	0.074
2	Theft of vehicles and motorbikes	556	0.423
3	Theft of bicycles	588	0.121
4	Car theft	1068	0.203
5	Other thefts in C_2	2504	0.180
Total:	C_2	$n_2 = 5681$	1
j	C_3^j: Economic crime	n_3^j	$\overline{d}_3^j$
1	Counterfeinting of money and documents	157	0.371
2	Credit card thefts	809	0.109
3	Thefts or destruction of electronic data	115	0.136
4	Embezzlement, fraud	808	0.273
5	Others in C_3	593	0.111
Total:	C_3	$n_3 = 2482$	1
j	C_4^j: Violent crime	n_4^j	$\overline{d}_4^j$
1	Murder	21	0.571
2	Assault	461	0.213
3	Theft	182	0.089
4	Abuse, oppression, restriction	381	0.081
5	Others in C_4	378	0.047
Total:	C_4	$n_4 = 1423$	1

Source Pavlík 2021

Table 6 The resulting size of risk in the second subgroup of crime

Type	j	1	2	3	4	5	Total
C_2	$p_2^j \cdot 10^{-5}$	80.957	46.645	49.329	89.598	210.069	$p_2 = 476.5 \cdot 10^{-5}$
	d_2^j		0.0309	0.1769	0.0506	0.0849	0.0753	$d_2 = 0.077$
	$R_2^j \cdot 10^{-5}$	2.517	8.221	2.517	7.634	15.772	$R_2 = 36.6 \cdot 10^{-5}$
C_3	$p_3^j \cdot 10^{-5}$	13.171	67.87	9.648	67.786	49.749	$p_3 = 208.2 \cdot 10^{-5}$
	d_3^j	0.2505	0.0736	0.0918	0.1843	0.0749	$d_3 = 0.122$
	$R_3^j \cdot 10^{-5}$	3.272	5.034	0.923	12.5	3.691	$R_3 = 25.4 \cdot 10^{-5}$
C_4	$p_4^j \cdot 10^{-5}$	1.762	38.675	15.269	31.963	31.712	$p_4 = 119.3 \cdot 10^{-5}$
	d_4^j	1.7782	0.6633	0.2772	0.2523	0.1464	$d_4 = 0.383$
	$R_4^j \cdot 10^{-5}$	3.1	25.67	4.279	8.054	4.614	$R_4 = 45.7 \cdot 10^{-5}$

Source Pavlík, 2021

Then, the respective probabilities of victimization of all examined subgroups of crime were calculated. The resulting values are shown in Table 6. Furthermore, the impacts of individual types of crime were examined. First, the impacts listed in Table 5 were determined by the Saaty's method.

Similarly, as in the previous chapter, it is necessary to multiply the values of the impacts $\overline{d}_2^j, \overline{d}_3^j, \overline{d}_4^j, j = 1, 2, \ldots, 5$ by suitable constants a_2, a_3, a_4, and to find the real impacts $d_i^j = a_j \cdot \overline{d}_i^j, j = 1, 2, \ldots, 5, i = 2, 3, 4$, which take into account the importance of the given types of crimes in the context of other crimes $C_1^1, \ldots, C_1^6$. Using the same procedure as in the previous chapter, it was calculated

$$d_2^j = 0.4182 \cdot \overline{d}_2^j, d_3^j = 0.6752 \cdot \overline{d}_3^j, d_4^j = 3.1142 \cdot \overline{d}_4^j, j = 1, 2, \ldots, 5. \quad (4)$$

The results obtained can now be summarized in Table 6.

Thus, the risk values for individual types of crime and their subgroups were calculated. Their graphic representation and commentary are the subject of Chap. 3.

2.5 Advantages of the Procedure Used

As already written in Chap. 2.1, when using Saaty's method directly to determine the impact of 26 types of crime, a total of $\frac{26^2 - 26}{2} = 325$ comparisons would have to be made. After dividing the problem into 5 parts, described in the previous chapters, the number of necessary comparisons decreased significantly. To determine the impacts of the 9 main types of crime, each respondent had to fill in $\frac{9^2 - 9}{2} = 36$ fields. In the analysis of the most common type of crime, respondents identified the impacts of 6 subgroups of crime. Here, each respondent had to make other $\frac{6^2 - 6}{2} = 15$ decisions.

Finally, in Chap. 2.4, each of the other three types of crime was divided into 5 subcategories of crime. Here, each respondent filled in $3 \cdot \frac{5^2-5}{2} = 30$ fields. Thus, each respondent made only a total of $36 + 15 + 30 = 81$ decisions. In total, only less than a quarter of the 325 they would have to fill out if all types of crime were studied in one Saaty's table. This reduction is very useful when studying large systems.

3 Results

The following section deals with the evaluation of results, their comparison and interpretation. Using the data from Tables 1, 4 and 6 in the previous chapter, it is possible to sort individual types of crime according to the size of their risk, from the most risky to the least risky.

The resulting risks were divided according to their size into four groups. This division took place on the basis of brainstorming in a group of 10 experts on the issue. The limits of the individual categories were set as follows:

- acceptable risks ($R \leq 10 \cdot 10^{-5}$),
- negligible risks ($R \leq 2 \cdot 10^{-5}$),
- adverse risks ($R \leq 200 \cdot 10^{-5}$),
- unacceptable risks $R > 200 \cdot 10^{-5}$.

Due to the obtained results and the set values separating the levels of individual risks, a risk map can be created. This is a graphical representation of Tables 1, 4 and 6, in which the probabilities p of the occurrence of a given crime are marked on the horizontal axis, and the magnitude of the impacts d of these crimes on the vertical axis. The curves that separate areas with different levels of risk are hyperbolas of equations

$$d = \frac{2 \cdot 10^{-5}}{p}, d = \frac{10 \cdot 10^{-5}}{p}, d = \frac{200 \cdot 10^{-5}}{p}$$

These curves connect points with the same amount of risk. The resulting risk map is plotted in Fig. 2.

The category of negligible crime is almost empty, from the investigated crimes it contains only graffiti taggers and theft or damage to electronic data. With the advent of digitalisation and electronisation of the world, it can be expected that the theft and damage to electronic data will gradually increase.

The acceptable category also includes the sum of all violent, property and economic crime, which, however, have a relatively high impact on a societal scale. In addition, property crime is the most frequent type of crime in the South Moravia Region.

The category of adverse risks includes burglary, embezzlement, assault causing bodily harm, and theft—others (i.e. all types of theft except pickpockets, theft of

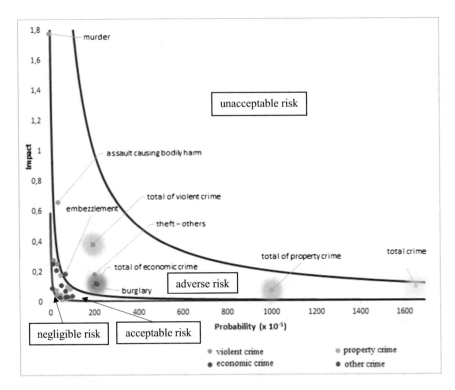

Fig. 2 The resulting map of risks

vehicles, bicycles and vehicle items). The statistics also show that burglary of various types (or theft by breaking into a building) is the most common crime in the South Moravian Region. The crime of murder is based on research as a criminal act that has an enormous impact on society and raises society's greatest fears of crime. However, due to the relatively low probability of victimization (the frequency for 2019 was only 21 cases in the SMR), murder is surprisingly one of the acceptable risks.

None of the examined types or subcategories of crime falls into the category of unacceptable risk. This fact reflects the relatively sustainable level of current crime with a slightly declining trend (Fig. 1).

The research question posed at the beginning of the chapter—*What kind of crime is the most risky for the South Moravian Region?* can be answered as follows:

The most risky type of crime is violent crime, and the most risky criminal offence is assault causing bodily harm. The value of these risks was calculated as the highest of all examined. The risk for violent crime is $R = 45.7 \cdot 10^{-5}$ and for assault $R = 26 \cdot 10^{-5}$ (Table 5, 6 and relations (2), (3), (4)).

4 Recommended Risk Mitigation Measures

The main activity of the security system in the area of crime is prevention and repression. In the area of prevention, I recommend a constant increase in the level of population education about current crime, the possibilities of securing their property and other preventive measures to reduce the risk of victimization. Preventive measures also include the frequency of police inspections or inspections by private security services in order to protect public order and property from crime. Unfortunately, the number of police inspections is connected with the economic costs of the region and the state, which ensure, among other things, the number of state and municipal police officers in active service.

Based on the brainstorming with an expert group on the topic of bulglary, the authors propose to increase the awareness of the population about modern security systems of buildings, e.g. increasing the level of passive security, camera systems, alarms, motion sensors, connection of the existing system to a central security desk managed by a private security agency etc.

Private security agencies are usually hired by private entities as part of preventive activities to protect their own property. Abroad, it is common practice that private security agencies also provide security for public buildings, this practice is partially beginning to be applied in the Czech Republic. This initiative is proving to be another possible extension of the region's security system in the fight primarily against property crime.

Unlike preventive measures, repressive measures are the sole responsibility of the public administration. However, the fear of repression is in itself a prevention, as no one wants to be punished. Penalties are always determined with regard to applicable legislation. Newly introduced alternative punishments have expanded the possibilities of repressive measures to include effective punishments for specific crimes (e.g. prohibition of activities, forfeiture, house arrest, community service, prohibition of entering public events, deportation of foreigners etc.), which act as another preventive stimulant.

The social area can also be described as a very important security area. The perception of crime by the population, but also the environment surrounding society, has a significant impact on the perception of crime. The current economic development of the regions in the Czech Republic has a positive effect on reducing the level of crime. However, regional development must perceive this issue and further actively participate in creating a safer environment, not only in physical form (e.g. wide and well-arranged public spaces, clean public spaces, landscaped parks), but also in security in the form of surveillance cameras and similar security systems, currently applied in the concepts of "Smart Cities".

Environmentalist factors, together with factors of social disorganization and elements of uncivilization, form an environment that plays in favor of further development of crime, development of social governance, as well as various nongovernmental organizations in the fight against this issue. The solution of the problem must focus on more problematic areas (e.g. ghettos) and actively involve the inhabitants of these areas in community life, and constantly work on restructuring and restoring the environment in which the community is located. The system must also actively educate children and influence their leisure activities with regard to their further personal development.

In conclusion, it is necessary to mention that as the most fundamental measure in the area of crime risk prevention, risk reduction in the area of potential sources of threat (crime) is recommended. Crime itself is not fully self-supporting, it is associated with a number of different perpetrators and reasons.

5   Conclusion

The aim of the study was to assess the regional security risks in the selected region and for the selected type of threat. The selected region for the research survey was the South Moravian Region and the chosen threat was crime.

The research survey included an analysis of nine types of crime, where selected types were divided into another 21 subgroups (usually specific crimes). The probability of occurrence was determined on the basis of data from police statistics, and the impact on society was determined using the Saaty's method. Subsequently, the risks posed by the individual types of crime studied were calculated. The result of the study are proposals for measures to mitigate the identified risks, which support higher efficiency of internal security of the territory.

Acknowledgements Irena Tušer would like to thank AMBIS College, Prague for its support and Jiří Jánský thanks for support of the project VAROPS, supported by the Ministry of Defence in the Czech Republic.

Supplement 1

See Table 7.

Table 7 Types of crime and numbers of individual criminal offences

C: Crime		Number of offfences		
C_1: Property—other	C_1^j	n_1^j	$n_1 = 6303$
	Burglary	2381	
	Fraud	943	
	Graffiti tagging	618	
	Evading alimony	91	
	Vandalism	1088	
	Others in C_1	363	
C_2: Property—simple thefts	C_2^j	n_2^j	$n_2 = 5681$
	Pickpocketing	965	
	Theft of vehicles and motorbikes	556	
	Theft of bicycles	588	
	Car theft	1068	
	Other thefts in C_2	2504	
C_3: Economic crime	C_3^j	n_3^j	$n_3 = 2482$
	Counterfeinting of money and documents	157	
	Credit card thefts	809	
	Thefts or destruction of electronic data	115	
	Embezzlement, fraud	808	
	Others in C_3	593	
C_4: Violent crime	C_4^j	n_4^j	$n_4 = 1423$
	Murder	21	
	Assault	461	
	Theft	182	
	Abuse, oppression, restriction	381	
	Others in C_4	378	
C_5: Addiction		$n_5 = 1208$	
C_6: Obstruction to authority's decision		$n_6 = 934$	
C_7: Road accidents		$n_7 = 537$	
C_8: Moral crime		$n_8 = 326$	
C_9: Others in C		$n_9 = 852$	

Source Ministry of the Interior ČR, 2020

References

1. Valadan M, Sharifi A, Ahmadi M (2003) Crime mapping and spatial analysis. M.Sc. Thesis, International Institute For Geo-Information Science And Earth Observation Enschede, The Netherlands
2. Ministry of the Interior ČR (2020) Report on the situation in the field of internal security and public order in the Czech Republic in 2019. Ministry of the Interior, Department of Crime Prevention, Prague. In Czech. Avaliable on: https://www.mvcr.cz/clanek/zprava-o-situaci-v-obl asti-vnitrni-bezpecnosti-a-verejneho-poradku-na-uzemi-cr-v-roce-2019.aspx
3. Pavlík O (2021) Regional security risk assessment. M.Sc thesis, AMBIS College
4. Police of the Czech Republic (2021) Crime statistical surveys for 2020. Police of the ČR © 2021. In Czcech. Avaliable on: https://www.policie.cz/statistiky-kriminalita.aspx
5. Saaty T, Sagir O, Mujgan S (2015) The rationality of punishment—measuring the severity of crimes: an AHP based Ordersof magnitude approach. Int J Inf Technol Decis Mak 14:5–16
6. Tušer I, Bekešienė S, Navrátil J (2020) Emergency management and internal audit of emergency preparedness of pre-hospital emergency care. Qual Quant. https://doi.org/10.1007/s11135-020-01039-w
7. Tušer I, Jánský J (2021) Security management in the emergency medical services of the czech republic—pre-case study. In: Soitu D, Hošková-Mayerová Š, Maturo F (eds) Decisions and trends in social systems. Lecture notes in networks and systems, vol 189. Springer, Cham, pp 409–423. https://doi.org/10.1007/978-3-030-69094-6_32
8. Tsohla SY, Polishchuk EA, Podsmashnaya IN (2020) Formal institutions in the development of crimean youth labor market. In: European proceedings of social and behavioural sciences. https://doi.org/10.15405/epsbs.2020.12.65
9. Vašková M, Náplavová M, Barta J (2018) Awareness and preparation of the population for emergencies, safety and reliability—safe societies in a changing world. In: Proceedings of the 28th international european safety and reliability conference. pp 45–52

Field Hospital Logistics Support System Risk Assessment

Zbynek Suchanek and Alena Oulehlova

Abstract The chapter deals with the process of logistics support system risk assessment of the field hospital of the Army of the Czech Republic, which performs tasks within health care provision of ROLE 2. Risk assessment was performed in three interrelated phases (risk identification, risk analysis and risk evaluation in accordance with ISO 31000 Risk Management—Guidelines). Semi-structured interviews with persons responsible for the operation of the field hospital were used to identify threats. Probability and impact levels were determined based on the threat and the vulnerability analysis which were further used for the risk assessment. A semi-quantitative approach was used for risk assessment. The results of risk assessment will be used in the following phase of risk treatment to propose measures to minimize unacceptable and undesirable risks.

Keywords Field hospital · Interviews · Logistic support system · Risk identification · Risk analysis · Risk evaluation · Risk assessment

1 Introduction

Growth of security threats at the regional, national and international levels requires a comprehensive readiness of the cooperating elements of the state security system to address them. Providing security and defence should be a fundamental commitment to the citizens in every state. Armed forces in each state represent one of the important elements contributing to the fulfilment of tasks in dealing with both non-military and

Z. Suchanek (✉)
Department of Military Medical Service Organisation and Management, University of Defence, Třebešská 1575, Hradec Králové 50001, Czech Republic
e-mail: zbynek.suchanek@unob.cz

A. Oulehlova
Department of Military Science Theory, University of Defence, Kounicova 65, Brno 66210, Czech Republic
e-mail: alena.oulehlova@unob.cz

© The Author(s), under exclusive license to Springer Nature Switzerland AG 2022
I. Tušer and Š. Hošková-Mayerová (eds.), *Trends and Future Directions in Security and Emergency Management*, Lecture Notes in Networks and Systems 257,
https://doi.org/10.1007/978-3-030-88907-4_13

military crisis situations, i.e. supporting the security preservation and its straightening. The Armed Forces of the Czech Republic, which the Army of the Czech Republic (ACR) forms its part, participate in providing both internal and especially external security within the framework of international obligations.

The Czech Republic, like other member countries of the North Atlantic Treaty Organization (NATO), relies on the principle of collective defence. NATO membership includes fulfilment of obligations and tasks, which include participation in all types of operations outside the territory of the Czech Republic. Field hospitals represent part of the destined forces to support operations under the leadership of NATO or the European Union or other organizations. Its task is to provide medical care to the wounded and sick during the operation, which is indispensable for the activities of troops. In addition to medical facilities, field hospital also consists of logistics facilities. Threats that endanger the functioning of the field hospital, staff and patients can be activated during the operation of all field hospital facilities. A well-conducted risk assessment can increase the quality of health services provided, rapidly reduce the risk of patients' deaths and increase the level of patients' return to fulfilment of their tasks.

2 The Theoretical Part

To determine the scope and context of the risk management of the field hospital logistics support process, it is necessary to understand its mission, structure, role, responsibilities, standards and guidelines.

2.1 Health Care Provision System in Operations

Activities of the ACR Medical Service in connection with NATO membership are governed by the Alliance's documents. Key standards are AJP 4.10, (C) Allied Joint Doctrine for Medical Support [8], AMedP-1.7, Capability Matrix [3], AMedP-1.8, Skills Matrix [4], AMedP-1.6, Medical Evaluation Manual [3].

According to the NATO AJP 4.10 doctrine [8], medical services provided should always meet the specific requirements of a particular foreign operation in relation to the task performed, size and structure of troops and, last but not least, the level of health risks that occur in the operation area.

The ACR medical evacuation system works similarly to those used by other Allied armies on the basis of the given doctrines such as AJMedP-1, Allied Joint Medical Planning Doctrine [4]. Another important document is AJMedP-2 Allied Joint Doctrine for Medical Evacuation [5].

The medical support system is divided into the so-called ROLEs in order to be able to meet the requirements effectively. Each ROLE has its capabilities, abilities and capacities with different degrees of providing continuous treatment, evacuation, replenishment of supplies and other functions necessary for the care of soldiers' health. Description of the capabilities of the individual modules is given in AMedP-1.7, [3].

Minimum requirements for individual levels of medical ROLEs are given in AJP 4.10, Allied Joint Doctrine for Medical Support [8].

The minimum requirements for individual ROLEs are generally specific to all higher roles.

ROLE 0

In connection with the lack of medical staff in smaller platoon-sized units, the system of self-help and mutual assistance is supplemented by the provision of first aid by soldiers who are trained to provide advanced first aid as a combat lifesaver (CLS).

These soldiers perform primarily tasks within their function in the unit and only if the commander orders it, or the situation allows, they provide first aid to the wounded.

Subsequently, the wounded are transferred to the Casualty Collection Point (CCP), where the platoon paramedic takes over and prepares them for further transport to the higher medical stage.

ROLE 1

The capabilities of this level are aimed at providing qualified emergency medical assistance to battalion-level units. These services are provided by the general practitioner (urgent procedures to save lives and stabilize vital functions, providing medication or oxygen), primary care, triage, resuscitation and patient stabilization.

This capability can be supplemented by a minimum temporary bed capacity, urgent dental care, basic laboratory tests and initial anti-stress therapy according to specific requirements. Task of the medical stage ROLE 1 is performed by the dressing station in the ACR [8], pp. 2–13.

ROLE 2

The capabilities of this ROLE are at a higher level than the previous ones. Main changes occur in the area of reception and triage of the wounded, resuscitation, treatment of shock conditions, which are also carried out in ROLE 1. However in this case they are provided to a greater extent. Damage control surgery (DCS) is also possible in emergencies to save lives, limbs and vital signs.

This level can be further divided according to capabilities and capacities into 2 main types—ROLE 2B (Basic) and Role 2E (Enhance). Medical stage ROLE 2 is a facility whose structure consists of individual modules. Numbers and character of the modules clearly define the possibilities and capabilities of individual medical stages. Required capabilities of medical stages depend on the character of the mission and the number of losses. Typical representatives of the medical stage of ROLE 2 are field hospitals in the ACR. Practical part of the article focusses on them [8].

ROLE 3

Compared to lower levels, it is equipped with facilities such as the production of medical oxygen, a module for hyperbaric medicine, a transfusion and haematology module, and it also contains diagnostic facilities.

In connection with the character of the operation and the assumption of the tasks performed, the given capabilities can be extended by specialized facilities such as those for thoracic surgery, burn care unit, etc. [8].

ROLE 4

This ROLE contains final surgical and non-surgical treatment and rehabilitation. It is performed within the ACR either in the Czech Republic or on the territory of the Allied state due to the fact that treatment at this stage is highly specialized and time-consuming.

For the needs of the ACR, it is provided by the Central Military Hospital Prague, military hospitals and the Institute of Aviation Medicine, Prague [8].

2.2 ACR Field Hospital

Field hospital is intended for the provision of health care, which can be provided depending on the character of the foreign operation. In terms of the building structures, it can be variably assembled with regard to the continuity of individual facilities, which are designed to provide professional medical care for moderately and severely injured, burned and sick. It is also intended to provide primary care to people affected by chemical or bacteriological weapons.

All health care activities must be provided within the framework of normative legal acts, in particular, in accordance with the Public Health Protection Act; state supervision over ionizing radiation, Act on Veterinary Care, Medicines and Related Substances Control Act, Act of Medical Devices, Act on Addictive Substances, etc.

At present, the ACR field hospitals consist of a so-called modular system, where the majority of professional facilities is located in containers and the main part—the corridor and other facilities in tent sections.

Containers and tent sections are interconnected and form a single unit, which can be assembled with variability depending on the requirements of a particular operation, local conditions and with regard to the functionality of the entire system.

It follows that the organizational structure and the location of the individual parts of the field hospital does not have its own fixed rules.

The basic structure of the field hospital consists of:

– Headquarters and staff, which control normal operation of the hospital,
– Professional sections belong to the practical part, which has the task of providing the reception, registration, triage and subsequently health care to the wounded and sick,
– Logistics units are determined to support all components of the field hospital,
– Support units.

It is necessary to specify requirements for medical provision from the point of view of fulfilment of the set tasks of the field hospital, so the construction of the entire unit and logistical support are further derived from these requirements.

Smooth functioning of the entire field hospital complex requires to further focus on other important parts, including provision of electricity and water supply, waste management, storage, repairs, material replenishment and revisions.

2.3 Medical Logistics in Foreign Operations

Medical logistics can be provided in several ways. The appropriate method was determined on the basis of efficiency, cost, time demandingness and required supplies availability, which limits the given operational task.

Individual methods of acquiring stocks are divided into own resources from home stocks, where, a logistics unit is deployed together with the dispatched units, which manages and coordinates the logistical support of national forces. This support may also be provided through the Host Nation Support (HNS) via the host country. The support itself is provided to NATO forces and NATO organizations present in or passing through the territory of the Host State. Another possibility is to request support from Allied forces that are present in the operational deployment area. The last option is direct purchase at the place of operation from international organizations, national commercial companies, or other entities that are authorized to provide medical services.

If the supplies are acquired from own resources of home supplies, this method can also be implemented in several ways, namely by the Medical Supplies Centre, or by military units and facilities of the military medical service. It is also possible to acquire them from other military units and facilities outside the scope of the military medical service. Another option is to obtain them from a department other than the Ministry of Defence, or by direct purchase from civilian sellers.

2.3.1 Ways of Implementing Logistic Support with Non-Medical Material

These ways of providing support are similar to those for medical supplies. Material can be acquired either for a fee or free of charge according to the military regulation Oper-1–5 Deployment of Forces and Resources of the Ministry of Defence in Foreign Operations [9]. Material that cannot be acquired in the area of operation is procured by own forces and resources. It is also possible to implement support in the form of Host Nation Support (HNS), a state with a specialized role in logistics (Logistic Lead Nation—LLN), a multinational integrated logistic unit and the Third-Party Logistic Support Services (TPLSS).

2.3.2 Storage of Medical Material Supplies

When storing material, it is important to distinguish what kind of stock it is. In the case of medicinal products, it is necessary to follow the procedures set out in the instructions for individual products, such as storage temperature, ambient humidity, etc. It is essential to record the expiration times of individual medications and, depending on it, to provide its timely replacement so that it was not used after the expiration date.

The entire storage process is monitored and all material must be recorded to prevent confusion, theft, loss, etc.

Storage of other non-medical stocks is based on stored commodities (e.g. food, fuel, spare parts). Standard operating procedures must be followed during their storage process.

3 Objective and Methods Used

The aim of the research described in the chapter was to assess the risks of logistic processes in the field hospital operation, whose individual steps were: assets and risks identification, adverse events impact determination and evaluation with regard to increasing the quality of services provided and the possibility of reducing the risk of patient death, threat determination and estimation of the probability of its activation, subsequent risk estimation including determination of the acceptability of individual risks.

Risk assessment process was carried out in accordance with ISO 31 000 Risk Management (2018) standard. Individual phases of the process are presented in detail in the practical part of the chapter. A qualitative method of risk identification and a semi-quantitative approach to determination of the threat activation probability, impact and risk estimation were used for the risk assessment.

Risk identification was performed by semi-structured interviews [1, 10–12]. Semi-structured interviews were used as they allow more freedom to identify risks than the structured interviews. The questions for the semi-structured interview were designed to be open and all focused on the area defined within the set scope of risk assessment. The semi-structured interview made it possible to ask additional questions for a deeper explanation of the answers and their understanding. Semi-structured interviews were also chosen since it was not possible to gather all the informants in the same place at the same time. Interviewed professionals were selected on the basis of their professional experience in the field of logistics processes of ACR field hospitals and the degree of management of these processes. These were both professional soldiers and civilian employees with extensive experience. A total of 8 interviews were conducted. Individual interviews lasted on average for 90 min.

Established semi-quantitative scales were used to determine the impact and the threat activation probability [13]. The risk was calculated as the product of the threat activation probability and impact.

4 The Practical Part

Practical part was divided into parts determining the scope, context, criteria, risk identification; risk analysis and risk evaluation in accordance with the ISO 31 000 [2] guidelines.

4.1 Determining the Scope, Context and Criteria

The scope of risk management was determined as the first step of the entire process so that risks could be assessed and risk treatment options subsequently determined. Comprehensive provision and field hospital operation use a large number of logistic processes. The logistic processes were divided into sub-processes for which the risk management process was implemented. The chapter presents a section of material acquisition and storage, which focuses on the operational part of these activities. The objective set at this stage was consistent with that of the present chapter. No risk management has been implemented for these activities so far, therefore a qualitative method of semi-structured interviews and semi-quantitative probability and impact assessment scales were used for risk identification.

Prior to the process initiation, research team got acquainted with the internal regulations governing the procedures for the acquisition and storage of material for the field hospitals. This formed part of the understanding of the internal environment of field hospitals to which risk management was applied. External context of field hospitals is quite variable depending on their operation and also on the chosen logistic method of material provision.

4.2 Risk Assessment

Risk assessment represents the overall process, which includes the risk identification, risk analysis and risk evaluation. Cooperation with representatives of the ACR, i.e. the key stakeholders, was used within its elaboration.

4.2.1 Risk Identification

Risk identification was implemented by two interrelated activities, in which risks and assets that are affected by risk sources were identified. Risk identification was based on semi-structured interviews. A total of 57 internal and external risks were identified based on them. The identified risks were mainly related to non-compliance with the set conditions for the acquisition and storage of material.

Field hospitals use a wide range of medical and non-medical material and logistics processes for their operation. For this reason, assets were grouped on the basis of the similar purpose of their use for the needs of the field hospital. The characteristics of grouped assets was performed. These were mainly tangible and intangible assets in the form of logistic processes. The following grouped assets were identified:

- Supplier/manufacturer,
- ACR storehouses,
- Purchase of material from other armies/local sources,
- Material storage
- Water provision,
- Electricity provision,
- Fuel provision,
- Waste management
- Field hospital operation.

The sources of individual risks, causes and events which may trigger the risk, vulnerability of individual assets and value of these assets from the point of view of the field hospital operation provision were taken into account (i.e. their price expressed in monetary units was not taken into account as this information was not available). The limitation of the implemented risk identification was mainly in the issue of the assessment of the external context change (especially the deployment of the field hospital, security environment development and deployment of the ACR in operations), limiting the availability of data as it represents classified or designated classified information whose detailed risk assessment and results cannot be published, and the time factor.

4.2.2 Risk Analysis

Risk analysis was performed as a semi-quantitative due to the limited data availability (source of information), the complexity and the interconnectedness of the logistic processes in the field hospital, which are affected by a large number of variables. The result of the risk analysis was to estimate critical risks of material provision for a field hospital. Countermeasures to minimize critical risks are going to be proposed in the subsequent phase of the risk treatment.

Risk analysis implementation, as well as risk identification, was influenced by the input information, reliability and opinions of individual evaluators.

The causes of risks activation and the probability of their activation were examined within risk analysis. Probability of activation was determined for each risk affecting the asset. Only limited historical data was available to determine the levels of risk activation, therefore the subjective probability of risk activation was determined on the basis of an estimate. The individual activation probability intervals are listed in Fig. 1.

Furthermore, the extent of impacts affecting the field hospital operation was considered. The impacts were assigned a value of 1–5 according to Fig. 2.

Fig. 1 Numerical and verbal risk occurrence probability assessment. *Source* [13]

Value	Verbal expression
1	Practically unlikely
2	Not very likely
3	Occasional
4	Likely to frequent
5	Very frequent

Fig. 2 Numerical and verbal impact assessment. *Source* [13]

Value	Verbal expression
1	Almost insignificant
2	Small
3	Significant
4	Very important
5	Unacceptable

The measures introduced so far to minimize the failure of the logistic process were considered throughout the process of determining probability and impacts determination.

Risk assessment was estimated by combining the assigned values after determining each risk activation probability and impact. Based on Figs. 1 and 2, the risk reaches values in the interval ⟨1; 25⟩. 5 levels of risk, where each level was implemented in relation to acceptance or non-acceptance and subsequent risk management activities, were determined with the emphasis on risk treatment or risk monitoring. Characteristics of individual risk levels are given in Fig. 3.

Interval	Risk	Description of the risk level
⟨1⟩	Negligible	Risks are acceptable and must be monitored.
⟨2;3⟩	Small	Risks are acceptable and must be monitored. It is not necessary to take measures to minimize risks, as the costs of minimization would exceed the value of the assets.
⟨4;9⟩	Medium	Risks are acceptable and must be monitored. Implementation of measures to minimize risks requires the application of the Cost-Benefit analysis and similar methods to assess the effectiveness of implemented countermeasures.
⟨10;16⟩	High	The risks are unacceptable. Implementation of adequate countermeasures on the basis of a developed risk minimization plan is required. Risks must be reviewed regularly.
⟨17;25⟩	Critical	The risks are unacceptable. They require immediate implementation of countermeasures, creation of a risk minimization plan, or cessation or interruption of activities until the risk is reduced.

Fig. 3 Characteristics of risk levels. *Source* author's own source

Risks	Risk source	Probability	Impact	Risk estimate
Quality of delivered material	Supplier/manufacturer	2	5	10
Material incompatibility	Supplier/client	2	5	10
Lack of material	Supplier/manufacturer	2	3	6
Higher price of material	Supplier/manufacturer	5	3	15
Time delays in delivery	Supplier/manufacturer	2	4	8
Material supply failure	Supplier/manufacturer	2	4	8

Fig. 4 Risk analysis affecting the process of purchasing material from other armies/local sources. *Source* author's own source

Risks	Risk source	Probability	Impact	Risk estimate
Expiration of stored material	Storekeeper	3	4	12
Material damage due to storage (temperature, dust, handling)	Storekeeper/Technical defect	2	4	8
Air conditioning malfunction	Storekeeper/Technical defect	2	4	8
Heating malfunction	Storekeeper/Technical defect	2	4	8
Electronic devices malfunction	Storekeeper/Technical defect	2	4	8

Fig. 5 Risk analysis affecting the material storage process. *Source* author's own source

Probability of activation and the impacts on the relevant assets were estimated based on the risk scenarios. Each risk was assigned a value from Figs. 1 and 2 and the product was used to estimate the risk. The potential source of threat was also examined within the risk analysis. An example of the risk analysis results is presented in Figs. 4 and 5.

4.2.3 Risk Evaluation

The estimated risks were, in this step, compared with the established risk characteristics in Fig. 3. The reference level that divided the risks into acceptable and unacceptable was level 9. The comparison showed that:

– 0 risks were negligible,

- 1 risk was small,
- 32 risks were medium,
- 22 risks were high,
- 2 risks were critical.

A total of 33 acceptable risks and 24 unacceptable risks were estimated. In accordance with the established characteristics of risk levels given in Fig. 3, it was recommended to only monitor and maintain existing countermeasures or improve them (e.g. stockpiling, increase in time reserve for material orders, increase and regularity of inspections), i.e. these were mainly organizational measures not requiring increased funding. For the two estimated critical risks, the lack of medication and the lack of blood derivatives, it was necessary to implement countermeasures. Both critical assets affect the operation of the field hospital. Lack of medicine would endanger the tasks performed by the field hospital in the mission and could potentially lead to the loss of human lives or the deterioration of patients' health. In this case, it is necessary to implement a countermeasure by creating a sufficient supply of medicine according to the assumption of its consumption, depending on the maximum passage of patients through the field hospital. For the second critical risk, the lack of blood derivatives, it was proposed to create a sufficient supply of blood derivatives and at the same time use the "walking blood bank" donation method.

5 Conclusion

Medical provision in operations represent a key logistical support for maintaining human life and health. Its activities are governed by national normative legal acts and, as in other Alliance states, are governed by NATO standards. The chapter characterized the health care system, including the provided medical assistance by individual ROLE levels, the field hospital, its possibilities, capabilities and ways of implementing logistical support.

Risk assessment of the logistic support processes of the field hospital operation was performed. Method of semi-structured interviews with professionals in the logistic processes of the field hospital was used to identify risks. A semi-quantitative estimate of the risk activation probability and impacts was performed within the risk analysis phase. The results of the risk assessment were compared with a reference value. Risk assessment found that two risks were critical and 22 unacceptable. Risk assessment was performed as a screening.

Despite the large number of estimated unacceptable risks, these risks can be relatively easily minimized throughout the system by implementing countermeasures. Possible countermeasures include, for example, revising or supplementing standard operating procedures or increasing the stock checks frequency. It follows from the above-mentioned facts that the cost of countermeasures is very low, the feasibility is very fast and it depends mainly on the management of individual parts of the field hospital.

References

1. IEC 31010 (2019) Risk management: risk assessment techniques, vol 2. International Organization for Standardization, Geneva
2. ISO 31000 (2018) Risk management—guidelines, vol 2. International Organization for Standardization, Geneva
3. NATO (2016a) AMEDP-1.7, Capability matrix. NATO standardization office
4. NATO (2016b) AMEDP-1.8, Skills matrix. NATO standardization office
5. NATO (2018a) AMEDP-1.6, Medical evaluation manual. Edition A Version 2. NATO standardization office
6. NATO (2018b) AJMEDP-1, Allied joint medical planning doctrine. NATO standardization office
7. NATO (2018c) AJMEDP-2, Allied joint medical doctrine for medical evacuation. NATO standardization office
8. NATO (2019) AJP-4.10, Allied joint doctrine for medical support. NATO standardization office
9. Oper-1–5 deployment of forces and resources of the ministry of defence in foreign operations (2018) In: Czech Oper-1–5: Nasazení sil a prostředků rezortu ministerstva obrany do zahraničních operací. Ministerstvo obrany, Praha
10. Ostrom LT, Wilhelsen CA (2012) Risk assessment: tools, techniques, and their applications, vol 1. John Wiley, Hoboken, New Jersey. ISBN 978–0–470–89203–9
11. Pritchard, CL (2015) Risk management: concepts and guidance, vol 5. Taylor and Francis Group, New York. ISBN 978–1–4822–5845–5
12. Rausand M (2011) Risk assesment: theory, methods, and applications, vol 1. John Wiley, Hoboken, New Jersey. ISBN 978–0–470–63764–7
13. Smejkal V, Rais K (2013) Risk management in companies and other organizations. In Czech: Řízení rizik ve firmách a jiných organizacích, vol 4. Grada, Praha. Expert (Grada). ISBN 978–80–247–4644–9

Risk Phenomena Prediction During the Deployment of the Czech Armed Forces Abroad

Zdenek Vostrel⬤, Alena Oulehlova⬤, and Jiri Jansky⬤

Abstract The development of world security under the influence of globalization causes a number of international problems and conflicts which, either directly or indirectly, threaten the security environment of the Czech Republic (CR). The deployment of the Czech Armed Forces abroad is a logical step to provide, support and create safe environment not only on our territory. All entities (combatants, civilians, stakeholders etc.) together with material, technology, and last but not least the environment, are threatened by the diversity of threats during these deployments. Protection of these assets must be provided inherently. The chapter deals with the prediction of the risk phenomenon of attacks on the armed forces during foreign activities depending on the time period. The calculation of the risk phenomenon was carried out in the paper based on historical data and using statistical methods. Statistical data processing showed that the prediction of the risk phenomena occurrence depended on time, especially on the time of the day. Prediction results can be further used to establish risk mitigation measures.

Keywords Armed forces · Foreign activities · Risk phenomenon prediction · Risk mitigation · Statistical methods · Time period

Z. Vostrel · A. Oulehlova (✉) · J. Jansky
Department of Military Science Theory, University of Defence, Kounicova 65, 662 10 Brno, Czech Republic
e-mail: alena.oulehlova@unob.cz

Z. Vostrel
e-mail: zdenek.vostrel@unob.cz

J. Jansky
e-mail: jiri.jansky@unob.cz

1 Introduction

A number of organizations operate in projects aimed at restoring, recovering, or building country or location infrastructure elements of various states within foreign activities, both with military and non-military focus. Prior to the actual deployment of any entity, i.e. with military or non-military-oriented projects and tasks, it is necessary to provide security and protection of each entity's individual assets.

The very prediction of risk phenomena, or its competence on the basis of available information, is an important element in the stages of preparation and planning, and during partial stages of each foreign operation/mission/project. Such prediction can significantly contribute to the protection of identified assets.

The decision-making process in the preparation phase of the operation or project is significantly influenced by predicting risk phenomena, when this information can be used to provide material aimed at risk mitigation or its complete elimination. These predictions help the adjustment of the protection means with respect to the new, changed conditions with regard to the development of the security situation in the following phases of the operation/project.

2 Theoretical Part

The North Atlantic Treaty Organization (NATO) is a key guarantor of security in the western part of the European continent. Negative development of the security environment in the recent period has been putting pressure to change the approach. The previously prioritized needs for high mobility of military forces, aimed primarily at their use in asymmetric conflicts, were re-evaluated after many years of building expeditionary capabilities of the military forces of NATO member states [4]. Attention of NATO returned to focus on the territorial defence, as it is not possible to automatically assume its absolute superiority in the areas such as electronic warfare, air domination, etc. [7].

The Czech Republic, as a member of the NATO, uses the following division of operations [2]:

- Combat operations;
- Crisis response:
- Measures against irregular activities (counter-insurgency, counter-terrorism, fight against crime);
- Military contribution to peace support;
- Military contribution to humanitarian aid;
- Military contribution to stabilization and reconstruction;
- Evacuation of non-combatants;
- Departure of forces;
- Sanctions and embargoes;
- Freedom of sea navigation and overflights.

Armed Forces of the Czech Republic perform tasks in the entire spectrum of the above-mentioned foreign operations, especially peace support operations within the framework of peaceful deployment. Czech Armed Forces are most often tasked with peacebuilding and humanitarian operations.

To provide examples of current operations of the Czech Armed Forces, it is possible to mention, for example, a training mission in Mali, Africa, where the task of Czech soldiers has been to train instructors for the needs of the Malian army. Then there is the Operation Inherent Resolve (OIR) in Iraq, where the key task has been to provide advice in the area of some military specialties. Moreover Afghanistan is an important and long-term destination where the Czech Armed Forces operate. Czech units are integrated into the International Security Assistance Force (ISAF) task force, where they perform various tasks in the field of medical support and in the field of protection of important persons [8].

Members of the Czech Armed Forces are also involved in the observation missions in Congo, Kosovo, Mali and the Central African Republic. Their main task is to monitor political security and military situation [9].

There is one common aspect of all these foreign operations and missions which is a post-conflict situation that requires humanitarian, medical, or military assistance and support on the basis of the United Nations Security Council resolutions. Given that security in these operations is a primary task, the deployment of members of the Armed Forces requires measures that minimize, as far as possible, the risks arising from the local threat identification.

Responsibility for risk decisions rests with the operation/mission commanders. The commander must consider risks that can occur across the range of operations, such as:

- In a wartime environment: Commander must consider the risk of collateral damage which may result in creating new adversaries.
- In a peacetime environment: Commander must consider political attitudes and previous actions of civilians in identifying risks to friendly forces and the population itself [3].

It is necessary to use risk management procedures both in the preparation and during the operation/mission for this reason. Prediction of risk activation helps in the military decision-making process to change activities so that the key/critical assets of the mission/operation are not adversely affected, or to adapt them to suit a specific situation.

The chapter is focused on the prediction of risk phenomena, which are based on the statistical processing of data on the number of incidents. These represent security incidents that were caused by a certain type of weapon or device and are further divided according to the identification of the persons who carried them out in terms of belonging to the armed forces or the civilian population. The data was obtained for the period from 5 February 2018 to 23 June 2019 in a foreign operation in Iraq. The data was taken from the document "Weekly report on security incidents", which was regularly updated as part of the weekly overviews of incidents at the place of the ACR unit deployment during its operation. The aim of the research described in the chapter was to use selected statistical methods to predict the risk phenomena occurrence and to confirm the hypothesis of the probability of a security incident occurrence in selected time periods according to recorded data.

3 Methods Used

Combination of statistical methods was chosen to predict risk phenomena. MAPLE software was used in their application.

The Poisson probability distribution [10] was used to estimate the number of attacks. It is an expression of the number of risk phenomena that occurred over a time period—interval, provided that they are independent of each other. The probability that the random variable X, expressing the number of attacks, acquires the values $X = m$ is given by the relation. Poisson probability distribution was calculated according to the formula (1):

$$P(X = m) = \frac{\lambda^m}{m!} e^{-\lambda} P(X = m) = \frac{\lambda^m}{m!} e^{-\lambda} \tag{1}$$

where

mm expresses the number of risk phenomena per time period,

$\lambda\lambda$ is the average number of phenomena occurrences per time period,

ee is Euler's number.

The mean value of the number of attacks, which is shown in Fig. 1, was calculated for the data processed in this chapter. One of the goals of the research described in this article was to approximate this value using the 95% confidence interval. According to Anděl [1], it results that the mean value of the number of attacks has $\chi^2 \chi^2$ (chi-square) probability distribution.

Formula (2) was used to calculate the confidence interval for the number of attacks in the intervals $-x_i, i = 0, \ldots, 23 x_i, i = 0, \ldots, 23$:

$$\frac{\chi^2_{0.05}(2x_i)}{2} < x_i < \frac{\chi^2_{0.95}(2x_i + 2)\chi^2_{0.05}(2x_i)}{2} < x_i < \frac{\chi^2_{0.95}(2x_i + 2)}{2} \tag{2}$$

where
$\frac{\chi^2_{0.05}(2x_i)\chi^2_{0.05}(2x_i)}{2}$ expresses the lower limit of the confidence band,

$\frac{\chi^2_{0.95}(2x_i+2)}{2} \frac{\chi^2_{0.95}(2x_i+2)}{2}$ expresses the upper limit of the confidence band.

Furthermore, it was verified whether the data had a uniform probability distribution. The Chi-square test of good agreement was used for this testing. This test uses the differences between the expected and observed values and is given by the Eq. (3):

Fig. 1 Total number of incidents in the monitored period during 24 h. *Source* modified source Incident summary report [5, 6]

Serial no.	From	To	Number of incidents
0	0:00	0:59	248
1	1:00	1:59	301
2	2:00	2:59	225
3	3:00	3:59	151
4	4:00	4:59	112
5	5:00	5:59	148
6	6:00	6:59	280
7	7:00	7:59	495
8	8:00	8:59	433
9	9:00	9:59	292
10	10:00	10:59	271
11	11:00	11:59	278
12	12:00	12:59	239
13	13:00	13:59	194
14	14:00	14:59	199
15	15:00	15:59	168
16	16:00	16:59	175
17	17:00	17:59	185
18	18:00	18:59	162
19	19:00	19:59	130
20	20:00	20:59	237
21	21:00	21:59	243
22	22:00	22:59	258
23	23:00	23:59	347
Total			k = 5,776

$$\chi^2 = \sum_{i=0}^{n} \frac{(x_i - kp_i)^2}{kp_i} = \frac{(x_0 - kp_0)^2}{kp_0} + \cdots + \frac{(x_n - kp_n)^2}{kp_n}$$

$$\chi^2 = \sum_{i=0}^{n} \frac{(x_i - kp_i)^2}{kp_i} = \frac{(x_0 - kp_0)^2}{kp_0} + \cdots + \frac{(x_n - kp_n)^2}{kp_n} \tag{3}$$

where $x_i, i = 0, \ldots, n x_i, i = 0, \ldots, n$ are the numbers of attacks in time intervals $i = 0, \ldots, n\ i = 0, \ldots, n$. Furthermore, $p_i p_i$ is the theoretical probability of a given event and $kp_i\ kp_i$ is the theoretical frequency of these events at times $i = 0, \ldots, ni = 0, \ldots, n$ and kk is the total value of the number of incidents for all time intervals.

The Least square method was used to approximate the data in which the periodic component was searched. The data function was interpolated as follows (4):

$$y = a + b \sin(\omega x) + c \cos(\omega x) y = a + b \sin(\omega x) + c \cos(\omega x) \tag{4}$$

where a, b, c a, b, c are the coefficients sought, $\omega = 52 \omega = 52$ is the value of the annual period in weeks.

The Turning point test generally verifies that the data is random. Formula (5) was used for the calculation.

$$U = \frac{5 - \frac{2n-4}{3}}{\sqrt{\frac{16n-29}{90}}} U = \frac{S - \frac{2n-4}{3} +}{\sqrt{\frac{16n-29}{90}}} \tag{5}$$

Where S is the number of turning points and nn is the value of the number of week intervals. U is the resulting value, which was compared with the relevant quantiles of the normal distribution. It was determined whether or not the data is random at a given level of significance.

The input data for the prediction evaluation was the total number of incidents recorded, including the time (Fig. 1) of their occurrence during 24 h. Figure 2 shows the data according to the number of incidents and the type of their execution (i.e. how the incident was committed and by whom), including the week time period.

4 The Practical Part

Practical part of the chapter deals with the analysis of the dependence of incidents on the time of day and the analysis of the number of attacks in individual weeks.

4.1 Analysis of the Dependence of Incidents on the Time of day

The analysis of the dependence of incidents on the time of day examines the development of the number of incidents during 24 h. The input data for the analysis are shown in Fig. 1. Figure 1 shows the hourly time intervals and the number of incidents that occurred in the given time period during the monitored period. Figure 3 shows the dependence of incidents on the time of day, including the determined bands of 95% confidence. Figure 3 shows that the highest occurrence of incidents takes part between 7:00 am and 9:00 am and then decreases. At night from 10:00 pm to 02:00 am the number of incidents increases again, and from 02:00 to 06:00 am it decreases. This fluctuation is probably due to the activity and movement of the local

Fig. 2 Types and numbers
of incidents per week
including source of risk.
Source Overview of security
incidents according to the
method of execution

Week	From	To	HG		IED		SAF	
			ISF	CIV	ISF	CIV	ISF	CIV
1	05/02/2018	11/02/2018	0	3	1	6	4	4
2	12/02/2018	18/02/2018	2	4	2	11	0	9
3	19/02/2018	25/02/2018	0	6	2	7	3	12
4	26/02/2018	04/03/2018	0	9	0	7	6	9
5	05/03/2018	11/03/2018	1	3	3	10	5	11
6	12/03/2018	18/03/2018	1	5	2	9	9	10
7	19/03/2018	25/03/2018	2	7	4	10	7	11
8	26/03/2018	01/04/2018	1	5	2	5	1	6
9	02/04/2018	08/04/2018	1	0	2	5	1	9
10	09/04/2018	15/04/2018	0	6	4	3	2	7
11	16/04/2018	22/04/2018	1	8	5	6	2	13
12	23/04/2018	29/04/2018	0	7	0	7	6	7
13	30/04/2018	06/05/2018	0	4	2	8	4	10
14	07/05/2018	13/05/2018	1	3	3	4	4	7
15	14/05/2018	20/05/2018	1	3	1	5	2	15
16	21/05/2018	27/05/2018	0	4	5	5	3	9
17	28/05/2018	03/06/2018	0	4	4	11	2	11
18	04/06/2018	10/06/2018	0	2	0	6	11	15
19	11/06/2018	17/06/2018	4	9	4	5	6	16
20	18/06/2018	24/06/2018	1	5	1	3	8	21
21	25/6/018	01/07/2018	1	7	7	8	6	16
22	02/07/2018	08/07/2018	0	10	3	9	7	18
23	09/07/2018	15/07/2018	1	8	6	5	3	15
24	16/07/2018	22/07/2018	1	9	7	7	5	18
25	23/07/2018	29/07/2018	2	5	4	12	5	21
26	30/07/2018	05/08/2018	0	5	2	6	10	20
27	06/08/2018	12/08/2018	0	3	2	0	8	4
28	13/08/2018	19/08/2018	0	0	12	11	13	29
29	20/08/2018	26/08/2018	0	0	4	1	3	6
30	27/08/2018	02/09/2018	0	0	0	0	0	0
31	03/09/2018	09/09/2018	1	2	1	4	2	4
32	10/09/2018	16/09/2018	0	3	4	11	0	13
33	17/09/2018	23/09/2018	0	8	8	4	2	8
34	24/09/2018	30/09/2018	0	0	1	5	0	2
35	01/10/2018	07/10/2018	0	0	4	6	3	9

Fig. 2 (continued)

36	08.10.2018	14.10.2018	1	2	2	17	4	14
37	15/10/2018	21/10/2018	0	4	6	4	4	3
38	22/10/2018	28/10/2018	0	1	1	4	2	4
39	29/10/2018	04/11/2018	2	5	3	8	9	5
40	05/11/2018	11/11/2018	0	4	0	6	0	3
41	12/11/2018	18/11/2018	1	4	1	2	3	12
42	19/11/2018	25/11/2018	0	3	1	0	2	5
43	26/11/2018	02/12/2018	0	4	4	3	2	12
44	03/12/2018	09/12/2018	0	3	0	3	0	7
45	10/12/2018	16/12/2018	2	2	0	1	0	3
46	17/12/2018	23/12/2018	0	5	0	2	4	3
47	24/12/2018	30/12/2018	1	0	1	1	0	5
48	31/12/2018	06/01/2019	0	0	1	1	0	1
49	07/01/2019	13/01/2019	1	2	4	7	3	11
50	14/01/2019	20/01/2019	0	2	0	1	1	2
51	21/01/2019	27/01/2019	0	3	4	8	7	11
52	28/01/2019	03/02/2019	1	3	1	3	3	4
53	04/02/2019	10/02/2019	0	8	1	6	2	10
54	11/02/2019	17/02/2019	0	5	4	9	4	6
55	18/02/2019	24/02/2019	0	2	0	3	1	2
56	25/02/2019	03/03/2019	0	3	3	4	6	8
57	04/03/2019	10/03/2019	1	1	3	4	6	6
58	11/03/2019	17/03/2019	0	6	2	8	2	11
59	18/03/2019	24/03/2019	1	2	2	4	3	7
60	25/03/2019	31/03/2019	0	3	2	1	6	5
61	01/04/2019	07/04/2019	3	1	4	6	8	9
62	08/04/2019	14/04/2019	2	4	11	9	8	16
63	15/04/2019	21/04/2019	1	2	2	5	2	3
64	22/04/2019	28/04/2019	1	2	1	4	8	14
65	29/04/2019	05/05/2019	0	0	4	5	4	6
66	06/05/2019	12/05/2019	0	2	3	1	1	7
67	13/05/2019	19/05/2019	0	3	0	8	3	6
68	20/05/2019	26/05/2019	0	11	6	3	5	15
69	27/05/2019	02/06/2019	0	4	5	7	7	12
70	03/06/2019	09/06/2019	0	3	2	3	8	10
71	10/06/2019	16/06/2019	0	0	1	0	1	2
72	17/06/2019	23/06/2019	0	1	1	1	0	3

Total			40	267	198	384	282	658

Meaning of abbreviations:

HG   Hand gun
IED   Improvised explosive device
SAF Small arms fire - weapons such as assault rifles, sniper rifles, machine guns, etc.
ISF   Iraqi Security Forces
CIV   Civilian population

Fig. 2 (continued)

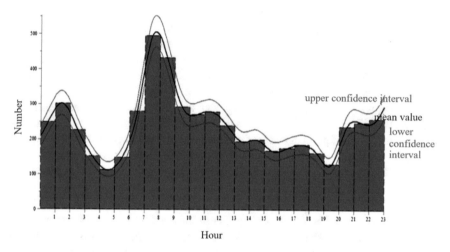

Fig. 3 Dependence of incidents on the time of day. *Source* author's own

security force units and the activity of the international units deployed in the operation, depending on the ordinary movement of the local population during the day time. The number of occurrences of independent phenomena that occur in a certain time interval is governed by the Poisson probability distribution. The probability of an incident is different at each time of day. The height of each column in Fig. 3 represents a point estimate of the mean value of the incident occurrence at a given time.

95% confidence bands, which are shown in Fig. 3, were calculated according to formula (2), using the values in Fig. 1.

The values of $\chi^2_{0.025}(2x_i)\chi^2_{0.025}(2x_i)$ and $\chi^2_{0.975}(2x_i + 2)\chi^2_{0.975}(2x_i + 2)$ were given in table values. At times $i = 0, 1, \ldots, 22, 23i = 0, 1, \ldots, 22, 23$; 95% confidence intervals were constructed for the number of incidents. In order to obtain estimates of the number of incidents at other times as well, the points $x_i, i = 0, 1, \ldots, 22, 23x_i, i = 0, 1, \ldots, 22, 23$ were interpolated with a cubic spline (indicated by a black curve in Fig. 1). Estimates of the upper limits $\chi^2_{0.95}(2x_i + 2), i =$

$0, 1, \ldots, 22, 23\chi^2_{0.95}(2x_i + 2), i = 0, 1, \ldots, 22, 23$ were interpolated with a cubic spline (indicated by a red curve above the black curve in Fig. 3). Subsequently, estimates of the lower limits $\chi^2_{0.05}(2x_i), i = 0, 1, \ldots, 22, 23\chi^2_{0.05}(2x_i), i = 0, 1, \ldots, 22, 23$ were performed and also interpolated with a cubic interpolation spline (indicated by a red curve below the black curve in Fig. 3).

It can be stated, based on the results of the determined 95% confidence interval, that if local conditions (long-term climatic, environmental, safety) do not change radically, then with a 95% probability, the number of incidents shall occur between these red curves at given times of day.

Chi-square test of good agreement was another statistical method used to analyse the dependence of incidents on the time of day. The analysed data from Fig. 1 was first tested whether the differences in the number of incidents that occurred at some times are statistically significant, or whether these are only random deviations in the implementation of a uniform probability distribution. The random variable indicating the time of the incident was denoted by X. Subsequently, hypothesis $H_0 H_0$: "Random variable X has a uniform probability distribution", i.e. given by the formula $p_i = \frac{1}{24} p_i = \frac{1}{24}$ for $i = 0 \ldots 23 i = 0 \ldots 23$ and $p_i = 0 \, p_i = 0$ for $x \notin \{0, 1, \ldots, 23\} x \notin \{0, 1, \ldots, 23\}$ was tested as opposed to the alternative hypothesis H: "Random variable X does not have a uniform probability distribution".

Chi-square test of good agreement can be used only under the assumption $kp_i > 5, i = 1, \ldots, n \, kp_i > 5, i = 1, \ldots, n$, where n is the number of intervals. In this case $n = 24$, where k represents the total number of attacks. Number 5 is a constant of the given formula.

It was necessary to verify the fulfilment of the assumption $kp_i = \frac{5776}{24} = 240.6 > 5 \, kp_i = \frac{5776}{24} = 240.6 > 5$ This assumption of the condition guarantees a sufficient number of theoretical frequencies, and thus meets the condition of performing the test. Furthermore, formula (3) was used to calculate the Chi-square test of good agreement. Calculation result was as following:

$$\chi^2 = \sum_{i=0}^{23} \frac{(x_i - kp_i)^2}{kp_i} = \frac{(x_0 - kp_0)^2}{kp_0} + \cdots + \frac{(x_{23} - kp_{23})^2}{kp_{23}} = 154$$

In the final phase of the test, the result was compared with the value $\chi^2_{1-\alpha}(n-1) = \chi^2_{0.95}(23) \cong 35 \chi^2_{1-\alpha}(n-1) = \chi^2_{0.95}(23) \cong 35$ t applies that the resulting value $154 > 35\,154 > 35$ at the significance level $\alpha = 0.05$ rejected the $H_0 H_0$ hypothesis that the random variable X has a uniform probability distribution.

It was found that the number of incidents depended on the time of day, respectively that the time of day, divided into hourly intervals, significantly affected the number of security incidents in a given location. Figure 4

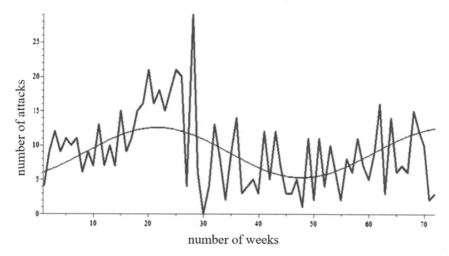

number of weeks

Fig. 4 Number of attacks in individual weeks with interpolation of the approximated function. *Source* author's own

4.2 Analysis of the Number of Attacks in Individual weeks

All types of attacks described in Fig. 2, respectively their data, was processed using methods listed in this part of the chapter and the computational procedure was performed identically for each type of attack, therefore only the procedure used in SAF—CIV data analysis is described in this chapter. In Fig. 2, the numbers of attacks $x_i, i = 1, 2, \ldots, n$ $x_i, i = 1, 2, \ldots, n$ $x_i, i = 1, 2, \ldots, n$ were plotted, where $n = 72$ $n = 72$ $n = 72$ is the number of weeks. For better clarity, the plotted values of x_i x_i were connected by lines.

The Least squares method was used for the data from Fig. 2. According to formula (4) it was substituted for $\omega = \frac{2\pi}{52}$ $\omega = \frac{2\pi}{52}$, in the denominator the period is 52 weeks, i.e. one year. It resulted in a modified formula (4) as follows $y = 8.9 - 3.1\cos(0.12x) + 1.9\sin(0.12x)$ · $y = 8.9 - 3.1\cos(0.12x) + 1.9\sin(0.12x)$.

If this interpolation was good, it was assumed that only the random component remained after subtraction from the data. This fact was tested using the so-called Turning point test. First, the values $\tilde{x}_t = x_i - y(i), i = 1, 2, \ldots n$ $\tilde{x}_t = x_i - y(i), i = 1, 2, \ldots n$ were calculated, where y_i y_i are the values of the function (4) for $i = 1, 2, \ldots n$ $i = 1, 2, \ldots n$ The Sign test proved that these values can be attributed to random noise. H_0 : H_0 hypothesis "$\tilde{x}_1, i = 1, 2, \ldots n$ $\tilde{x}_t, i = 1, 2, \ldots n$ are independent and identically distributed random variables (iid)" was tested against alternative hypothesis H_1 : $\tilde{x}_t, i = 1, 2, \ldots n$ H_1 : $\tilde{x}_l, i = 1, 2, \ldots n$: are not independent and identically distributed random variables (iid) For this purpose, formula (5) was used, where $S = 44$ $S = 44$ is the number of turning points and a $n = 72$ $n = 72$ is the number of weeks. This statistic had a normal distribution of N(0,1) N(0,1). The hypothesis that it can be attributed to random noise was accepted at the significance level α, if $|U| < u_{1-\frac{\alpha}{2}}$ $|U| < u_{1-\frac{\alpha}{2}}$, where $u_{1-\frac{\alpha}{2}}$ $u_{1-\frac{\alpha}{2}}$

Type of attack	Approximated function
HG-CIV	$y = 3.6 - 0.25\cos(0.12x) + 0.84\sin(0.12x)$
HG-ISF	$y = 0.52 + 0.06\cos(0.12x) + 0.16\sin(0.12x)$
IED-ISF	$y = 2.7 - 1.1\cos(0.12x) + 0.4\sin(0.12x)$
IED-CIV	$y = 5.2 - 0.2\cos(0.12x) + 0.52\sin(0.12x)$
SAF-ISF	$y = 3.7 - 0.94\cos(0.12x) + 1.2\sin(0.12x)$
SAF-CIV	$y = 8.9 - 3.1\cos(0.12x) + 1.9\sin(0.12x)$

Fig. 5 Approximated functions of the individual types of attacks. *Source* author's own

is the appropriate quantile of the normal distribution. For the above-mentioned data $U = -0.75\,U = -0.75$ was calculated and for $\alpha = 0.05$, the inequality $|-0.75| < 1.96 |-0.75| < 1.96$ was obtained, and thus for the sequence, the hypothesis that the sequence $\tilde{x}_t, i = 1, 2, \ldots n\tilde{x}_l, i = 1, 2, \ldots n$ is random was accepted. It was shown that the approximation of attacks using the modified formula (5) was good. The computational procedure described above was also used for the calculations performed for other types of analysed attacks. The results of the calculations are shown in Fig. 5.

It can be predicted that the numbers of attacks of the given type will follow the functions from Fig. 5 assuming that the conditions will not change or that they will remain similar [11, 12].

5 Conclusion

Current world is very diverse in terms of security. New conflicts, which are caused by various reasons, persist or continue to arise. It is desirable to predict their development to provide security, which gives a significant advantage, especially in the preparation phase of a project or operation. Risk phenomena prediction can, in particular, significantly affect protection of assets of an operation or project.

The chapter used data from one of the operations abroad, where the Czech Republic has been engaged with breaks since 2003. The data was used to model the risk phenomena prediction, which were attacks by certain types of weapons or devices as well as persons executing the attack.

Probability calculation by chosen methods confirmed that the data on the numbers and types of attacks was not random and it is likely that their numbers correspond to the time interval. This statement is conditioned by the fact that no significant change in conditions (cultural, political, security, environmental) will take place. The results of the research described in the chapter provide relevant information that can be used in further preparation of the operation/project in the corresponding country for the purpose of risk mitigation, or its complete elimination.

References

1. Anděl J (2005) Basics of mathematical statistics. In Czech: Základy matematické statistiky. Matfyzpress, Praha. ISBN 80–86732–40–1
2. Czech Army Doctrine (2019) 4th edn. Ministry of Defence of the Czech Republic, Department of Communication and Promotion, Prague
3. Field Manual No 100–14 (1998) Risk management. Department of the Army, Washington
4. Hlavizna P (2020) The development of electronic warfare in the Czech Armed Forces providing combat support in the electromagnetic environment during operations in the near future. In Czech: Rozvoj elektronického boje Armády České republiky zajišťující bojovou podporu v elektromagnetickém prostředí operací blízké budoucnosti. Dissertation, Univerzita obrany, Brno, p 452
5. Incident summary report 2018 (2018) [internal document]. Ministry of Defence, Prague
6. Incident summary report 2019 (2019) [internal document]. Ministry of Defence, Prague
7. Stojar, Richard et al (2017) Security environment 2017. In Czech Bezpečnostní prostředí 2017. Univerzita obrany, Brno. [online]. Available online at: https://www.unob.cz/cbvss/Documents/publikace/20180213%20-%20Bezpe%C4%8Dnostn%C3%AD%20prost%C5%99ed%C3%AD%202017.pdf
8. Ministry of Defence (2020) Current deployments. [online]. Available online at: https://www.army.cz/scripts/detail.php?pgid=450
9. United Nations (2020) Where we operate. [online]. Available online at: https://peacekeeping.un.org/en/where-we-operate
10. Vose D (2009) Risk analysis: a quantitative guide, vol 3. Wiley, Chichester. ISBN 978–0–470–51284–5
11. Weekly report on security incidents (2018) [internal document]. Ministry of Defence, Prague
12. Weekly report on security incidents (2019) [internal document]. Ministry of Defence, Prague

Identification and Analysis of Risks Related to PESCO Project in the Electronic Warfare Area

Alena Oulehlova⬤, Petr Hlavizna⬤, and Radovan Vasicek⬤

Abstract The European Union member states have decided to enhance their defence and security cooperation with emphasis on increased operational readiness, cooperation capabilities and mutual interoperability. The purpose of the Permanent Structured Cooperation (PESCO) projects is to strengthen Europe's defence capabilities. The Czech Republic has become the leading nation responsible for the management and execution of the Electronic Warfare Capability and Interoperability Programme for Future Joint Intelligence, Surveillance and Reconnaissance (JISR) Cooperation project. The chapter describes the methods of identification and analysis of the risks related to this project. Due to the fact that this was the initial assessment with limited data input, the preliminary hazards analysis method was applied for the hazard identification. The identified hazards then provided the basis for the hazard register. Simultaneously, the project assets were also identified. The resulting findings were used for analysis of the assets and hazards which established the baseline for the risk assessment.

Keywords Assets · Capabilities · Hazard register · Electronic warfare · PESCO · Preliminary hazard analysis · Risk analysis · Risk identification

A. Oulehlova (✉)
Department of Military Science Theory, University of Defence, Kounicova 65, 662 10 Brno, Czech Republic
e-mail: alena.oulehlova@unob.cz

P. Hlavizna · R. Vasicek
Department of Intelligence Support, University of Defence, Kounicova 65, 662 10 Brno, Czech Republic
e-mail: petr.hlavizna@unob.cz

R. Vasicek
e-mail: radovan.vasicek@unob.cz

1   Introduction

The European security environment is affected by a broad spectrum of internal and external hazards. Each of these hazards is unique and distinctive in terms of the intensity of its influence, frequency and duration in relation to assets which shape the European security environment. Not only the European Union (EU) as a whole, but also its individual member states must be ready to reduce the impacts and conduct effective risk management. In the ideal case scenario, relevant risk management controls must be implemented well in advance before the assets are physically exposed to the actual hazard. As a response to the changes in the security environment, several European countries have recently increased their defence spending. This is particularly valid for those EU member states that also have obligations resulting from their simultaneous membership in NATO, as they are moving towards their pledge of dedicating 2% of their GDP to defence.

In order to meet the EU top defence priorities, i.e. peace support, peace keeping, crisis and military conflict prevention, it is crucial to develop and strengthen the international security and defence. Therefore, the EU member countries and their partners need to share their knowledge and experience. As this cooperation should be based on agreed rules of operational and technical interoperability, the member countries shall build and sustain essential effective defence capabilities to satisfy their obligations related to the mutual assistance and solidarity [2]. Such a transparent and unified approach towards partnership will increase preparedness and resilience of the participating states, while supporting the collective security on the EU and NATO level.

The convergence of the EU defence capacities is based on the build-up of adequate capabilities, which require concurrency of personnel, financial and material resources. This process provides long term benefits as it accelerates the defence industry, supports sharing of exclusive capabilities of the participating nations and facilitates participation in the development of the EU security architecture. The PESCO projects are a tangible challenge for integrity of common approach of the EU nations, as they represent a real European initiative and basis for increase of credibility, readiness, reduction of instability, cooperation or key position in the European defence industry.

2   The Theoretical Part

The European Council established the PESCO with reference to the Lisbon Treaty on the European Union (article 42.6, 46 and Protocol 10) in December 2017. The purpose of the PESCO initiative has been "*raised cooperation on defence among the participating EU Member States to a new level*" [9]. As of now, 25 out of 27 EU member states have joined 47 PESCO projects, which are currently in various stages of progress. According to their focus areas, the projects can be divided into seven main categories [9]:

- training, facilities,
- land, formations, systems,
- maritime,
- air, systems,
- enabling, joint,
- cyber, C4ISR (Command, Control, Communications, Computers, Intelligence, Surveillance and Reconnaissance),
- space.

Each of the projects is unique. The current status of their progress depends on the specific approach chosen in order to achieve the objectives, and activities that respective EU stakeholders have conducted so far. Despite existing general rules of the project management and availability of multiple project management methodology portfolios, it is always necessary to thoroughly consider whether the implementation of the generic approach solution is suitable for a particular PESCO project. This especially applies to those PESCO projects, which are focused on the comprehensive military capabilities development where specific factors must be taken into account.

The Electronic Warfare Capability and Interoperability Programme for Future Joint Intelligence, Surveillance and Reconnaissance (JISR) Cooperation, here abbreviated as the PESCO EW, is one of such projects. *"The primary objective of the project is to produce a comprehensive feasibility study of the existing EU electronic warfare (EW) capabilities and the gaps that need to be filled. The findings of the feasibility study should potentially lead to the adoption of the joint EW concept of operations (CONOPS). The CONOPS may include joint training of EW experts and, if agreed upon by the member states, the establishment of a joint EW unit."* [10].

The main idea of the PESCO EW project originates from undisputable dependence of any actors of contemporary or future military conflicts, including military forces, on the electromagnetic environment (EME). The ability to gain and maintain superiority in the EME can be not only instrumental for achievement of the planned end state, but it should also be seen as an effective factor to increase the combat effectiveness and combat power of own and cooperating military forces. The PESCO EW project deals with EW, generally perceived as a combat activity conducted in the EME, in connection with EW activities, measures and tasks. Furthermore, it elaborates on the relationship between military services as well as the EW role and contribution to combat support to other military capabilities in the battlefield. In addition, part of this combat support provided by EW can be described in relation to JISR; a set of intelligence and operational capabilities synchronized and coordinated in a manner which facilitates planning and execution of joint military operations by multiple services of military forces.

3 The Methods

Prior to the risk management process of the PESCO EW project it was necessary to choose an appropriate hazard identification method. The criteria for the method selection were as follows:

- the method must be applicable in the early stage of the project solution,
- the method must be able to work with a limited amount of input information,
- the method must also be applicable to a system, which has not been subjected to the risk identification and analysis yet,
- the method must be usable for multiple types of activities conducted during II Follow-on Phase,
- the use of the method must be uncomplicated not only for risk management experts,
- it must be applicable in a small team (up to ten members).

The quantitative method of hazard identification was deemed unsuitable due to the required criteria for applicability at the early stage of the project and a limited amount of input information. Therefore, one of the qualitative hazard identification methods, the Preliminary Hazard Analysis (PHA), was chosen instead.

The PHA method was introduced in the U.S. military through the Military Standard System Safety Program Requirements 882-C in 1984 [5]. The reason why it was chosen was its applicability at the early stage of the project, when there is not enough information on the concept details or processes required in order to achieve the project objectives. The implementation of the method was a suitable starting point for further study, and it provided information needed for specification of the system proposal [4], in this case the PESCO EW project. Furthermore, it also met the remaining criteria including a limited number of the PESCO EW working team members, their specific expertise, existing limits in the risk management area or flexibility provided by a broad scope of the PHA method applicability.

The PHA is a simple inductive method of analysis. Its main purpose is to identify hazards and hazardous situations which will inflict adverse consequences on a system or activity [4, 8]. The method enables not only to identify hazards, but also to determine probability of hazard activation, its impacts and to estimate the risk. Thus, the PHA has extensively been used in various disciplines [12]. There are multiple advantages of the PHA method (the source amended [12]):

- due to its application at the early stage of development, it is possible to improve critical areas of the evaluated system, and in this way to increase its standards and contribution,
- controls are adjusted so that risks can be managed effectively,
- if risks cannot be treated through minimization to an acceptable level, the method offers controls which will at least bring those risks down to an tolerable level at least.

Tasks and requirements involved in preparing the PHA should include the following [8]:

- For purpose of the analysis establish:

    - boundaries between the system, any system with which it interacts, and the domain,
    - an overall system structure and functionality.

- Identify:

    - a detailed list of hazards of the system based on preliminary hazard list report and the requirements,
    - an update hazards list,
    - accidents to the most practicable extent,
    - events of accident sequence and those that can be discounted,
    - a record in a hazard list.

- Assign:

    - each accident a severity categorization and each accident sequence a predicted qualitative/quantitative probability,
    - each hazard a preliminary random and systematic probability target.

- Document:

    - any safety features that are to be implemented during the design and development phase.

    Rausand [11] names seven PHA main steps:

- plan and prepare,
- identify hazards and hazardous events,
- determine the frequency/probability of hazardous events,
- determine the consequences of hazardous events,
- suggest risk-reducing measures,
- assess the risk,
- report the analysis.

The comparison of the content of activities conducted in individual steps described by the above mentioned authors [8, 11], reveals that tasks and procedures are almost identical, and only their division differs, as Ostrom does not mention the assessment of applied controls and risk evaluation.

Due to the extent of the PESCO EW risk management process, the chapter describes only the identification phase and the risk analysis phase. This corresponds to [11] Steps 1–4 and Ostrom's [8] Steps 1–3.

Prior to the actual application of the PHA method, it was necessary to set up the PESCO EW working team, consisting of competent personnel. For this process, the requirements for PHA team members, formulated in the U.S. Occupational Safety and Health Administration [7], were adapted to the context of the PESCO EW project as follows:

- the PHA is performed by a team possessing EW knowledge, expertise and experience at the national and international level,
- at least one of the team members is knowledgeable in the risk management.

Due to the limited availability of personnel resources, the permanent facilitator and recorder were not appointed, which partially deviated from the rules of the PHA team composition [1, 6]. In order to compensate for such a deficiency, all the team members regularly took turns in the facilitator and recorder's roles. All of them tried to respect the facilitator's positive attributes and reduce negative attributes in order to maintain a friendly and creative atmosphere [1]. The team became truly international, as it consisted of national EW and risk management experts from the countries participating in the project. The EW subject matter experts (SME's) benefitted from their long-term experience and also from the fact that they knew each other. This was really important as their interpersonal relationships positively contributed to the team's performance. At the beginning, the risk management expert introduced the process and the PHA method to the EW SME's. Brainstorming was frequently used for stimulation of creativity, generation of the list of risks and description of impacts.

4   The Practical Part

The risk management of the PESCO EW project was initiated in accordance with the ISO 31000 risk management procedure and the PHA specific tasks and requirements [8, 11]. In order to provide more detailed description of the process, [11] approach was chosen for the risk identification and analysis phase.

4.1   Step 1: Plan and Prepare

The first step, Plan and Prepare, is important for the quality and applicability of the risk identification and analysis phase.

First, the purpose of the risk management process was determined in relation to the objective and execution of the PESCO EW project. The Risk Assessment I was produced in support of the PESCO EW project feasibility study. Based on the authorized decisions, taken in I B and II A Phases, the next step of the PESCO EW project will involve the risk treatment (II B) which will be included in II Follow-on Phase. The risk treatment will be performed in order to minimize risks and employ measures against unacceptable risks. The risk evaluation will facilitate targeted reduction of risk activation probability, impacts or their combination. This process will increase the quality and the level of detail of expected formulated outcomes of II Follow-on Phase, shown in Step II B (see Fig. 1).

Electronic Warfare Capability and Interperability Programme for Future JISR

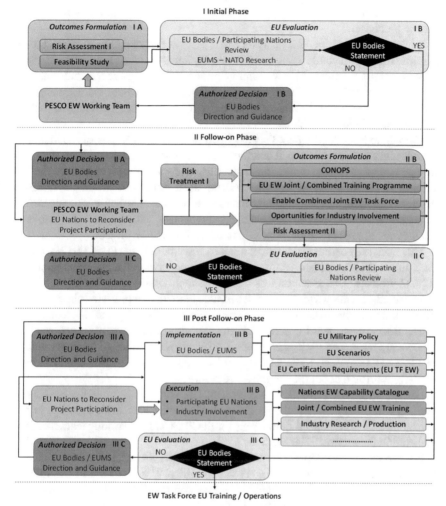

Fig. 1 The concept of the PESCO EW solution (Risk Assessment I covers the areas outlined in red). *Source* Own

The PESCO EW feasibility study represents the comprehensive product initiating the determination of the EU bodies direction and guidance (see II A) for execution of II Follow-on Phase. For this reason, it is also the basis for the risk identification and analysis process.

The boundaries of the risk analysis process were defined with respect to the declared purpose of the management process. The analysis covered Step I B (the EU Evaluation and Authorized Decision taken by the EU bodies) and the whole II Follow-on Phase (II A and II B). The Fig. 1 shows the boundaries of the risk identification and analysis outlined in red. The risk assessment II will be conducted within Outcomes Formulation (Step II B) in order to assess the risks for III Post Follow-on Phase.

It is expected that the risk analysis carried out at this early stage of the project solution will significantly contribute to the higher quality of II Follow-on Phase outcomes (i.e. CONOPS, EU EW Joint / Combined Training Programme, Enabling of the Combined Joint EW Task Force, Opportunities for Industry Involvement), and at the same time it will provide a solid basis for an appropriate decision making process in the follow-on phases.

The whole Risk Assessment Process I will provide an overview of the identified hazards and assets. Its additional outcomes will include the hazard register, scales for the assessment of hazard activation probability and their impact as well as the risk estimation. The format of the outcomes will be based on the PHA hazard identification qualitative method and semiquantitative form of the risk estimation outcomes. The Risk Assessment I results are presented simultaneously with the feasibility study so that they are applicable to the next steps of the PESCO EW project.

There was only a limited amount of input data available to the PESCO EW working team due to a lack of historical data and sparse coverage of regulations related to PESCO projects. This situation was one of the main reasons why the PHA method was selected, as it allows for work with a limited amount of input data, while enabling establishment of the level of impact, risk activation probability and risk level.

The Risk Assessment I is conducted in support of processing the outcomes of II Follow-on Phase. The Risk Assessment I results can be exploited for every step of the PESCO EW project. National stakeholders from the participating countries were regularly informed about the progress and results.

For the purposes of the Risk Assessment I production, the initial PESCO EW working team was augmented by a Czech MoD professional with expertise in risk management. Professional requirements for the PESCO EW working team members are listed in the section "The Methods."

Prior to the second step, it was necessary to identify the assets for the established boundaries of the system (see Fig. 1) which will play a vital role in the PESCO EW project solution, and whose damage or loss would significantly affect successful achievement of the project's objectives. Due their complexity, the individual assets were merged into seven key ones:

1.   A country participating in the project. The asset comprises the armed forces of a project member state, its Ministry of Defence, national legislation (law, internal regulations, standards, etc.), budget and personnel, military infrastructure and foreign policy.
2.   The Concept of Operations (CONOPS[1]) represents the key outcome of II Follow-on Phase. It will describe approaches and procedures for operations as well as the training concept. CONOPS will be classified.

---

[1] A clear and concise statement of the line of action chosen by a commander in order to accomplish his given mission [3].

3. The training programme. It will include training on both operational and tactical levels, training concept, requirements for personnel and material and technical support for training.
4. Enabling of combined joint EW Task Force. The outcome will specify the format (template) of the capability catalogue and requirements for administration of core material.
5. Relevant EU institutions. The asset involves EU civil and military authorities responsible for the project execution.
6. The working team. It includes specialists and subject matter experts responsible for the project solution and other members nominated by respective countries.
7. The project (II Follow-on Phase). The asset compiles achievements of goals of II Follow-on Phase.

4.2 Step 2: Identify Hazards and Hazardous Events

The purpose of this step is to conduct a survey on hazards which may affect activities within the boundaries of the Risk Assessment I. This step is one of the most crucial ones out of the whole risk management process, because any failure to identify hazards would result in their omission in the following steps. Since the PESCO EW project is the first, and actually unique, EU project the purpose of which is to provide favourable conditions for military forces to achieve superiority in the electromagnetic environment, the hazard register is also an initial product which can be modified and amended. As all military forces conduct activities in the EME, either deliberately or unintentionally, the PESCO EW hazard register can potentially be used even outside the framework of the actual PESCO EW project.

Using the PHA method, the international working team, consisting of EW, JISR and risk management experts, has identified 65 internal and external hazards related to the PESCO EW project. There are several examples listed below:

- setback or change in the EU support to PESCO projects
- the EU dismissive stand against the outcomes of II Follow-on Phase,
- loss of national political support in project member states,
- restriction of national political support in project member states,
- resignation from the project by a member state,
- abandonment of the working team by its member,
- change of the working team manning,
- the specified content of the EU Task Force EW specific training on tactical-operational level does not reflect real-life requirements,
- insufficient definition of material and technical support requirements to enable EU Task Force EW specific training on tactical-operational level,
- failure to determine requirements for the personnel in charge of EU Task Force EW specific training on operational level and instructors of the EU Task Force EW specific training on tactical-operational level (language and professional skills),
- differing perceptions and attitudes towards the project's OPSEC.

The team investigated what, where and when individual hazards, as well as their combination, would occur in the future. During brainstorming, the team considered not only individual hazards, but also combinations of their mutual effects on the assets. In Step 3, this hazard definition was then used for description of hazard activation probability. In order to visualize whether an asset is affected by a hazard, the hazard register was compiled. A cross in the register represents a relationship between the threat and asset. All identified relations were then included into the risk analysis and evaluation (see Fig. 2).

As shown in Fig. 2, each hazard affects a different number of assets. The maximum number of risks, resulting from interaction of 65 hazards and 7 assets,

Hazard	Assets						
	Country	CONOPS	Training programme	Enabling of combined joint EW Task Force	EU	Working team	Project
Set back or change in EU support to PESCO projects	x	x	x	x	x	x	x
Loss of national military support in project member states	x	x	x			x	x
Restriction of national military support in project member states	x	x	x			x	x
Resignation from the project by a member state	x	x	x	x	x	x	x
EUMS disapproval of procedures for execution of II Follow-on Phase	x	x	x	x	x	x	x
Abandonment of the working team by its member	x	x	x	x		x	x
Change of the working team manning	x	x	x	x		x	x
Insufficient level of expertise of the working team members		x	x	x		x	x
Failure to keep to the time frame of the project solution	x	x	x	x	x	x	x
Insufficient funding of the working team	x					x	x

Fig. 2 An example of the assembled hazard register. *Source* Own

	Number of affected assets						
	1	2	3	4	5	6	7
Number of hazards ($\Sigma = 65$)	0	1	1	0	16	17	30

Fig. 3 The total number of assets affected by hazards. *Source* Own

	Assets						
	Country	CONOPS	Training programme	Enabling of combined joint EW Task Force	EU	Working team	Project
Number of hazards	56	52	61	60	40	64	64
Position	5	6	3	4	7	1–2	1–2

Fig. 4 The number of hazards affecting a single asset. *Source* Own

is 455 (65 × 7). Figure 4 shows that a total of 397 risks were identified in the PESCO EW project.

The PESCO EW working team analyzed effects of the hazards on the assets. Figure 3 shows the number of various hazards jointly affecting single assets. It also includes summary information on the number of hazards. It indicates that none of the hazards affects only one asset. Two or three assets are affected by one hazard. However, the most hazards affect all seven assets.

It was also necessary to identify the assets affected by the highest number of hazards in order to apply controls to increase resistance of the assets identified. Figure 4 presents the assets and the number of specifically related hazards.

As shown in Fig. 4, the most affected assets include the working team and the project. These two assets were affected by all but one out of 65 hazards, which was the discrepancy between the project member states national legislation, including internal regulations of the armed forces, and the project outcomes (national caveats). The least affected asset is the EU, because the risk analysis is primarily intended for internal needs of the PESCO EW project, while the EU participates only in its specific phases (approval, statements, etc.; see Fig. 1).

4.3 Step 3: Determine the Frequency/Probability of Hazardous Events

In this step, the working team identified and discussed the causes of hazard activation in order to determine the probability of hazard activation. As there was neither historical data nor data from other PESCO projects available, the PESCO EW working team applied its own hazard activation probability levels, based on its experts' subjective assumptions for Risk Assessment I. Figure 5 contains the index values and their verbal description.

The value 4 was assigned to none of the hazards. All the values of the hazard activation probability were recorded in the table.

Index value of likelihood	Likelihood	Description
1	Low	no indication of activation and hazard impact on asset vulnerability in upcoming period
2	Medium	occasional activation and hazard impact on asset vulnerability
3	High	frequent activation and hazard impact on asset vulnerability
4	Critical	very frequent activation and hazard impact on asset vulnerability

Fig. 5 The scale of the hazard activation probability. *Source* Own

Index value of impact	Impact	Description
1	Negligible	damage of the asset partially affects achievement of the objective, it is not necessary to replace or repair the asset
2	Low	damage of the asset affects achievement of the objective, the asset can be easily replaced or repaired
3	Medium	damage of the asset impacts the achievement of the objective, its replacement or repair is necessary and requires additional resources
4	High	damage of the asset is crucial for achievement of the objective, its replacement or repair is almost impossible/unfeasible
5	Critical	damage of the asset is critical for achievement of the target, replacement or repair is impossible/unfeasible

Fig. 6 Verbal and index evaluation of the impacts. *Source* Own

4.4   Step 4: Determine the Consequences of Hazardous Events

In Step 4, conducted simultaneously with Step 3, the PESCO EW working team identified and evaluated the consequences of hazardous events. During this process, the worst conceivable case of hazard impacts on the assets was always considered. In order to describe impacts on the assets, the team created general characteristics of impacts, and the values were recorded in Fig. 6.

5   Conclusion

Although it is a sovereign right of each of the EU member countries to secure its own defence, the complex challenges posed by the current dynamic development of the security environment call for extensive cooperation, interoperability and experience sharing. Without such a partnership, it will be impossible to develop and increase combat effectiveness of national armed forces. The purpose of the PESCO projects is to encourage and strengthen mutual cooperation at the EU level. The PESCO EW has become the first PESCO project with the Czech Republic as the leading nation. The feasibility study, which is the most important outcome of the project's initial phase, proposes processes and solutions for II Follow-on Phase.

A lack of information is specific to this early stage of the project. The PHA method was selected as the most suitable for the Risk Assessment I, because it met most criteria required. The method was applied in seven steps. Nevertheless, due to the extent of the chapter, only the first four steps were presented. Steps 5 to 7 of the PHA method, following [11], will be released in future publication.

Within the outcome of the PHA Step 1, the scope of the PESCO EW project was specified, the working team with EW and risk management expertise was established, and assets were identified. The assets were then aggregated by their purpose or common characteristics into logical groups. A total of seven assets were defined. The working team and the project represented the most affected assets. In Step 2, several brainstorming sessions identified 65 hazards. They were implemented into the hazard register, which indicated whether a particular hazard would affect the assets. In total, 397 out of 455 maximum potential interactions were discovered. Steps 3 and 4 were conducted simultaneously. At this stage, the PESCO EW had to define and describe the levels of the hazard activation probability and impacts. The PESCO EW working opted for a semiquantitative scale (a combination of index and verbal description). The probability of activation was allocated to each hazard, and the level impact was determined for the respective asset.

The comprehensive results of the PHA method will be used for application of controls during II Follow-on Phase which will contribute to minimization of hazard activation and impacts, or the combination of both. The outcomes of Risk Assessment I will reduce potential deficiencies related to II Follow-on Phase, thus improving its outcomes.

References

1. Baybutt P (2015) Competency requirements for process hazard analysis (PHA) teams. J Loss Prevent Process Ind 33:151–158. ISSN 09504230.https://doi.org/10.1016/j.jlp.2014.11.023
3. CONOPS (2020) NATO Term: the official NATO terminology database [online]. NATO, Brussels [cit. 2020–08–27]. https://nso.nato.int/natoterm/Web.mvc
2. European Union (2016). Shared vision, common action: a stronger Europe: a global strategy for the European Union´s foreign and security policy, vol 1. Publications Office of the European Union, Luxembourg. ISBN 978-92-9238-369-5
4. ISO 31000:2018 Risk management—Guidelines, 2018, vol 2. International Organization for Standardization, Geneva
5. MIL-STD-882D System Safety (2012) Environment, safety, and occupational health, risk management methodology for systems engineering, 2012, vol 3. Military and Government Specs & Standards, Depatrment of Defence, Washington

6. Montewka J (2020) Kul-24.4230 safety and risks of marine traffic: L7—preliminary hazard analysis. MyCourses [online]. Aalto University, Espoo, 2020 [cit. 2020–09–27]. https://mycourses.aalto.fi/pluginfile.php/213175/mod_resource/content/2/Kul-24.4230% 20-%202016%20-%20L7%20-%20Preliminary%20Hazard%20Analysis.pdf
7. OSHA (1992). Process safety management of highly hazardous chemicals, 29 CFR Part 1910.119, U.S. Department of Labor, Occupational Safety and Health Administration
8. Ostrom LT, Wilhelsen CA (2012) Risk assessment: tools, techniques, and their applications, vol 1. Wiley, Hoboken, New Jersey. ISBN 978-0-470-89203-9
9. PESCO (2020) Member States Driven. PESCO [online]. PESCO Secretariat, Brussels [cit. 2020–09–10]. https://pesco.europa.eu/
10. PESCO Project (2020) Electronic warfare capability and interoperability programme for future Joint Intelligence, Surveillance And Reconnaissance (JISR) cooperation. PESCO [online]. PESCO Secretariat, Brussels 2020 [cit. 2020–09–10]. https://pesco.europa.eu/project/electr onic-warfare-capability-and-interoperability-programme-for-future-joint-intelligence-survei llance-and-reconnaissance-jisr-cooperation/
11. Rausand M (2011) Risk assesment: theory, methods, and applications, vol 1. Wiley, Hoboken, New Jersey. ISBN 978-0-470-63764-7
12. Yan F, Kaili XU (2019) Methodology and case study of quantitative preliminary hazard analysis based on cloud model. J Loss Prevent Process Ind 60:116–124. ISSN 09504230. https://doi.org/10.1016/j.jlp.2019.04.013

Cyber Risks

Why Human Firewall Fails in the Battle with Sophisticated Spear Phishing Campaigns

Miroslav Čermák

Abstract The aim of this chapter is to familiarize readers with the results of long-lasting research that was conducted in several organizations in the Czech Republic during the last four years. The integral part of this research was an experiment examining the cyber security resilience of these organizations and their employees to sophisticated cyber-attacks, and test a hypothesis regarding human firewall as a main countermeasure to social engineering techniques, especially targeted phishing used by a skilled attacker. The results of this experiment revealed that although the security awareness and training significantly decrease the click rate after the first run, it is rather impossible to reach a score below a few percentage units in consequent years. The results and recommendations stated in this chapter can help the organizations to set up their security awareness, training and testing, as well as security metrics.

Keywords Phishing · Human firewall · Human error · Security awareness · Testing · Security metrics

1 Introduction

For many years, security experts have argued that any system is as secure as its weakest link, which is the user. At the same time, however, they are convinced that in the fight against phishing it is not possible to rely only on security measures of a technical nature and that the most effective security solution is the so-called human firewall. Here, however, it is possible to see a certain contradiction. On the one hand, the user is the weakest link, and on the other hand, an informed user should be the most effective measure in the fight against sophisticated phishing.

Is it even possible? If so, how effective is such a human firewall? Does it make sense to invest in it? Is it possible to rely on a sufficiently educated user not so

M. Čermák (✉)
Department of Management and Informatics, Police Academy of the Czech Republic in Prague,
Lhotecká 559/7, 143 01 Praha 4, Czech Republic
e-mail: cermak.miroslav@pacr.eu

© The Author(s), under exclusive license to Springer Nature Switzerland AG 2022
I. Tušer and Š. Hošková-Mayerová (eds.), *Trends and Future Directions in Security and Emergency Management*, Lecture Notes in Networks and Systems 257,
https://doi.org/10.1007/978-3-030-88907-4_16

easily becoming a victim of social engineering and not clicking on an attachment in an e-mail or a link leading somewhere on the Internet, not giving information to an unauthorized person by phone (vishing) or allowing an unauthorized person to access protected areas (piggy backing)?

It should be noted that phishing uses basic social engineering techniques, such as inducing urgency, time pressure, and exerting a position of authority, which are relatively effective. Especially in the case of phishing, which we will deal with in this chapter, and which is also the least risky for the attacker, because the attacker does not have to respond promptly to any questions from the employee or worry about his detection and arrest.

But first, let's look at what phishing is. Since phishing is defined in several NIST standards and has evolved in the meantime, we can come across several different definitions.[1] However, they all describe phishing as a form of social engineering in which an attacker posing as a trusted authority aims to redirect email recipients to a similar website that looks like a legitimate site, retrieve sensitive personal information from it, and misuse it.

However, the definition does not cover the case where the attacker's goal is only to get the victim to click on a link or attachment in the e-mail, followed by a compromise of the terminal in which no theft of sensitive data may occur. However, this terminal can be integrated into the botnet, misused to attack other targets, or it can receive any malware such as adware, dialer, ransomware, bankware, or cryptominer, etc.

Therefore, in this post, we will consider phishing as an activity that exploits social engineering techniques in order to lure the recipient of an email or a website visitor to click on a link or attachment, which will lead to his subsequent redirection to a fraudulent site or to the download of malicious software code, or both. In practice, however, there is usually either a download of malicious code or a redirect to a fraudulent site.

Knowbe4 (according to Gartner considered a leader in Security Awareness Training) states[2] that an untrained user is involved in a phishing campaign in more than 35% of cases, and that after three months of training the campaign can be reduced to 14% and approached after a year of training to the 5% threshold.

In the context of the above, we asked ourselves the question of whether a human firewall is really effective enough to protect organizations from sophisticated spear phishing campaigns not only today but also in the near future, and whether it makes sense to further develop it. Finally, whether it would be possible to reach a value close to 0% after a few years. The following null hypothesis (H) is also related to this research question.

H: *Once all employees of the organization are familiar with the cyber-attacks towards them and trained in phishing detection for several years, it will no longer be possible to penetrate the organization in this way, because phishing will be correctly detected by them.*

---

[1] NIST [4]

[2] KNOWBE4 [3].

If this hypothesis was confirmed, it would be possible to state that the human firewall is the most effective security measure. In order to find the answer to the research question and verify the validity of this null hypothesis formulated above, we decided to carry out a long-term experiment.

As part of this experiment, which took place between 2016 and 2019 and consisted of repeatedly sending test phishing e-mails of various qualities to employees of selected organizations in the Czech Republic, we were interested in how these employees would respond to them. Specifically, how they will react to phishing immediately after completing the relevant security training, and whether there will be any significant change in their behaviour, which was the subject of our observation.

2 Characteristics of the Environment in Which the Experiment Took Place

The individual organizations in which this survey was conducted operate in different sectors of the national economy and have different numbers of employees, ranging from hundreds to hundreds of thousands, so it is a fundamentally homogeneous but externally very heterogeneous group of organizations which have in common the fact that they are aware of the threat of a cyber-attack and their lack of cyber resilience, and try to address it through security training for their employees and planned phishing tests to verify the effectiveness of the training. For the sake of completeness, it should be added that they do not rely only on the human firewall, but also run various antispam solutions that can filter out most spam and thus phishing.

In addition, it is typical for these evaluated organizations that all employees must attend introductory training as part of on-boarding, where they are trained in information and cyber security, among other things, and then must undergo mandatory e-learning training focused on information and test-terminated cyber security. However, if the employee does not pass the test, he/she will have to complete the relevant training again and undergo the final test again.

The e-learning courses and training itself have always been prepared in close collaboration with the organization's security experts to take into account applicable security policies and processes to achieve the desired change in behaviour. That is, in order for the employees to react correctly when they receive a phishing e-mail, someone will call them, contact them via a social network, or otherwise contact them and ask them to provide internal information or perform a certain action.

The security policy in the organizations is set so that the employee must complete the e-learning training culminating in an online test every year, with control questions generated for each participant at random from the database, so it is unlikely that two employees will receive the same set of questions or the employee would receive the same questions as in the previous year.

In addition, the questions are continuously updated and supplemented by the organization's security experts to reflect current developments in cyberspace threats and to enable the required level of security awareness to be achieved. The training does not test knowledge of the concepts, but the employee answers how to proceed in the event of a situation, learns how to recognize an ongoing attack using social engineering techniques, and how to respond to phishing, vishing, piggy backing, etc.

The emphasis in this security program is primarily on phishing and spear phishing, which these organizations encounter most often. As part of the training, employees are presented with various e-mails and they must decide whether it is phishing or not, and indicate in the e-mail what they consider to be suspicious, and on which basis they decided so.

Given that the cybersecurity manager was aware of the fact that an employee could correctly answer test questions and pass the test does not mean that he will always, and under all circumstances, respond correctly even when exposed to such a situation, because he may find himself under time pressure, he may be pressured by his superior authority, he may be negatively overwhelmed by emotions in general, and he will not be able to think clearly and will not always behave properly. For this reason, the management has its employees and members of the top management tested repeatedly.

3   The Course of the Experiment

In collaboration with the company's management, it was decided that an experiment would be conducted to examine how employees would react to the actual ongoing attack. For this purpose, a set of e-mails in various qualities was prepared, according to the severity classification of phishing e-mails from low danger to critical,[3] which is used to assess the severity of the HOAX server, and which informs de facto about all widespread phishing campaigns in the CR.

Two to three such phishing campaigns take place within each organization each year. In total, up to 9 phishing campaigns have taken place in each organization over the past 3 years. As part of these campaigns, several thousand e-mails were sent and it was evaluated how individual employees responded to them. That is, whether they click on a link in an email or an attachment.

Only the company's top management, security manager and Data Protection officer were informed of the launch of this campaign, and all were asked not to disclose information about ongoing testing.

Later, the IT security team learned about the ongoing campaign at the time when this fact was reported to them by the user who became the recipient of the phishing e-mail, correctly assessed it as phishing and reported this fact in accordance with the security policy.

The security team would normally update the appropriate rules and not pass another test phishing email to the organization. However, in this case, they set only passive rules by agreement, which allowed them to monitor who received the phishing

---

[3] HOAX [2].

e-mail of the given type and whether they clicked on the attachment or link in the e-mail.

Each click on the link led to the launch of the browser and the generation of an entry in the log, as well as the click on the attachment led to the launch of the associated application and writing to the log. The log stores information about who and when clicked on the attachment or link in the e-mail. After the end of each campaign, it was then possible to compare the number of e-mails sent against the number of records in the log.

In this way, it was possible to compare not only the development over time, but also the success of individual campaigns, i.e. which e-mail reached a higher click-through rate. Needless to say, at the time of the actual attack, clicking on an attachment or link in an e-mail would not only write information to the log, but trigger any malicious code that could do essentially anything.

The campaign used a pre-arranged email link that led to a site that was completely under the experimenter's control, so there was no risk that the site might contain actual malicious code. Prior to launching the campaign, it was verified that the checksum of the email attachment matched the code submitted to the Cyber Security Manager for review.

As for the actual content of the e-mail, according to Steves et al. [5], the more the content of the e-mail relates to their personal or professional life, the greater the chances that the recipient of the e-mail will click on the link in it or the attachment.[4] In other words, context is a key factor in deciding whether a user can detect phishing. And the more the content of the email relates to the recipient's life and profession, the less likely the recipient is to label the email as phishing. We've taken this into account when compiling the phishing email.

The emails were sent in several waves, and various obfuscation techniques such as ZeroFont, Ropemaker, Punycode, MetaMorph, Base striker, OCR homograph or EML attachment were used to prevent the antimalware/antiphishing solutions used by these companies from being detected.

As for the domain name, the same or a very similar domain name was used as the organization the sender impersonated, or the domain was located on another top-level domain. A non-existent.cs domain or an existing.com domain was used instead of the.cz domain. The FROM field has also been changed to include an email address located on the same domain.

However, the primary goal of this experiment was not to verify the effectiveness of the security solutions used. It has been found out that most techniques known for more than a year are still applicable to bypass antiphishing filters, especially when they are properly combined, and the resulting form does not correspond to other previous and known cases, based on which the rules of antiphishing solutions are formed.

---

[4] Steves et al. [5].

Year	Round	Level 1	Level 2	Level 3	Level 4
2016	1	33			
	2	6			
	3	3			
2017	4	1	20		
	5	2	5		
	6		6		
2018	7		4	33	
	8			23	
	9			31	
2019	10				30
	11				25
	12				35

Fig. 1 Phishing campaigns in years. *Source* Own

It should be noted that most solutions work on the basis of IoC or IoA, which are passed to an algorithm working on the principle of machine learning, so they are able to detect a global phishing campaign, but not local, conducted in another language, in this case Czech.

4 Results of Experiment

Based on the results of a repeated experiment and using induction, it can be stated that the success of the attack itself depends on the quality of phishing. Furthermore, as the quality of phishing increases in consequent years, so does the success of the attack. This can be measured as the percentage of those who clicked on a link or attachment in an email out of the total. See Fig. 1 for phishing campaigns in years that evaluate the trend in click rate.

This success rate of the attack then ranges from 1 to 35% in the organization where the training and testing took place. If we look at the results of these tests, we can also conclude when using the inductive method that further repeated training cannot achieve a significantly better result, only to maintain the status quo. This is confirmed by the results of previously performed experiments where researchers also concluded that security awareness training is not a silver bullet in phishing defence and click rate will not go to zero but will vary according to the phish level.[5]

The results of the experiment were also compared with the conclusions of other security experts who test employees to order. It turned out that they came to similar

---

[5] Greene et al. [1].

conclusions and that for untargeted "apparently phishing" messages (Level 1), the success rate ranges from about 5–20%. The message without context (Level 2) has a success rate of between 5 and 30%, and for more sophisticated/more targeted campaigns (Level 3 and Level 4), the success rate is usually between 15 and 30%.

We see that although different teams are using their own methodology and are testing different companies, they all have achieved similar results in their phishing test campaigns.

In other words, when high-quality phishing e-mail is distributed, almost every third recipient clicks, while in the case of general phishing, it is only every twentieth. By simple deduction, we can conclude that it is not so important what percentage of e-mail recipients click, because it is enough to click only once, and then we can talk from the attacker's point of view of success, because the attacker basically does not care who clicks and whom he compromises.

It also follows from the above that in the case of generic phishing, it is necessary to send a larger volume of e-mails in order to find someone who clicks on the link or attachment in the e-mail, and staff training significantly reduces this click rate. In the case of targeted, context-oriented phishing, it is usually sufficient to send only a few such e-mails, and employee training does not significantly reduce this click rate.

The fact that for a successful generic phishing it is necessary to reach a larger number of recipients, leads to the fact that such a campaign is then detectable by antispam solutions, so the human firewall does not have to get the word out in such a case. This is also confirmed by statistics from the antispam solution used in individual organizations, according to which up to 90% of e-mails are marked as SPAM and discarded at the entrance to the organization and are thus not delivered to the mailboxes of end recipients at all.

Based on the performed experiment, it was possible to reject the hypothesis because it was not confirmed, and to answer the research question about the potential of the human firewall.

It has been proven, on the basis of results of the above mentioned experiment, that security education and training positively change the attitude of employees and their ability to detect a low level phishing campaign. However, in case of sophisticated phishing attacks it is necessary to consider human firewall rather as a measure which effectiveness is very limited and does not prevent endpoint compromise. Continuous improvement is not possible and further training and testing cannot reduce the click rate below the percentage units, which may vary from organization to organization.

5 Partial Findings and Recommendations for Further Research

During the experiment, it was observed that as soon as one of the employees detects a phishing e-mail, he asks other colleagues if they also received it and then informs them that they are probably being tested, and they should be careful. The question is whether other research that has already taken place in this area has encountered this phenomenon and how they dealt with it. In this way, there is a certain bias in the evaluation of phishing tests, which then show significantly better results. However, on the other hand, if it were not a test and it was real phishing, it can be assumed by analogy that the person would probably behave the same way, because he would not know whether it is a test or not.

The question is how to send e-mails as part of testing, whether to find out who is sitting with whom and to send e-mails to people from different departments, where it can be assumed that they do not sit next to each other, and thus prevent the dissemination of information about an ongoing test.

When sending typical e-mails in waves, they are sorted alphabetically or randomly, but in such a case it cannot be ruled out that the same e-mail will come to two employees sitting next to each other and they will tell each other about it and immediately inform the others.

In such a case, the incident management process and early warnings are tested rather than the ability to detect phishing.

Another question is how long the testing should take. The fact that the recipient did not click on the e-mail may be due to other factors. Several such factors have been noted, e.g. the employee is on vacation, a retrospective analysis is performed by antispam solution, and the e-mail is moved on the server to the SPAM folder.

6   Conclusion

An experiment conducted in many different organizations from 2016 to 2019 confirmed that the human firewall has its limits and cannot be completely relied on, and must always be deployed in combination with other security measures and be the integral part of a defence-in-depth security solution. The null hypothesis was not confirmed and it was possible to reject it based on the performed experiment.

While it must be acknowledged that with every phishing campaign an organization's resilience increases, it is impossible to achieve a situation where no one would click on the attachment or link in the e-mail. There are always a number of people who click on an attachment or link in an email, and even if it's just one person, the organization can be compromised this way.

It turns out that employees are relatively reliably able to identify low level quality phishing, but as its quality increases, their ability to detect it decreases rapidly, especially when the e-mail is spelled correctly, comes from a seemingly trusted address, relates to what the employee is dealing with in his private or professional life, and in addition acts on emotions, for example when reacting to current events.

Given that our second survey, conducted from January to May 2020, which was attended by 93 organizations in the Czech Republic and aimed to find out what attacks these organizations actually face, showed that the sector, size and criticality of the system operated does not have a significant impact whether or not phishing will be delivered to the organization. We can say that the conclusions we came to in the experiment described above should also be valid for other organizations. At least for those whose approach to cyber security is similar to the organizations in which this research took place.

By simple deduction, it can be concluded that organizations that do not address security education in this area and do not engage in training and testing of employees, are significantly more vulnerable, because existing security solutions are not able to detect targeted phishing.

References

1. Greene K, Steves M, Theofanos M (2018) No phishing beyond this point. Computer [online]. roč. 51, č. 6, s. 86–89. ISSN 0018-9162, 1558-0814. https://doi.org/10.1109/MC.2018.2701632
2. HOAX. Methodology for hazard assessment. In: Czech Metodika hodnocení míry nebezpečnosti [online] [vid. 8. červen 2020]. https://hoax.cz/cze/metodika-hodnoceni-miry-nebezpecnosti/
3. KNOWBE4. Security awareness training. KnowBe4 [online] [vid. 5. červenec 2020]. https://www.knowbe4.com
4. NIST. Phishing—glossary. CSRC. Phishing [online] [vid. 3. červenec 2020]. https://csrc.nist.gov/glossary/term/phishing
5. Steves MP, Greene KK, Theofanos MF (2019) A phish scale: rating human phishing message detection difficulty. In: Workshop on usable security: proceedings 2019 workshop on usable security [online]. Internet Society, San Diego, CA [vid. 3. červenec 2020]. ISBN 978-1-891562-57-0. https://doi.org/10.14722/usec.2019.23028

Scams Committed Solely Through Computer Technology, Internet and E-Mail

Miroslav Čermák and Vladimír Šulc

Abstract The main aim of this chapter is to describe how information technology is, in combination with social engineering, misused to lure a big sum of money from internet users just via e-mail, how the money transfer is performed, who is a victim, who is an attacker, what are the possibilities to investigate such a case, how to protect yourself, and what developments can be expected in this area of cybercrime.

Keywords SCAM419 · SPAM · Fraud · Crime · Samples

1 Introduction

Economic profit is given as the difference between costs and revenues, while always taking into account the risk that the entrepreneur is taking. This is no different in the case of SCAM419 fraud involving the sale of a non-existent product, whether goods or services, where the offender has minimal costs, high returns and minimal risk, which is why the number of these crimes is not declining and on the contrary, it is only growing with the expansion and availability of the Internet.

As of the SCAM419, it should be noted that this is not a completely new type of fraud, but a fraud centuries old. Its roots go back to the sixteenth century, when someone got the idea to start pretending to be a prisoner who has a treasure hidden somewhere, saying that he does not have enough money to pick it up.

In a letter written with a quill on paper, he addresses the recipient with a request for help, which should consist in sending a certain amount of money to a related person, who should pay him for the help after collecting the treasure. Needless to

M. Čermák (✉)
katedra managementu a informatiky, Policejní akademie České republiky v Praze, Lhotecká 559/7, P.O. Box 54, 143 01 Praha 4, Czech Republic
e-mail: cermak.miroslav@pacr.eu

V. Šulc
Ambis College, Lindnerova 575/1, 180 00 Praha 8-Libeň, Prague, Czech Republic
e-mail: lada.sulc@seznam.cz

© The Author(s), under exclusive license to Springer Nature Switzerland AG 2022
I. Tušer and Š. Hošková-Mayerová (eds.), *Trends and Future Directions in Security and Emergency Management*, Lecture Notes in Networks and Systems 257,
https://doi.org/10.1007/978-3-030-88907-4_17

say, the person who believed the letter and sent the money to the address never saw it again.

Later, by sending these letters for decent money, real prisoners came to France in the eighteenth century and also in England in the nineteenth century. In the twentieth century, these frauds began to occur all over the world. In the twenty-first century, with the development of the Internet, this type of fraud accelerated.

In principle, however, ~~only~~ this crime has shifted to cyberspace. Instead of written letters on paper and sending them by regular mail, massive e-mails are sent to a huge number of e-mail addresses.[1] Computer technology allows the offender to achieve significant economies of scale, as they do not have to buy paper, quills and ink and just write one letter, translate it into many different languages using the online Google translator and then send it to thousands of addresses at minimal cost.

Compared to the previous period, we also see a significantly lower level of quality of the language used, because if the perpetrator wanted to write a letter, he had to master the language at least at the elementary level. Here, however, we encounter machine translation, in which the quality of the message suffers considerably, for example when translating from English to Czech. On the other hand, according to some security experts, this may be the intention, as such a letter will seem much more credible and the perpetrator may assume that if the recipient of the email does not stop and think about the level of communication, then he will be less cautious in other stages of this fraud.

When it comes to withdrawing money from the banking system, the simple transfer of money between bank accounts and the use of white horses to withdraw cash still prevail. The total damage within one fraud ranges from several thousand to millions of crowns. Until 2014, Western Union and MoneyGram companies were actively used to transfer money. In the following years, various payment gateways and bitcoin wallets were used, because in their case it is very difficult to find their owner.

It is obvious that the perpetrator does not act under his own name, but only pretends to be someone, and as a rule it is not possible to verify his identity in any way or even meet him. The moment the offender is offered a personal meeting, he refuses it under some pretext, eg with reference to the fact that he is currently abroad, he has to deal with an urgent family matter, with regard to his or your safety, with the fact that he is monitored or that the transfer would be endangered, etc.

Its sole purpose is to lure money from the victim under various pretexts, from corruption of officials, payment of a lawyer, insurance, customs warehouse fee, a transport company, etc., threatening to report their activities to the authorities, or terminating communication with the victim and shutting down completely. He cancels the e-mail, phone and can no longer be contacted, and the victim will never see their money.

In addition to individuals, entire communities and gangs make a living from this type of fraud. False identities, documents, certificates, official documents, company websites are also created, which contain contact details and telephone numbers that can be called.

---

[1] Džubák [2].

In some cases, the user's email box is compromised, which the perpetrator impersonates or which confirms a fictional story. However, communication takes place not only via email, but also on social networks and various communication platforms, such as Skype, VoIP, SMS.

2 Types of SCAM

Over the last ten years, we have evaluated several thousand e-mail messages using automated tools and subsequent visual inspection, and we have come to the conclusion that there are countless variants of this fraud. In principle, however, these scams have certain features in common, so we tried to subsume individual SCAMs into the following categories:

- **Charity**—The perpetrator is under the auspices of an organization, stating that he is raising money to help orphans, protect endangered animals, stop logging, burn the rainforest, clean the ocean of waste, etc. In fact, all the money ends up in fraudsters accounts, so the victim will help, but to someone completely different than they wanted.
- **Favourable investment**—The perpetrator of the victim writes about the offer of a profitable investment, always abroad, in a country far enough away. Then he demands to pay some entry fee, acquiring a share in the form of shares, etc.
- **Request for assistance**—The offender turns on behalf of the person whose identity he appropriated with a request for assistance with payment, debt, medical expenses, departure from the country. It can be both a compromised account and a fake account of the victim's friends on Facebook or another social network.
- **Inheritance**—The perpetrator pretends to be a disgruntled bank employee who finds that one of the accounts he manages contains a large amount of money and that no one has registered for one year, and that you could declare yourself an heir and help him withdraw money from the bank. And since he realizes that you also take some risk when you decide to help him, and you incur some costs, he offers you a commission of up to several tens of percent of the transaction. The moment a victim joins the game and provides his or her identification and banking details, he or she becomes blackmailed because he or she obviously wants to take part in the crime of fraud, tax evasion or withdrawal from the state budget, which is affected by Nigerian legislation and its law no. 419, which also gave rise to the name of this fraud.
- **Persecuted dissident**—The perpetrator pretends to be a victim of persecution and prosecution in his country plagued by civil war, he has assets in a bank in a secret account, but he needs to transfer it to another account abroad, which he cannot do without the help of someone outside. And since he realizes that you, too, take some risk when you decide to help him, and you incur some costs, and you may be acting in violation of the law, he offers you a share of that hidden

property. Sometimes it happens that the war is long over and what he says clearly doesn't make sense.

- **Fictitious sale of goods**—The perpetrator on advertising servers offers, for an advantageous price, renting an apartment, selling a car, computer, animal or goods that are not so easy to find in a given country, or are offered at a significantly higher price. The perpetrator claims to be abroad and does not want to return back because of this and requires a deposit to be made to the intermediary. The perpetrator then charges a variety of fees.
- **Offer of goods**—The perpetrator responds to an advertisement offering certain goods and expresses an interest in purchasing them, but requires sending the goods, such as a laptop abroad, paying both postage and the price of the goods themselves.
- **Demand for goods**—The offender responds to an advertisement requesting certain goods and claims that he has the goods and sends them to a courier. He then changes his mind and demands payment in advance.
- **Unexpected prize and lottery**—The perpetrator addresses the victim on the grounds that he has been drawn, and that he can collect the prize, but only has to pay a fee, etc.
- **Participation in the selection procedure**—The offender addresses the victim with an offer of work in another country and the need to appear in person for an interview and reserve a place in a specific hotel, etc.
- **Bride over the Internet**—The perpetrator pretends to be a citizen, usually a woman from Russia, Ukraine, the Philippines, and claims that she has an education and that she wants to get out of her country because of the poverty there. For this purpose, she establishes a profile on social networks and addresses potential partners by e-mail, which, after a certain period of communication with them, asks for help with paying for the ticket, insurance, transportation of goods, medical examination, treatment, etc.
- **Threatening Attack**—The perpetrator threatens devastating long-lasting DDoS, which will have an impact on the functioning of society. In reality, however, it does not have the computational power to carry out an attack at all.
- **Extortion**—The perpetrator threatens to publish a video that captures the victim masturbating or reports it to the police, that he has stolen SW, video, audio, illegal pornography, etc. on his computer.

3 Research Questions

As part of our research, we asked ourselves several research questions about the perpetrators and victims of these crimes which are answered in this subchapter.

Where do the perpetrators of these crimes come from?

The perpetrators of these crimes are found practically all over the world, but in general most of them can be found in most African states, and some states of the former Union of Soviet Socialist Republics. It is typical for these countries that they are plagued by civil unrest, there is high unemployment and the associated significant crime, violent crimes are committed, there are restrictions on fundamental human rights and freedoms, and we can also encounter very poor law enforcement, a dysfunctional health and social system with a short life expectancy, a dysfunctional infrastructure, high internal and external indebtedness and considerable pervasive corruption, where the black and grey economy forms a significant part of the otherwise relatively low gross domestic product.

Who are the perpetrators of these crimes?

The perpetrators of these crimes are usually citizens of the above-mentioned troubled states who are on the brink of economic poverty and the proceeds of this crime allow them to survive in difficult conditions. This activity is also tolerated in their country and accepted by the society as a way of livelihood. These are often people who do not even have a basic education, but on the other hand have a relatively good command of the English language and information technology needed to commit these scams. They connect to the Internet via a mobile data connection or from public Internet cafes, which makes it difficult to catch them. These attackers work completely independently or as a part of smaller organized groups.

Even after being convicted of a lie after long communication, they insist on their version. In some cases, they state that the difficult economic situation lead them to commit these frauds.[2]

Who are the victims?

The victims are exclusively citizens of states and possibly continents other than the perpetrator himself, which is no coincidence, but intention. This is mainly because you cannot meet the offender, and also because the offender assumes that it is easier for the victims to believe what he writes, because they will not be able to verify it and will not want to risk their own health and life by a personal visit, especially if civil unrest or a military coup is taking place in the country, in which opponents are being liquidated. The victims who were caught in this way were men and women of all ages and backgrounds. Even a university degree does not seem to guarantee that the person will detect the fraud in time and not become another victim.

What is cooperation with law enforcement authorities?

Authorities active in criminal proceedings in the Czech Republic receive reports from citizens regarding these crimes in the form of notifications of the crime of fraud, when the citizen was misled as a victim and a considerable amount of funds were fraudulently lured from them. However, not every such fraud is reported, because the moment the victim realizes that they have been caught, they feel embarrassed and afraid of the reaction of those close to them. In some cases, they are afraid of accusations of being involved in organized crime and money laundering.

---

[2] Edwin [3].

At the moment when law enforcement authorities turn to their partner abroad in the framework of international cooperation, in the vast majority of cases, information is passed on and taken over, but the perpetrators never manage to be caught, charged and convicted. This is partly due to the fact that, of all criminal activities, this is the least harmful from their point of view, because the victim is not their citizen, but a citizen of another state. In addition, it is very difficult to trace the device from which the communication took place and who operated it at the time.

It is very difficult to get the necessary support from the state where the offender is located, in many cases it is not clear exactly who to turn to in a given country. To obtain this information, it is necessary to contact the consulates of other countries. Cooperation can in some cases be enforced under the United Nations Convention against Transnational Organized Crime. However, for more effective cooperation, it is necessary to conclude a bilateral agreement on legal aid in criminal matters. It takes up to several years to conclude such an agreement.

Even if the identity of the perpetrator can be established, attempts to solicit it always end in failure in virtually all African countries, with the exception of the northern states such as Morocco, Tunisia, Egypt and South Africa. From other states, no sent request was ever returned processed or its acceptance for processing was never confirmed.

What is the damage?

The proceeds from this crime are in the order of tens of millions of dollars a year, the damage starts at tens of thousands, but also reaches millions of Czech crowns in individual cases.

4 Real Samples with Comments

In this chapter are stated few real examples of email messages we analysed during our research.

Example No. 1—the oldest type of letter, which has been spread in various variations in the Czech Republic for more than 20 years. It should be noted here that this is a machine translation, which according to some security experts may give the recipient a certain impression of authenticity.

From: Kate Guadorres <velikovaa1948@gmail.com>
Sent: Saturday, February 1, 2020 11:30 AM.
Subject: Hello and how are you?
Hello and, how are you?
I am very sorry for this sudden contact. I received your e-mail address from the address book and will contact you personally because I needed your urgent help. My name is Kate Guadorres. I am 19 years old and I am the only daughter of my parents who were murdered by insurgents during the political crises in my country, Côte d'Ivoire. When my father was alive, he worked for an oil and gas company and invested some money, three million five hundred thousand euros (3,500,000,000

euros) with his name in his bank, before he died. I would like you to help me transfer this money to your bank account in your country for investment and also to help me come to your country to continue my education.

I am asking for your urgent help, because after the death of my parents, my evil uncle wants to kill me and collect his inheritance money from me by force, because I have no other relative who can help or defend me here in my country. I reported my evil uncle to the local police here in my country, but there has been nothing positive since then. I am writing this report from a local hotel, where I am currently hiding behind my safety, and I am willing to offer you 30% (EUR 1,050,000.00) of my inheritance funds as compensation for your post-transfer assistance. Please answer me urgently to send you more details …

Regards,

Kate.

Example No. 2—Another variant of a previous e-mail that refers to the relationship between the deceased and the recipient of the e-mail. In some cases, the sender also took the job of stating the name directly in the e-mail to increase credibility. These were automatically generated e-mails that used a general address and were again characterized by very poor Czech, which was the result of machine translation.

From: Lia Ahil <lahiali361@gmail.com>

Sent: Saturday, February 1, 2020 7:50 PM.

Subject: How are you today?

Dear friend

My name is Mission Lia Ahil (Attorney) I am from Manila in the Philippines. I need your help to raise funds for my late client ($ 8,500,000 in the US) and 250 kg of Gold deposited in the Bank Security Department for safe management. Keep in mind that your last name is similar to that of my late client's family, so I contacted you and helped me obtain unsolicited funds because the bank gave me an official mandate to submit the beneficiary's name to claim or they would seize and cancel the account if none answer.

If you would like more information about this mutual profit transaction, please contact me urgently. Our sharing ratio is 50% - 50%. He will send me your information.

(1) Full name ——————

(2) Skype address ————

(3) Mobile number ————

Regards

Lia Ahil.

Email …. (liaahil802@gmail.com).

Example No. 3—It should be noted here that high-quality Czech was used, and that the sender uses the awareness of the social network Facebook and its CEO Mark Zuckerberg, as well as the current coronavirus situation.

—Original Message—

From: Nadia Senhadji <Nadia.Senhadji@osu.cz>

Sent: Monday, May 11, 2020 9:11 AM

Subject: Good day..

FACEBOOK INC.
1601 WILLOW ROAD MENLO PARK, CA 94,025.
www.facebook.com
Dear Facebook user:
We would like to inform you that your Facebook account with e-mail received financial support in the amount of EUR 2,800,000.00 (two million eight hundred thousand thousand EUR) for the promotion of Facebook for the 2020 edition.
Congratulations on being one of the people you choose.
We encourage you to process your claim by clicking below on this direct link on Facebook and filling out the following information. If you can't click, copy and paste the address into your browser:
https://bit.ly/2LizPCR
Congratulations!! Once again.
For security reasons, we recommend that all winners keep this information confidential from the public until your request is processed and your prize is released to you. This is part of our security protocol to avoid double claims and to misuse the benefits of this program by non-participating or unofficial employees.
STAY HOME AND STAY SAFE. We can all fight coronavirus (COVID 19) TOGETHER.
President's Office.
CEO of Facebook.
Mark Zuckerberg.
Example No. 4—Here the e-mail is distributed in English, as the sender can address virtually anyone at minimal cost and it can seem more credible, as the e-mail looks like it is coming from a foreign company. Another advantage of using English is the fact that the sender usually speaks it, and often better than the recipient of the e-mail.
From: Clara González [mailto:cg4087494@gmail.com].
Sent: Thursday, April 23, 2020 3:38 PM.
Subject: Re:
Dear benefactor,
Final Notice of Payment of Unused Prize Money. We would like to inform you that the Spanish law firm has commissioned our law firm as a legal advisor to process and pay prize money credited in your name and has not claimed for more than two years.
The total amount you are entitled to is currently 3,540,225.33 EUROS.
The original prize money is 2,506,315.00 EUROS. This sum has now been invested in profit for more than two years, hence the increase to the above amount. According to the Office for Unclaimed Prize Money, this money was deposited with them as a non-claimed profit from a lottery company to be managed and insured on their behalf.
According to the lottery company, the money was awarded to them after a Christmas campaign. The coupons were bought by an investment company.
According to the lottery company, they were registered at that time to inform you of this money, but unfortunately no one had registered to claim the prize by the

specified date. This was the reason why the money was earmarked for administration. According to Spanish law, the owner must be informed of his current profit every two years.

If the money is not used again, the profit will be reinvested for another two years through an investment company. We were therefore instructed by the office of the unused prize money to write it down.

This is a notification for the use of this money.

We would like to point out that the lottery company checks and confirms whether their identity is identical before the money is paid out to them. We advise you on how you can make your claim. Please contact our German-speaking lawyer DR. Rodrigo De Barros TEL: + 35 130 881 2400.

E-MAIL:—info.rodrigodebarros01@gmail.com is responsible for withdrawals abroad and supports you in this matter. The claim should be submitted before 05-012-2020, otherwise the money would be reinvested.

We look forward to hearing from you and assure you of our legal support.

Please fill out the attached form and send it back to me to forward your data to the bank for processing.

Sincerely yours.

Clara González.

It is strange that although there were used obvious linguistic patterns [4] well known for couple of years,[3] those emails were not flagged as a SPAM and were delivered. Despite the provided examples, we see a certain shift in SCAM campaigns, which are becoming more and more sophisticated, of course not all of them, but in very specific cases. This also leads to a blurring of the boundaries between SCAM and BEC, where the difference lies only [1] in whether the email comes to a private or corporate email address and whether there is an effort to pump users for their own or corporate money.[4]

5   Conclusion

As part of our research, we have analysed SPAM identified as SCAM419 that was not detected by the SPAM filter and was delivered to end users' e-mail boxes. We have identified common features of these scams, and subsumed them.

As part of our research, we further identified the common signs of the perpetrators of these crimes and verified the possibilities of cooperation at the international level in detecting and apprehending the perpetrators of these crimes, which proved to be very limited. Given that this crime is tolerated in the countries where the perpetrators themselves are located and that there has been no significant shift in this area over the last decade, we believe that further research will need to focus more on technical measures to detect and block this type of spam.

---

[3] Schaffer [4].

[4] Čermák [1].

As for the frequency of occurrence of individual types of scam, they vary in different periods, however, the survey showed that some users receive a certain type of scam repeatedly. Therefore, in further research, it would also be appropriate to focus on the interests of the user and the website visited.

References

1. Čermák M (2020) Business email compromise. CleverAndSmart management consulting [online] [vid. 29. červenec 2020]. ISSN 2694-9830. https://www.cleverandsmart.cz/business-email-compromise/
2. Džubák J (2020) HOAX. Scam 419 [online]. https://www.hoax.cz/scam419/
3. Edwin G, americký voják z Bagdádu (2019) investigace.cz [online]. https://www.investigace.cz/gregg-edwin-americky-vojak-z-bagdadu/
4. Schaffer D (2012) The language of scam spams: linguistic features of "Nigerian Fraud" e-mails. ETC Rev Gen Semant. roč. 69, č. 2, s. 157–179. ISSN 0014–164X

Ransomware Attacks on Czech Hospitals at Beginning of Covid-19 Crisis

Jan Kolouch, Tomáš Zahradnický, and Adam Kučínský

Abstract The chapter describes cyber-attack at the Rudolph and Stephanie Regional Hospital in Benešov, authors' analysis of the attack, situation in smaller hospitals, and calls for a minimal cyber security standard. The attack and its consequences are described, along with actions taken by the Czech National Cyber and Information Security Agency. The chapter provides a qualitative analysis of the attack and issued measures in smaller hospitals. The level of ICT throughout hospitals was found very uneven mostly because of their technology debt and also because there is no minimum ICT security level they must meet, unless they are part of the critical infrastructure. Authors propose to establish a minimal cyber security standard for all essential service sector organizations, be they part of the critical infrastructure or not.

Keywords Critical infrastructure · Cyber security incident · Emotet · Ransomware · Ryuk · TrickBot · Cyber crime

Electronic supplementary material The online version of this chapter (https://doi.org/10.1007/978-3-030-88907-4_18) contains supplementary material, which is available to authorized users.

J. Kolouch (✉)
Department of Security and Law, Ambis College, Prague, Czech Republic
e-mail: jan.kolouch@law.muni.cz; jan.kolouch@ambis.cz

CyberCrime and Critical Information Infrastructures Center of Excellence (C4e), MUNI, Brno, Czech Republic

T. Zahradnický
Department of Systems Analysis, Faculty of Informatics and Statistics, University of Economics, Prague, Czech Republic
e-mail: tomas.zahradnicky@vse.cz

A. Kučínský
Department of Cybersecurity Regulation, The National Cyber and Information Security Agency, Brno, Czech Republic
e-mail: a.kucinsky@nukib.cz

1 Introduction

From December 2019 on, we witnessed a series of ransomware attacks on hospitals in the Czech Republic. The first attack began on December 11, 2019 at the Rudolph and Stephanie Regional Hospital in Benešov (HBEN). The IT infrastructure was recovering for 3 weeks and there was a loss of more than 2 million EUR. According to the hospital, there were, fortunately, no human casualties. Besides the HBEN, next victim, the St. Anne's University Hospital Brno (HBRNO) was paralyzed in March 2020, while other hospitals such as the University Hospital Olomouc, University Hospital Ostrava, or Hospital Pardubice Region resisted the attack. There is at least one case known in which a cyber-attack resulted in a human casualty [8]. The following sections describe the most probable attack vector and malware discovered during the analysis: Emotet, TrickBot, and Ryuk. Following the description, the attack experience on HBEN will be described. Next, a relationship of the HBEN, the Cybersecurity Act, and their relationship to the National Cyber and Information Security Agency (NCISA) will be discussed. Then a set of publicly known recommendations will be summarized. Finally, the main purpose of this chapter will be presented, a view on the measures taken at the operative level. It is evident that it is yet impossible to unify measures within a single sector or branch, but the purpose of the chapter is to present a critical analysis of processes applied during incident resolution. Based on the analysis, authors will present their conclusions de lege ferenda, which would contribute to a faster and more efficient handling of similar situations.

1.1 The Initial Attack

The incident at HBEN was preceded with a massive spear-phishing campaign aiming to plant a banking Trojan Emotet [14, 15]. Targets of the campaign received an e-mail with an attached macro-enabled Microsoft Office document. The e-mail (i) originated from a person the target might have known because of the distribution mechanism described below, and (ii) was written in good language. Both of these two facts raised credibility of the e-mail and increased probability of opening the attachment by an unsuspecting user. Once the attached document was opened, Microsoft Office application launched, showing a security warning "Macros have been disabled." allowing the user to override the warning and execute the macro, unless configured otherwise in Microsoft Office's Trust Center [17, 18] or with a group policy after installing Administrative Templates files (ADMX/ADML) and Office Customization Tool for Microsoft 365 Apps for enterprise, Office 2019, and Office 2016 [17, 18]. When the user overrode the warning by hitting the "Enable Content" button, the malicious macro within the document ran further executing a PowerShell script which in turn downloaded Emotet from the Internet, running it, and starting off the first stage of the infection. There are other possibilities for Emotet installation, such as running a stand-alone infected script or downloading its executable directly by

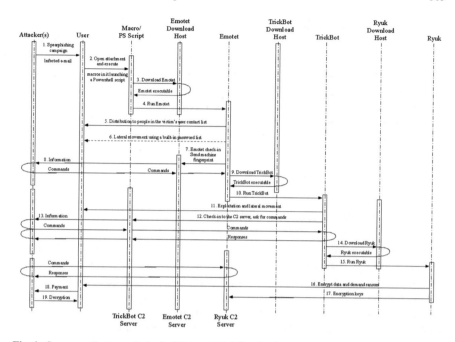

Fig. 1 Sequence diagram of a typical Emotet-TrickBot-Ryuk attack. The attack starts with a spear-phishing campaign (1) aiming at users to open an infected Microsoft Office document. Once the malicious macro inside the document gets executed (2), Emotet is downloaded from the Internet (3) and launches (4), distributing itself to any users' contacts (6) and trying to move laterally on the network (7). Emotet also sends a machine fingerprint (8) to the C2 server waiting for instructions. When instructions to plant TrickBot come from the Emotet command and control (C2) server, TrickBot is downloaded from another server (9) and run on the machine (10). TrickBot moves laterally using exploitation (11) and check-ins to its C2 server (12), asking for commands. In this phase TrickBot harvests internet history and credentials, and opens a network for remote access and hacking. Communication with the C2 server is performed in a loop (13). Ultimately, the TrickBot C2 decides to deploy Ryuk and downloads Ryuk from the Internet (14) and runs it (15). Ryuk encrypts user data (16) and drops a ransom notice to the user. Encryption keys are sent to the Ryuk C2 server (17). If the user decides to pay the ransom (18), the attacker provides him with a decryption tool (19). Note the attacker column is discontinuous, as there may be different cybercriminal groups controlling each of the C2 servers

accessing a malicious link in e-mail. The attack on HBEN was carried out with Emotet, TrickBot, and Ryuk malware cascade, as depicted in Fig. 1.

1.2 Emotet

Emotet [14, 15, 20, 27] is a Windows banking Trojan originally created to steal banking information from the victim's computer. It evolved over time to spread itself via malicious e-mail. Emotet distributes itself to people in the victim's contact list through e-mail sent from the victim's hijacked account. It also tries to use around a thousand common passwords [16] to log-in into network services accessible over LAN to spread laterally. Emotet also collects a list of running processes and other machine information [11] and sends it to a command and control server (C2) using RSA/AES-128 encryption. The C2 server decides which other malware to deploy and instructs Emotet to download and install it [14, 15]. In the HBEN case another Trojan called TrickBot was downloaded and installed.

1.3   TrickBot

TrickBot [21] is a Windows banking Trojan, often deployed by Emotet. It is capable of antivirus deactivation, credentials and data theft, lateral movement, providing remote access, establishing connection to a C2 server, installing other malware, and maintaining its persistence in the attacked system [2, 29]. TrickBot accomplishes its functionality with a set of modules [27], each with a unique purpose. Center for Internet Security [2] divides these modules, having form of dynamically linked libraries, into 4 main classes: (i) Banking information stealers, (ii) System/Network reconnaissance, (iii) Credential and user information harvesting, and (iv) Network propagation.

Banking information stealers. Modules in this class use various injection techniques to attack a user's web browser and get into the Man-in-the-Browser (MitB) position. The user can be attacked by a web injection, e.g. Trickbot adds extra fields or pop-ups to the pages being browsed, be redirected to a similarly appearing website provided by the attacker, or have added some extra client code to responses from the server. Harvested information is then submitted to an external IP address [2]. TrickBot is also capable of tunneling user's web connection through a local SOCKS proxy, or provide a VNC service for the attacker to enable him user observation capability through its VNC module [27].

System/Network reconnaissance. Modules in this class aim at harvesting information from the attacked system itself. This includes information queried by using Windows Machine Instrumentation (WMI) queries, information about the current user, amount of RAM, and a list of all running processes [29].

Credential and user information harvesting. Modules in this class gather data from installed web browsers including web form autofill information, cookies, and browsing history. They also try to harvest credentials from domain controller, Microsoft Outlook, VNC, Remote Desktop, PuTTY, and from the Windows Local System Authority SubSystem (lsass.exe) process using the Mimikatz [7] tool. Mimikatz is a tool capable of extracting plain text passwords, hashes, PIN codes, and Kerberos tickets from lsass.exe by examining its memory image.

Network propagation. These modules provide capability for multiplication and for lateral movement over the network. The Server Message Block (SMB) and Lightweight Directory Access Protocol (LDAP) are used for this purpose. SMB access is supported with abuse of the SMB protocol version 1 (SMBv1) EthernalBlue vulnerability MS17-010 [1, 19].

Similarly, like Emotet, TrickBot is also capable of installing other malware. In the HBEN case the Ryuk ransomware was installed, triggering encryption of computer hard drives throughout the network, causing its disruption. Although it is not known to the author how long it took in the HBEN case for the TrickBot C2 server to deploy Ryuk, [12] estimates the average time might have been 2 weeks. During this period, until Ryuk was deployed, the network was accessible for remote access, manual hacking, data theft, and lateral movement. Generally, we could accept that the attacker's motive could have been qualified as financial—either to steal and market potentially sensitive data and/or to collect ransom by deploying Ryuk.

1.4 Ryuk

Ryuk [9, 22, 34] is a Windows ransomware, often deployed after previous infection with TrickBot. Ryuk is capable of process injection where it copies and runs its payload into other running processes, making it difficult to remove once it completes. Next, Ryuk performs process and service termination of processes and services matching a built-in list disabling, for example, antivirus software and backup software services (cf. [34, Appendixes C and D] or [9] for complete lists). Another capability is to delete data shadow copies, making data recovery more difficult. The ransomware also establishes its persistence by writing itself into the Run registry key under the current user. To move laterally over the network, Ryuk reads the host's ARP table and uses MAC addresses found there to send them a Wake-on-LAN datagram. If a device receiving this datagram is configured to do so, it wakes up from the sleep and becomes another possible target Ryuk can spread onto laterally. Ultimately, Ryuk iterates all local and mapped network drives and encrypts all files located there not matching a list of file names and an extensions blacklist. Network drives on the LAN are tried to be accessed with user's privileges using a UNC path consisting of an IP address and a drive letter followed by a dollar sign. Encryption is performed using a combination of RSA and AES algorithms and the key is uploaded onto the C2 server.

Ryuk also leaves a file at the user's Desktop containing an e-mail address, typically registered at @protonmail.com or @tutanota.com for further decryption instructions. Despite the decryption ratio being 93% according to [3], organizations like The National Cyber and Information Security Agency (NCISA) do not recommend to pay it [5], advising the user to prefer recovery from backup.

1.5 Infection at Rudolf and Stephanie Regional Hospital Benešov

According to information shared by the hospital at a workshop [31], which has one of the authors attended, the Ryuk encryption started on December 11, 2019 at 2:50 a.m. with a malfunctioning application. Minutes later, a ransomware attack was suspected. The network infrastructure started to shut down to a bare minimum, including the domain controller. The hospital's crisis management team meeting took place analysing the situation, contacting the Governmental CERT (operated by NCISA), ESET spol. s r.o., an antivirus company, and the Police of the Czech Republic. There was no ransom note found, only an e-mail contact. Both ESET and NCISA sent their expert teams to help and to acquire data samples. After performing analysis of data samples, they provided a set of recommendations to the hospital. In subsequent days, 600 PCs and servers were reinstalled and their data restored from backups. On December 18, first systems returned to their normal operation and on December 31, all systems were back functioning again. The hospital operations were paralyzed for almost 3 weeks and the loss was estimated to be 60 million CZK (2.4 million EUR).

2   Recommendations, Measures, and Offers as a Respond to Ransomware Attack

In HBEN's case, NCISA provided a list of recommendations to the hospital after an analysis of the samples, but did not issue any warning or any other measure until the attack on a OES-hospital in March. After the attack on the St. Anne's University Hospital Brno on March 12, 2020 NCISA issued a reactive measure (RM) and a methodology on March 17, 2020, followed by a recommendation and another methodology distributed to select hospitals on March 18, 2020, not only the ones in OES. Though it is not possible to describe either of these documents in detail, RM defined 4 sets of tasks to be performed immediately, in 2 days, in a week, and in 2 weeks. A summary of tasks in the RM [32]:

1. To be performed immediately:

    a.   Block systems intercommunication except for necessary cases
    b.   Block communication to the Internet except for necessary cases
    c.   Isolate the network of medical devices from rest of the network
    d.   Change passwords of privileged accounts
    e.   Report IP address ranges used by the hospital to NCISA.

2. To be performed in 2 days:

   a. Move backups to off-line, verify their functionality
   b. Check business continuity management plans and move them off the systems
   c. Avoid deletion of data about security incidents
   d. Check specified indicators of compromise (IoC)s
   e. Warn employees about the risk of phishing.

3. To be performed in a week:

   a. Check that backups are separated so even a privileged administrator could not delete them
   b. Ban unsigned macro use, if possible
   c. Check network segmentation and intersegment controls
   d. Harden security policies of endpoints (ban unapproved applications, unsigned PowerShell scripts, ...)
   e. If there is no business continuity management, create it at least for key systems
   f. Assure a vulnerability scan (NCISA can perform the scan).

4. To be performed in 2 weeks:

   a. Deploy antivirus software on all relevant devices
   b. Consider deploying upgrades, test, and deploy them.

   Recommendation distributed on March 18, 2020 can be divided into two sets—administrative and technical measures. A summary of these:

1. Technical measures

   a. Segment the network isolating core services provisioning systems from other systems. Intersegment traffic shall be controlled and evaluated.
   b. When not necessary, do not connect systems to the Internet.
   c. For medical devices, modalities, create a special network segment and isolate it from rest of the infrastructure.
   d. Have a privileged account use policy. Avoid using privileged accounts for common system operation and restrict their use to the minimum.
   e. Make backups, test their functionality, and keep backups offline. Production systems shall not be able to access backups even with the highest privileges.
   f. Ban execution of unapproved applications.
   g. Ban execution of unsigned PowerShell scripts.
   h. Regularly make a vulnerability scan of applications accessible from the Internet.
   i. Use antivirus protection and update it regularly wherever possible.
   j. Keep your systems up to date, wherever possible.

2. Administrative measures

    a. Have business continuity management plans calculating with a loss of information systems' availability. Each system should have its priority assigned based on its importance for operations and have recovery plans.
    b. Discover new threats and vulnerabilities and make appropriate measures.
    c. Train employees in cybersecurity. Focus especially on phishing and Microsoft Office macro use.
    d. Avoid using unsigned macros.

On March 27, 2020 NCISA offered vulnerability monitoring services for select hospitals and e-learning classes for hospitals in OES. CZ.NIC offered a free Turris Mox router to help with the network segmentation recommendation, including training and deployment assistance.

On April 16, 2020 NCISA issued a warning [24, 25] of imminent attacks aimed at hospitals and other Czech significant targets. The warning was followed on April 17, 2020 with a recommendation [24, 25] and recalled on May 20, 2020. The recommendation aimed to avoid or mitigate impact of an imminent cyber security incident and contained the following:

1. Essential information to the recommended processes in the warning

    a. Alert users about the spear-phishing threats and encourage those who opened suspicious attachments to contact their infrastructure administrators and alert users about masking of executable files used in phishing such as "picture.png.exe", "text.txt.exe", or "document.pdf.exe";
    b. Immediately block remote infrastructure accesses and block services opened to the public network, unless required;
    c. Immediately create offline backups and backup data based on their importance for the organization;
    d. Check consistency of existing backups;
    e. Update antivirus solutions in the infrastructure;

2. Additional recommendations that can be performed to prevent or mitigate impact of a cyber security incident

    a. Change passwords for privileged accounts. Check/Set appropriate policy for privileged account use;
    b. Check and assure that the backup system is isolated from other systems so that backups of a system cannot be deleted, even with the highest privileges to that system;
    c. Block access and interconnection in between organization-vital systems and systems and networks that are not important for service provisioning or system security;
    d. Check network segmentation and controls, evaluate the situation, and take appropriate measures to establish at least elementary segmentation;
    e. Consider updating all used systems under the condition that the update was tested. If an update is tested and functional, perform it;

f.   Check existing business continuity plans and emergency plans related to systems and check their validity, efficacy, and usability especially with request to possible inaccessibility of these systems;

g.   Assure business continuity and emergency plans related to systems operations are stored outside of systems for which they are made for;

h.   If the business continuity or emergency plans are inaccurate or do not exist, create these plans at least for the most critical service provisioning systems;

i.   Do not delete any data about a cyber security incident without authorization of the Police of the Czech Republic or NCISA, and educate all administrators and all relevant security and IT operation roles;

j.   For health care providers only: Isolate medical devices network – modalities (CTs, X-Rays), from rest of the network;

k.   Generic NCISA recommendations for administrators.

3   Analysis of Health Care ICT

The conclusions in this section are based on authors' own analysis of the HBEN incident, evaluation of NCISA's recommendations and measures, applicable law, and by consulting experts in cyber security, from the Governmental CSIRT, and ICT managers in select hospitals. Based on these analyses and consultations, we define the following areas forming real cyber security risk for the vast majority of healthcare providers.

3.1   ICT

1.   Often obsolete and unsupported ICT equipment;

2.   uneven ICT level throughout health care providers, yet it is unreal to harmonize it in the near future with the current approach;

3.   lack of investments into ICT development and maintenance further contributes to an already very high technical debt;

4.   insufficient redundancy in backup systems (data centers, server rooms, etc.);

5.   black boxes (appliances provided for laboratory work, X-Rays, CTs, etc.) that must be connected to the network but cannot be managed by the organization nor their connection to the Internet restricted to avoid losing support or warranty from their vendors;

6.   data mining from internal security surveillance systems and the ability to share with others through a common platform;

7.   reactive measure, not only in case of healthcare providers, often "hits" a barrier in an inability to create off-line backups, while off-line backups are one of the standard data security requirements. Backing data up to cloud typically operated by Amazon, Google, or Microsoft is not real, even in case of encrypted data.

3.2 Administration

1. Missing administrative directives;
2. unclear definition of roles in case of an ICT incident;
3. rules for system information/knowledge sharing;
4. lack of fast communication;
5. business continuity management;
6. other recommendations mentioned in Sect. 4. Recommendations, Measures, and Offers.

3.3 Human Resources

1. Lack of IT staff employees trained for health care;
2. low computer literacy of medical staff (doctors, nurses, laboratory staff) especially in computer security.

4 Evaluation of the Measures and Recommendations

According to the FBI's Internet crime report for 2019 [10], expenses spent for recovery from cyber-attack inflicted damages (either by performing recovery from a backup or by paying a ransom) topped 3.5 billion USD in 2019 a billion more than in 2018. For HBEN, the loss was estimated to be 60 million CZK/2.6 million USD/2.4 million EUR. We see that the number of attacks grows each year, and from HBEN's case we can expect a loss of more than tens of millions CZK per a non-OES hospital. Medical care providers as well as other sector organizations can be therefore expected to be under a higher load of attacks in the future and should prepare while they still can.

The NCISA's can act only when law allows it. Their competence is strictly defined by the Cybersecurity Act and the Decree no. 82/2018 Coll. on Security Measures, Cybersecurity Incidents, Reactive Measures, Cybersecurity Reporting Requirements, and Data Disposal (the Cybersecurity Decree). This explains why the RM could have been issued after the attack on HBRNO (part of OES), but not after the previous attack on HBEN (not part of OES) 3 months earlier. Despite that, HBEN still received assistance and recommendations from NCISA shortly after the attack.

The measures and recommendations summarized above were meant with good intention yet they may be difficult to implement. When we look at the recommendations more closely, some of them may be difficult to implement within a very tight time frame set on by the RM. For this reason, even few of the hospitals in OES did not implement all of them [32]. Hospitals in OES have to fulfil stringent security requirements laid on by the Cybersecurity Act and are likely to implement the RM faster than non-OES hospitals. Non-OES hospitals, who have received the recommendations, are not required to meet any minimal security standard and would need help with the implementation of measures to bring their infrastructure into the recommended state.

Mr. Dušan Chvojka, a head of ICT in Hospital Bulovka proposes the state could establish a common security center to oversee and protect cyberspace of the hospitals. He compares the proposed center to the Air Navigation Services of the Czech Republic operating the aerospace. Hospitals would not have to purchase security operations centers on their own as demanded by the Cybersecurity Act but could rent its services in a similar way as they rent physical security employees now (Hospodářské noviny 2020).

It is a question whether the proposed solution is conceptual or it is a temporary solution intended to overcome a long unresolved ICT situation in the health care sector. The authors believe that the proposed solution would encounter major difficulties accounting for specifics of each health care provider, mostly because of their technology debt.

We should also ask which other sectors could demand a similar solution, with the same reasoning. Using a common center with ICT security outsourcing could paradoxically lead to a situation where there will be no ICT security staff in the hospital capable of resolving the incident as they could only wait for intervention from the center.

With the above claims, authors do not try to call for nor approve reduction of ICT service outsourcing possibilities. In case of solid ICT architecture, *reasonable* outsourcing can significantly increase security of the system and data within. ICT services whose outsourcing could benefit the hospitals are, for example:

1.   Honeypots
2.   IPS and IDS systems
3.   Content Management Systems
4.   Certified VPNs.

An appropriate approach appears to define a minimal standard that could be used to build and maintain ICT throughout sector organizations. Rules within the standard would contribute to reduction of probability of a successful cybersecurity attack.

Minimal cybersecurity standard. While hospitals in OES have to meet security requirements of the Cybersecurity Act in addition to other requirements, non-OES hospitals have no such obligation. Their obligations originate only from requirements defined by the Ministry of Health, government, and national and international healthcare legislation and standards. There is no minimal cybersecurity standard (MCS) that would be mandatory for healthcare providers. The same applies in other sectors named in the Decree no. 437/2017 Coll. Annex 1. When a subject does not qualify in essential service criteria set on by the Decree, it does not need to meet any cybersecurity standard. If there is no MCS assumed by sector subjects, recommendations and measures such as the RM above may be difficult to implement, as they may require larger changes to network infrastructure. Naturally, if there is a cybersecurity standard in the Cybersecurity Act (CACS), we could ask if the existing standard shall apply also to non-OES subjects, not just hospitals but all sector organizations, as if they were part of the OES. Consider now a situation that CACS were mandatory for subjects not in the OES. On one hand NCISA could have assumed a common cybersecurity level throughout the sector industry and it would be easier to regulate the industry since possible heterogeneity would be reduced. On the other hand, non-OES subjects would have to bear additional significant expenses to the ICT to meet the level required by the CACS. The same situation would apply if the essential service criteria were loosened, as there would be much more subjects under CACS. Could NCISA with their current capabilities still regulate the subjects if their number grew 10 or more times? The regulator would have to expand. For this reason, the MCS appears to be a part that would put all sector organization's ICT at a common minimal level without the necessity of the regulator to expand nor to put significant expenses into the ICT.

Clear and transparent MCS one for each sector or a common one for all sectors could help since they would enable:

1. easier implementation of technical and administrative measures;
2. easier and systematic information/knowledge sharing and application.

In the case of an acute case, respect the prepared rules, but not always formally adhere to them (A strictly formal approach can often make the resolution of an entire incident worse). In case of a cyber security incident, it holds that faster and ideally coordinated communication is more appropriate than communication performed later with precision and without errors.

5 Conclusion

After analysis of the cyber security incidents at Czech hospitals occurring during the COVID-19 pandemic and discussion with medical care cyber security experts that were unaffected by the attack, we have to conclude that there is no "collective intelligence". There is no platform health care providers (and other sectors) could share data on collectively to learn from and to apply experiences of others.

As an appropriate collective intelligence solution, authors propose to create a cyber-attack information coordinator, perhaps per sector, within NCISA or the National CSIRT team. It is a question whether these organizations are understood as a trustworthy partner for the hospitals and other sector organizations, as trust can be built by active approach to share data about attacks and by presentation of appropriate measures, not by using directive approach.

Another long-term underestimated and underfinanced area beside health care is education. NCISA prepared a good educational campaign about spear-phishing and how to defend against it [6]. Authors believe that NCISA should continue in the same direction and use the current attacks as a use-case to prepare further educational material.

From our own experiences, we can tell that e-learning does not work very well. For the user to learn a lesson, he needs to be attracted by the material and/or become part of it. It is thus necessary to present real attacks and their real impact, or simulate attacks so the users can experience them on their own and feel the consequences. Mistakes committed by the victims can be then used to give recommendations on how to behave in similar situations in the future.

There is no golden rule how to deal with a cybersecurity incident or with technology malfunction with service unavailability as a consequence. Due to this there are various regulations, suggestions, or advice from various subjects that may be opposing one another. The solution to this is cooperation, communication, and endeavor to resolve the problem.

Acknowledgements The authors would like to express their gratitude to the Rudolf and Stephanie Regional Hospital Benešov for sharing their sad experience with the attack at Ransomware Ryuk & Nemocnice Benešov 12/2019 Workshop.

This research/work/article was supported by ERDF "CyberSecurity, CyberCrime and Critical Information Infrastructures Center of Excellence" (No. CZ.02.1.01/0.0/0.0/16_019/0000822).

References

1. Avast (2020) What is EternalBlue and why is the MS17-010 exploit still relevant? [2020-08-18]. https://www.avast.com/c-eternalblue
2. Center for Internet Security (2019) Security primer September 2019—TrickBot [2020-08-18]. https://www.cisecurity.org/wp-content/uploads/2019/03/MS-ISAC-Security-Primer-Trickbot.pdf
3. Coveware, Inc. (2019) Decrypt Ryuk ransomware. How to recover Ryuk encrypted files [2020-08-20]. https://www.coveware.com/blog/decrypt-ryuk-ransomware-guide-to-recovery-encryp ted-files
4. Czech Ministry of Health (2016) Traumacentrums. In: Czech: Traumacentra [2020-08-24]. https://www.mzcr.cz/traumacentra/
5. Czech National Cyber Security Center (2019) [2020-08-21] Emotet-TrickBot-Ryuk threat warning. In: Czech: Varování o hrozbě Emotet-TrickBot-Ryuk. https://www.govcert.cz/cs/inf ormacni-servis/hrozby/2721-varovani-o-hrozbe-emotet-trickbot-ryuk/
6. Czech National Cyber Security Center (2020) Spear-phishing and how to defend against it. In: Czech: Spear-phishing a jak se před ním chránit [2020-09-15]. https://nukib.cz/cs/infoservis/ doporuceni/1514-spear-phishing-a-jak-se-pred-nim-chranit/
7. Delpy B, Le Toux V (2020) Mimikatz [2020-08-19]. https://github.com/gentilkiwi/mimikatz/ releases
8. Deutsche Welle (2020) German police probe 'negligent homicide' in hospital cyberattack [2020-09-18]. https://p.dw.com/p/3ieQl
9. Elshinbary A (2020) Deep analysis of Ryuk ransomware [2020-08-20]. https://n1ght-w0lf.git hub.io/malware%20analysis/ryuk-ransomware/
10. Federal Bureau of Investigation (2020) 2019 Internet Crime Report [2020-09-15]. https://pdf. ic3.gov/2019_IC3Report.pdf
11. Fox N (2019) Emotet deep dive analysis [2020-08-19]. https://neil-fox.github.io/Emotet-Deep-Dive-Analysis/
12. Ilascu I (2020) Ryuk ransomware deployed two weeks after Trickbot infection [2020-08-20]. https://www.bleepingcomputer.com/news/security/ryuk-ransomware-deployed-two-weeks-after-trickbot-infection/

13. Kolouch J, Basta P, Kropacova A, Kunc M (2019) Cybersecurity [2020-08-21]. https://knihy. nic.cz/files/edice/cybersecurity.pdf
14. Malwarebytes, Inc. (2018) Malware analysis: decoding Emotet, part 1 [2020-08-17]. https:// blog.malwarebytes.com/threat-analysis/2018/05/malware-analysis-decoding-emotet-part-1/
15. Malwarebytes, Inc. (2018) Malware analysis: decoding Emotet, part 2 [2020-08-17]. https:// blog.malwarebytes.com/threat-analysis/2018/06/malware-analysis-decoding-emotet-part-2/
16. McLellan M (2018) Lazy passwords become rocket fuel for Emotet SMB Spreader [2020-08-19]. https://www.secureworks.com/blog/lazy-passwords-become-rocket-fuel-for-emotet-smb-spreader
17. Microsoft Corp. (2020) Administrative templates files and office customization tool for Microsoft 365 Apps and enterprise, Office 2019 and Office 2016. English version [2020-08-17]. https://www.microsoft.com/en-us/download/details.aspx?id=49030
18. Microsoft Corp. (2020) Enable or disable macros if Office files [2020-08-17]. https://support. microsoft.com/en-us/office/enable-or-disable-macros-in-office-files-12b036fd-d140-4e74-b45e-16fed1a7e5c6
19. Microsoft Corp. (2017) Microsoft Security Bulletin MS17–010—Critical [2020-08-18]. https:// docs.microsoft.com/en-us/security-updates/security-bulletins/2017/ms17-010
20. Mitre Corp. (2020) Emotet [2020-08-19]. https://attack.mitre.org/software/S0367/
21. Mitre Corp. (2020) TrickBot [2020-08-19]. https://attack.mitre.org/software/S0266/.
22. Mitre Corp. (2020) Ryuk [2020-08-19]. https://attack.mitre.org/software/S0446/
23. National Cyber and Information Security Agency (2019) Emotet-Trickbot-Ryuk warning. In: Czech: Varování o hrozbě Emotet-Trickbot-Ryuk [2020-08-24]. https://nukib.cz/cs/infoservis/hrozby/1478-varovani-o-hrozbe-emotet-trickbot-ryuk/
24. National Cyber and Information Security Agency (2020) Cyber attack warning against attacks at hospitals and other significant targets in the Czech Republic. In: Czech: Varování před hrozbou kybernetických útoků na nemocnice a jiné významné cíle ČR [2020-09-10]. https:// nukib.cz/download/uredni_deska/Varova-ni_NUKIB_-2020-04-16.pdf
25. National Cyber and Information Security Agency (2020) Recommended security measures to the warning from April 16, 2020. In: Czech: Doporučená bezpečnostní opatření k varování ze dne 16. dubna 2020 [2020-09-10]. https://nukib.cz/download/uredni_deska/Doporuceni_k_varovani_2020-04-17.pdf
26. Nemocnice Rudolfa a Stefanie Benešov, a.s. (2019) Annual report for business year January 1, 2019–December 31, 2019. In: Czech: Výroční zpráva za hospodářský rok 1. 1. 2019 – 31. 12. 2019 [2020-08-24]. https://www.hospital-bn.cz/wp-content/uploads/2020/05/V%c3%bdro%c4%8dn%c3%ad-zpr%c3%a1va-rok-2019.pdf
27. Pinkas N, Rochberger L, Zatz Matan (2019) [2020-08-21]. https://www.cybereason.com/blog/triple-threat-emotet-deploys-trickbot-to-steal-data-spread-ryuk-ransomware
28. Polčák R (2013) Explanatory memorandum to the Cybersecurity Act, p 70 [2020-09-09]. https://www.govcert.cz/download/legislativa/container-nodeid-708/nbu-zkb-navrh-130415-duvodzprava.pdf
29. Robinson M (5/2019) TRICKBOT—Analysis [2020-08-18]. https://www.sneakymonkey.net/2019/05/22/trickbot-analysis/
30. Robinson M (10/2019) TRICKBOT—Analysis Part II [2020-08-18]. https://www.sneakymonkey.net/2019/10/29/trickbot-analysis-part-ii/
31. Spolek pro ochranu osobních údajů (2020) Ransomware Ryuk & Hospital Benešov 12/2019 Workshop. In: Czech: Ransomware Ryuk & Nemocnice Benešov 12/2019 Workshop
32. Tate International—Metamorfosa (2020) Obligate steps to safe health care industry—X days after cyber attacks at hospitals at Benešov, Brno, Ostrava, and others. In: Czech: Nutné kroky k bezpečnému zdravotnictví – X dní po kyberútocích na nemocnice v Benešově, Brně, Ostravě a dalších [2020-06-11] Workshop
33. St. Anne's University Hospital Brno (2020) Annual report for year 2019 [2020-08-26]. https:// iweb3.fnusa.cz/wp-content/uploads/VZ_2019-2.pdf
34. Zeligson A, Rubinfeld A (2020) Ryuk revisited—analysis of recent Ryuk attack [2020-08-19]. https://www.fortinet.com/blog/threat-research/ryuk-revisited-analysis-of-recent-ryuk-attack

Cybersecurity Insurance as the Method of Protection Against Cyber-Risks

Petr Dobiáš

Abstract This chapter is dealing with use of insurance as the method for mitigation of the cyber risks. The significance of the cyber insurance as the protective measure against cyberattacks is recently underestimated. Business circles are still not aware of the danger connected with the utilisation of the unsecured modern computer networks and technologies. Author of the chapter is analysing development and contents of the cybersecurity insurance.

Keywords Cybersecurity · Insurance · Mitigation · Protection · Risk

1 Introduction

Cyber security insurance is currently of interest not only to insurance companies and policyholders from the business and non-business community, but also to governmental and non-governmental international organisations. This proposition can be demonstrated on documents issued by European Union institutions (The European Insurance and Occupational Pensions Authority—EIOPA and The European Union Agency for Cybersecurity—ENISA) and the German Insurance Association (hereinafter the "GDV"), which will be part of the analysis conducted in this chapter. At the same time, it applies that it can be a combination of property, legal protection, liability and financial loss insurance. This is not surprising, given that in the event of damage to machinery at an entrepreneur's production plant, data from the entrepreneur's computer equipment may also be stolen, his reputation in the eyes of other entrepreneurs may be damaged, as well as the loss of existing contracts for the supply of goods. An overview of the types of cyber incidents and the resulting losses is provided in the report issued by the Organization for Economic Co-operation and Development entitled Enhancing the Role of Insurance in Cyber Risk Management [1]. An entrepreneur can insure himself against these insurance perils by taking out a suitable insurance, whereby he must first examine the exclusions from the

P. Dobiáš (✉)
Department of Private Law, CEVRO INSTITUT, Jungmannova 28/17, Prague 1, Czech Republic
e-mail: petr.dobias@vsci.cz

© The Author(s), under exclusive license to Springer Nature Switzerland AG 2022
I. Tušer and Š. Hošková-Mayerová (eds.), *Trends and Future Directions in Security and Emergency Management*, Lecture Notes in Networks and Systems 257,
https://doi.org/10.1007/978-3-030-88907-4_19

insurance cover and the maximum amount of the insurance benefit. In the case of household insurance, the extent of property damage may not be as large as in the case of entrepreneurs, but defamatory effects may arise in the event of theft and misuse of sensitive data (cybergrooming—contacting a person through channels such as social networks in order to arrange a meeting, at which the victim is usually subjected to manipulation or abuse, cyberbullying—harassment, intimidation or ridicule of another person using information technology, cyberstalking—monitoring and pursuing a certain person, which usually increases in intensity, etc.—cf. [2, p. 309]. A specific feature of cyber risk insurance is the fact that a cyber-attack can be carried out from a place that is far from the place where the effects of the cyber-attack manifest themselves (cf. [3, p. 32]). The fact that not even the European Union is underestimating the risk of cross-border cyber-attacks and their impact on Member States is evidenced by its recent legislative activities [cf. Commission Recommendation (EU) of 13.9.2017 on Coordinated Response to Large Scale Cybersecurity Incidents and Crises, C/2017/6100, Official Journal of the European Union L 239, 19.9.2017, pp. 36–58] and the activities of Europol and ENISA [4]. In cases well known from the media (e.g. cyber-attacks conducted by the Carbanak group against banks in various countries), it is possible to prove that the victim of a cyber-attack may be located in a different state than the computer from which the cyber-attack was carried out. In this context, it should be noted that terrorist cyber-attacks are usually excluded from insurance cover in the insurance terms and conditions of domestic and foreign insurers. This conclusion can also be supported by statistical data [5, p. 7].

Cyber insurance is provided by multinational insurance companies, which provide their products under similar insurance terms and conditions through their subsidiaries, branches and intermediaries in several different countries, where they are able to fulfill their obligations under the insurance contract in the event of an insured event. The problem is that insurers usually offer the same insurance terms and conditions for cyber risk insurance in different countries, although, in practice, this type of insurance requires adapting the conditions to the individual needs of the policyholder and the insured. This situation is to some extent surprising, because as Kalinich correctly states and proves on theoretical models, it is in this area that an insurer who can offer its customers a tailor-made offer should have a competitive advantage [6, p. 153]. Nevertheless, when negotiating, it can be assumed that insurers will be willing to offer individual insurance terms and conditions in the case of the insurance of large risks.

In view of the growing number of cyber-attacks that are occurring worldwide, cyber risk insurance is one of the most important ways to protect the modern information society. The importance of insurance for minimising the consequences of cyber-attacks is also confirmed by the brief of the project of the Organization for Economic Co-operation and Development [7], which states that: "Cyber risks pose a real threat to society and the economy, the recognition of which has been given increasingly wide media coverage in recent years. Cyber insurance is one of the risk transfer mechanisms to address the financial costs that arise from cyber-attacks, assisting in the recovery of those affected. In addition, cyber insurance can support risk reduction by promoting mitigation and prevention measures." The opposite view is held by Pavlík, who believes that policyholders reduce their costs of protection against cyber-attacks because they rely on insurance cover [8]. However, Berdykulova states that cyber risk insurance is an important tool for protection against cybercrime for many companies without sufficient financial resources [9]. The importance of cyber risk insurance in the Czech Republic is also beginning to be reflected in the negotiation of framework insurance contracts, which can be demonstrated on the example of the Czech Chamber of Authorized Engineers and Technicians in Construction, who prepared a framework insurance contract for their members in cooperation with the brokerage firm OK GROUP a.s.

Most insurance companies on the Czech insurance market offer cyber risk insurance (sometimes referred to as Internet risk insurance) as part of their other products (e.g. household insurance or supplementary insurance of electronic and mechanical equipment housed in properties), or allow you to take out insurance for these insurance risks (note: risk means the degree of probability of the occurrence of an insured event caused by an insurance peril. We define insurance risk as a possible cause of an insured event. In some legal regulations and in the insurance terms and conditions, these terms are incorrectly confused.) In the following text, the **term Internet insurance** will include the insurance of individuals and possibly also persons living with this individual in a common household (Note: according to the currently repealed §115 and §475 paragraph 1 of Act No. 40/1964 Coll. and established case law—cf., for example, the judgement of the Supreme Court of the Czech Republic of 10 October 2012, file number 21 Cdo 678/2011 and the case law cited in this judgement—a common household was understood to mean the cohabitation of two or more individuals, who live together permanently, and who cover the costs of their needs jointly). The term **cyber risk insurance** will be understand to mean the insurance of legal entities, particularly companies. Separate general insurance terms and conditions are publicly presented on their websites by ČSOB pojišťovna [10], Chubb European Group Ltd, branch office and Colonnade Insurance S. A, branch office [11] (Note: AXA pojišťovna, a.s. also offers a separate product, cyber risk. Insurance products in this area are also offered by the insurance intermediaries RENOMIA, a.s. and INSIA a.s.). As part of the analysis of Internet risk insurance, we will not deal with insurance against terrorist attack [Note: Insurance against terrorist attack could be arranged as part of travel insurance, life insurance and international cargo insurance. In the case of Internet risk insurance, unlike cyber risk insurance, the insurance terms and conditions do not state the motive of the attack as being the decisive fact for the provision of an insurance (the exception to this rule is Article 7(4) of the AXA 2020 insurance terms and conditions)] and insurance for damage caused by cyber risks. The cover of both mentioned insurance perils is arranged on the basis of special insurance terms and conditions (Note: for example, according to Article 2, paragraph 1, letter h) of the [12] insurance terms and conditions, this insurance does not cover damage caused as a result of cyber risks. However, insurance companies usually make it possible to adjust the insurance terms and conditions for professional personal liability insurance of an entrepreneur, provided that the amount of the insurance premium is also determined individually). With regard to the limited scope of the chapter, the scope of insurance cover of Internet and cyber risks offered by insurance companies in the Czech Republic will be analysed using the method of descriptive, analytical, synthetic, comparative and literary research.

The chapter is focused on the analysis, comparison and evaluation of the content of cyber risk insurance in the Czech Republic with a focus on general insurance terms and conditions. The chapter will also make a partial comparison with the insurance terms and conditions of selected insurance companies used in the Federal Republic of Germany and the United Kingdom. Furthermore, an analysis will be conducted of recent activities of the European Union in the field of cyber security insurance. In conclusion, with the help of the synthesis of knowledge gained during the analysis of insurance terms and conditions and the documents of the European Union, an evaluation of cyber risk insurance as one of the tools for protection against cyber-attacks is performed.

2   Systematic Classification of Internet and Cyber Risk Insurance

From the point of view of public law, insurance of Internet and cyber risks is included as part of non-life insurance and from the point of view of private law it is included as part of loss insurance. Internet and cyber risk insurance, as a rule, consists of elements of several non-life insurance branches according to Annex I (Classes of non-life insurance) Directive 2009/138/EC of the European Parliament and of the Council of 25 November 2009 on the taking-up and pursuit of the business of Insurance and Reinsurance (Solvency II) and Annex 1 (Part B) of Act No. 277/2009 Coll., on Insurance. According to §2811 of Act. No. 89/2012 Coll., of the Civil Code "in the case of loss insurance, the insurer shall provide an insurance benefit which, to the extent agreed, compensates for the loss of property resulting from the insured event". The purpose of loss insurance is therefore to compensate for damage as a result of an insured event.

3   Cyber Risk Insurance in the Czech Republic

In the Czech Republic, insurance companies usually offer separate insurance of Internet risks for households and insurance of cyber risks for entrepreneurs. Some insurance products also include in the insurance cover claims for compensation of damage caused by the unauthorised handling of personal data, although they also separately offer insurance protection in the case of a breach of personal data protection by acting in violation of the GDPR regulation (e.g., Colonnade insurance, S.A. branch office, ČSOB Pojišťovna, a.s., or the insurance intermediary MARSH s.r.o.).

3.1 Household Insurance

Internet risk insurance is arranged in accordance with Article 2 of the [10] Insurance Terms and Conditions as insurance of legal protection and financial losses that may arise to the insured:

(a) by the purchase of goods via the Internet,
(b) by the misuse of the insured's identity via the Internet,
(c) by the unauthorised use of the insured's payment card and other electronic means of payment,
(d) by damage to the reputation of the insured via the Internet or social networks.

A similar definition of the scope of insurance is offered on its website by UNIQA pojišťovna a.s., which also includes the provision of assistance in the event of a home computer or mobile device being infected by a hacker or a malicious virus (cyber assistance). Kooperativa pojišťovna a.s. As part of home insurance, Kooperativa pojišťovna a.s. Vienna Insurance Group allows, as part of its household insurance, supplementary insurance for the "freezing" of monitoring and security equipment, which occurred as a result of a hacker attack, when it is necessary to reinstall software or replace hardware ([13], Article 7(3)). Given that Kooperativa pojišťovna, a. s. Vienna Insurance Group undertakes to pay up to a sublimit of CZK 30,000 within the limit of the insurance benefit with a co-participation of 10% of the total amount of the insurance benefit, it can be concluded that this limit may be acceptable for household insurance. UNIQA pojišťovna, a.s. and Kooperativa pojišťovna, a.s. Vienna Insurance Group thus also offer cyber security insurance for households abroad as well (Note: Internet risk insurance tends to be part of packages for household insurance abroad. An example is the insurance terms and conditions of [14] offered in Switzerland, which include assistance services in the information technology field). ČSOB Pojišťovna, a.s. tends towards providing insurance against the harmful consequences of the non-delivery of goods, and damage to reputation on social networks by the misuse of identity and authorisation data. Compared to ČSOB Pojišťovna, a.s., AXA pojišťovna, a.s. offers a narrower scope of insurance, whose insurance only covers the misuse of the insured's identity and the purchase of goods via the Internet [15], Article 6).

Of the main importance **for cyber security as part of this** type of insurance is the misuse of the insured's identity on the Internet, which includes the provision of legal protection and compensation for financial losses caused by the misuse of identification and authorisation data. The insurance in question includes legal protection as well as compensation for damage in the event that a third party misappropriated the identification or authorisation data of the insured, which enabled it to wrongly collect funds from the insured's account.

In the case of internet risk insurance, it is necessary to realise that this type of insurance covers the whole world from the territorial point of view, with the exception of the purchase of goods via the Internet. For example, according to Article 3 of the [10] insurance terms and conditions, this insurance only covers a total of 37 states, of which 30 are European and seven are non-European states. The decisive factor is whether the seller has a registered office or residence in one of those states. Non-European countries do not include, for example, China, from which some buyers from the Czech Republic order consumer goods.

An insured event includes a factual or legal fact that results in the need to protect and enforce the rights of the insured, as well as the damage that has been suffered by the insured as a result of such facts.

3.2 Business Insurance

The insurance terms and conditions of cyber risk insurance have a significantly more extensive structure and regulate in more detail the rights and obligations of the insurer and the policyholder in comparison with the insurance terms and conditions of Internet risk insurance. The content of these general insurance terms and conditions can also serve as a case study. We can demonstrate the content of these conditions on the insurance of cyber risks offered by ČSOB Pojišťovna, a.s.—[16], Colonnade—[11] and the conditions of Chubb European Group—Cyber Enterprise Risk Management [17].

When concluding an insurance contract, it is necessary to examine not only the scope of the insurance cover (i.e., the range of insurance perils) and the limits of insurance benefit, but also exclusions from insurance cover. The OECD report states, based on the conducted survey, that 50% of entrepreneurs cited insufficient insurance cover as a reason for not taking out cyber security insurance. Only 3% of respondents stated that they were not at all aware of the possibility of protection against cyber-attacks via insurance cover [18, p. 41].

Insurance terms and conditions of ČSOB Pojišťovna, a.s. are a typical example of insurance terms and conditions intended for entrepreneurs, which demonstrates the range of insurance perils covered, in respect of which it is possible to conclude:

(1) financial loss insurance (e.g. financial losses arising in connection with data loss and the need to recover them)—some insurance companies also cover other losses typically caused by cyber-attacks [e.g. losses related to network failure, which include a reduction in the net profit of the insured company for the period from the expiration of the waiting period until the restoration of the computer system, which the insured company would have generated in the absence of a network outage (and which corresponds to a loss of revenue), less related savings—cf. [11] insurance terms and conditions, Article F],

(2) legal protection insurance (e.g., costs associated with legal representation before state authorities),

(3) liability insurance. (e.g., damage caused to a third party by a data leak as a result of a cyber incident).

The first two points are referred to in the insurance terms conditions as property damage to the data of the insured, while the last point is liability insurance for damage caused to a third party. However, both types of insurance are subject to a large number of exceptions. It should be emphasised that, according to Article 2(1)(a), this insurance does not cover damage caused as a result of the deliberate negligence of the insured or of the policyholder. Therefore, if the insured or the policyholder knows that it may cause a certain consequence, but without reasonable reason assumes that it will not occur, such a loss event will not be covered by the insurance. This is a significant difference compared to the insurance terms and conditions of Pojišťovna VZP, a.s. These conditions, which are designed universally and can therefore be used for the insurance of entrepreneurs as well as households, apply to negligent conduct and the subject of the exclusion is only intentional conduct. Intentional conduct of employees is covered by this insurance only if they are not employees who "were, are or will be the statutory body of the insured or its member, general or executive director, risk manager, technical or operational director, including head of IT department or head of IT security department, finance director, head of legal department, or another management employee of the insured." The insurance terms and conditions of [11] are stricter, as they exclude from the insurance cover both intentional as well as deliberately negligent conduct of the authorised officer, member of the statutory or supervisory body of the insured company. From the point of view of the form of the fault of the damage caused, the insurance terms and conditions of Pojišťovna VZP, a.s. are thus significantly more advantageous than the conditions of ČSOB Pojišťovna, a.s. In addition, it is appropriate to also mention the slightly different structuring of the conditions of Pojišťovna VZP, a.s. (property insurance, liability insurance and insurance of fines incurred by the supervisory body and costs of administrative proceedings).

Cyberplus [11] insurance forms the basic cover, which can be extended variably by modules of optional extending cover. The basic scope of insurance consists of:

(a) claims for compensation for damage caused by unauthorised handling of personal data and confidential information to the insured or its subcontractors,
(b) claims against the insured due to a breach of network security,
(c) costs of regulatory proceedings [According to Article 3.24, regulatory proceedings are to mean "any proceedings against the Insured or an investigation or audit of the Insured conducted or carried out by the Supervisory Authority (i) due to the use or alleged misuse of Personal Data; or (ii) for the purpose of verifying the procedures for the management and processing of Personal Data; or (iii) arranging such processing through a Subcontractor, to the extent regulated by the Personal Data Protection Regulations."] and
(d) costs of professional services (cyber experts and independent consultants in the fields of law, media strategy, crisis management and personal relations).

Insurance cover can be extended by the following areas:

(a) publishing digital content via multimedia,
(b) blackmail via a computer network,
(c) network failure.

The insurance conditions of Cyber Enterprise Risk Management 2016 contain the scope of insurance cover, which in principle corresponds to the [11] insurance. It is therefore insurance relating to:

(a) unauthorised handling of data,
(b) liability for breaches of network security,
(c) liability in connection with the media,
(d) cyber blackmail,
(e) loss or corruption of data; and
(f) interruption of operations.

According to Article 3.11, Cyber Enterprise Risk Management 2016 insurance also applies to a cyberterrorist attack, while [11] insurance does not, according to Article 4.11, cover any losses resulting from or otherwise related to war and terrorism (for a definition of cyberterrorism, cf. Blanco 2009, p. 202) [19].

The insurance benefit limits are not specified in the insurance terms and conditions of Cyberplus or Cyber Enterprise Risk Management, which can be considered logical with regard to the fact that insurers will arrange this type of insurance according to the individual needs of the policyholder.

In the event of an insured event, the extent of the costs paid by the insurance company will also be a crucial issue. Here, too, there are significant differences in what costs and under what conditions they will be covered by the insurance company. For example, pursuant to Article 10 of the Insurance Terms and Conditions of the insurance company Pojišťovna VZP, a.s., this insurance company will, as part of its legal protection insurance, cover the costs of the insured's legal representation at all levels. However, the insurance terms and conditions of Cyber Enterprise Risk Management 2016 issued by Chubb European Group Limited, branch office, stipulate that only the reasonable and necessary costs of a lawyer approved by the insurer in advance in writing will be covered. The costs associated with the appeal (guarantees or bonds) will be paid by the insurer at the insurer's discretion and there is no claim for their payment. It is clear from the above that the insurance conditions of the branch office of Chubb European Group Limited set significant limits on the amount of insurance benefits provided in connection with the costs of legal representation and defense of the policyholder's rights.

4 Cyber Risk Insurance in the Federal Republic of Germany

In the Federal Republic of Germany, the GDV (*Gesamtverband der Deutschen Versicherungswirtschaft e.V.*) responded to the problems associated with the non-uniform form and content of the terms and conditions of cyber insurance creating standard conditions entitled "General Insurance Conditions for Cyber Insurance" (*Allgemeineversicherungsbedingungen für die Cyberrisiko-Versicherung*, April 2017). Standard conditions are divided into two parts. The first part contains key provisions for defining the scope of insurance cover (general provisions, insurance benefit, liability insurance and insurance of damage caused to the insured). The second part defines the commencement of the insurance cover, the payment of the insurance premium and the obligations of the policyholder in the event of an insured event. In addition, GDV also seeks to address the issue of the difficulty of understanding cyber insurance terminology for some entrepreneurs, as well as the evaluation of their insurance risks. To this end, a relatively recent "Non-binding questionnaire for risk assessment in cyber insurance for small and medium-sized enterprises" has been developed (*Unverbindlicher Muster-Fragebogen zur Risikoerfassung im Rahmen von Cyber-Versicherungen für kleine und mittelständische Unternehmen, December 2019*). As part of the standard conditions for liability insurance for damage caused by operating or performing a profession (Allgemeine Versicherungsbedingungen für die Betriebs- und Berufshaftpflichtversicherung, May 2020), GDV seeks to avoid, to some extent, the problem of silent cyber risk, which will be mentioned in the following subchapter, when these conditions in Article A1-6.13.1 exclude from insurance cover the deletion, retention, rendering unusable or modification of third party data via viruses and other malicious codes. Although the GDV standard documents are non-binding, they are recognized and employed by insurers operating in the Federal Republic as templates for their insurance conditions. The insurance conditions of individual insurers tend to be shorter and simpler, because of the fact that the standard conditions drafted by GDV endeavour to cover all elements and possible variants of cyber insurance in a comprehensive manner. German insurers, including Allianz, seek to provide comprehensible information on the scope of insurance cover and indemnity limits to parties interested in concluding an insurance contract, which is an obligation imposed on them by the German Insurance Contract Act 2007 [20].

5 Cyber Risk Insurance in the United Kingdom

Insurers in the United Kingdom and some other states are addressing an issue referred to as silent cyber risk. The problem lies in solving the question of whether property and liability insurance taken out at a time when cyber risks were not so significant and were not subject to regulation in special insurance conditions, also covers insurance perils associated with current cyber-attacks. Lloyd's of London has taken a step to prevent this "unforeseen" insurance cover by requiring all insurance intermediaries and underwriters operating in the Lloyd's insurance market to inform customers (those interested in the insurance) whether the insurance covers cyber risk or not, and to explicitly state it in the contract [21].

The insurance market in the United Kingdom is still statistically one of the most important in the world, which is also reflected in the range of insurance products. This can be demonstrated by taking as an example AIG UK, which has several special sets of insurance terms and conditions for cyber insurance (interruption in network connectivity, discount vouchers in case of unavailability of services provided to the customer via the Internet, cyber theft, unauthorized access to the phone, computer system failure, provision of services outside the company, cyber blackmail and breach of privacy), which make it possible to extend the scope of insurance protection provided when negotiating the basic variant of Cyberedge insurance.

In the case of English insurance companies, the products offered are further divided according to the turnover of the company taking out the insurance. Another approach can be found in the insurance terms and conditions of Angel Cyber Liability 2018, the introductory section of which contains a clear summary of the scope of the insurance cover and contact details for insurers and bodies responsible for resolving any complaints.

To compare the insurance cover, we will select the 2018 insurance terms and conditions of the insurance company Hiscox Limited for cyber protection and data protection. Although it is an insurance company operating in the United Kingdom, the scope of insurance cover is essentially the same. Insurance cover includes loss of connection, interruption of business activities, damage caused by hackers, cyber extortion, protection of personal data and liability in connection with the media. Similar insurance conditions [22] are offered in the United Kingdom by Royal & Sun Alliance Insurance plc. This is not a surprising finding, as the insurance companies AIG and the Chubb European Group operate essentially worldwide and therefore know the insurance conditions of other insurance companies in the field of cyber risk insurance. The difference between the Czech Republic and some states lies in the length of experience with cybercrime, which is also evident in the insurance conditions of English insurance companies. Statistics are already available abroad in this field and procedures have been tested on how to proceed in the event of a cyber-attack [23].

6   Documents Drawn up Within the European Union in the Field of Cyber Security Insurance

In response to the growing risks of cyber security breaches, reports were produced within the European Union entitled Commonality of Risk Assessment Language in Cyber Insurance—Recommendations on Cyber Insurance [24], which continued on from a study entitled Cyber Insurance—Recent Advances, Good Practices and Challenges [25], and Understanding Cyber Insurance—A Structured Dialogue with Insurance Companies [26]. Although both reports are of a recommendatory nature, they also mention the relevant provisions of Regulation (EU) 2016/679 of the European Parliament and of the Council of 27 April 2016 on the protection of natural persons with regard to the processing of personal data and on the free movement of such data, and repealing Directive 95/46/EC (General Data Protection Regulation—GDPR) a Directive (EU) 2016/1148 of the European Parliament and of the Council of 6 July 2016 concerning measures for a high common level of security of network and information systems across the Union. In addition, the conclusions set out in these reports are based on questionnaires and consultations with large insurance companies, which also increases their relevance and informative value.

The ENISA report of 2017 identifies as a fundamental problem in the field of cyber insurance the inconsistency of standards, that would allow for the harmonisation of insurance companies' terminology and offers. This conclusion is correct from the point of view that insurance terms and conditions sometimes use different designations for the same term and that certain terms are defined differently in different insurance terms. On the other hand, the definition of terms is within the contractual autonomy of the policyholder and the insurer. The harmonisation of terminology would make sense in that it could make it possible to define the scope of insurance cover with a greater degree of precision. The scope of insurance cover is, in addition to the scope of insurance benefits, one of the most common reasons for disputes between the parties to an insurance contract. The recommended adjustment of insurance terms and conditions at the European Union level could be of importance during the judicial and extrajudicial resolution of disputes arising from cyber insurance. The authors of the 2017 ENISA report also focused on the importance of terminology in terms of the policyholder's ability to correctly understand the scope of the insurance cover defined on the basis of the insurers' questionnaire and insurance conditions. In terms of the scope of insurance cover, it was found that the scope of insurance cover in cyber insurance can be divided into liability insurance for damage caused to the insured, liability insurance for damage caused to third parties and other services and costs paid by the insurer (e.g. services provided by IT or law firms required to deal with the consequences of a cyber-attack). I consider it important to mention that the conclusion reached in the cited report is that terminological problems are caused, inter alia, by insufficient legislation and terminology used by policyholders from various sectors, who have to meet different conditions for ensuring cyber security.

The EIOPA report of 2018 considers the lack of specialized insurers and insurance intermediaries to be a major problem [26, p. 3]. In the case of specialized insurers, their deficiency cannot, in my opinion, be considered a fundamental problem because it is essential that insurers offer well-formulated insurance conditions to their customers and are able to offer "tailor-made" insurance contracts. It is unsatisfactory if the cyber risk insurance forms an addendum to the general terms and conditions of liability insurance, which is not compatible with these general terms and conditions, or contains gaps and is incomplete. It is therefore crucial that the insurance terms and conditions of cyber insurance are formulated by cyber security and insurance experts in cooperation with customers of insurers who need to reduce the level of risks they face in connection with cyber-attacks. The EIOPA Report of 2018 makes use of the results of a questionnaire survey conducted in the ENISA Report of 2017 [26, p. 7], on the basis of which it is clear that individual insurance companies provide different levels of insurance cover, which only coincides in the fact that cyber insurance always applies to the insured's liability insurance. It is further argued that insurers primarily target business corporations with their products and largely ignore the specific insurance needs of individuals. This conclusion clearly corresponds to the current situation on the insurance market, as evidenced by the publicly available insurance terms and conditions of the insurance companies listed below.

7 Conclusion

The insurance market is facing a diverse range of insurance products, which may not be comprehensible to those interested in taking out insurance. With regard to the autonomy of will when concluding an insurance contract, it does not seem problematic to negotiate different insurance conditions with different policyholders, but ambiguities regarding the scope of insurance cover provided by individual insurers, which is due to inconsistent terminology, does seem problematic. A recommendation document (guide, handbook) issued by a governmental or non-governmental international organization could contribute to the unification of terminology. The first attempt at such a publication was a report called Preparing for Cyber Insurance, co-produced in 2018 by the Federation of European Risk Management Associations (FERMA) [27], Insurance Europe (member of the Global Federation of Insurance Associations) and the European Federation of Insurance Intermediaries (BIPAR). Although their publication is instructive, it is relatively brief and focused on communication between insurers and insurance intermediaries. ENISA is currently also planning to issue such a recommendation document at European Union level.

Based on the analysis performed in this chapter, cyber risk insurance has been identified as belonging to non-life loss insurance, given that it may include property insurance (e.g., loss related to breaches in network security, network outages, hacker attacks, or cyber extortion), and liability insurance (breach of privacy, confidential information and personal data; media liability). In the insurance of cyber risks there is a close connection with the legal regulation of the protection of personal and sensitive data according to Act No. 110/2019 Coll., on the processing of personal data, and the GDPR regulation, because cyber risk insurance also covers the unauthorised handling of personal data by the insured and his subcontractors.

Based on the content comparison of the insurance terms and conditions of Internet risks and cyber risks insurance, it was found that:

(a) Internet risk insurance is usually arranged as part of household insurance or insurance of electronic and mechanical equipment housed in properties,
(b) cyber risk insurance is usually concluded with entrepreneurs, with insurance companies having separate conditions prepared for this type of insurance.

The scope of Internet risk insurance has a common basis as far as the insurance terms and conditions are concerned, but individual insurance companies differ in that they provide cover for certain insurance perils that are not covered by other insurance companies, or that they do not state these insurance perils within the scope of the insurance cover.

The insurance of cyber risks offered by insurance companies in the Czech Republic is identical in their basic elements. The differences lie in the scope of insurance cover. The insurance indemnity limits are negotiated individually according to the needs of the policyholder or the insured. The scope of cyber risk insurance is similar also in comparison with selected insurance conditions used by insurance companies operating in other states.

When arranging cyber security insurance, the policyholder must assess whether the insurer offers insurance cover that meets its needs (property insurance, liability insurance, or legal protection). Furthermore, it is necessary to evaluate the range of exceptions from insurance cover. As a rule, the larger the scope of insurance cover, the more expensive the insurance. Likewise, negotiating a smaller range of exemptions from insurance cover means that the insurer sets a higher premium. The amount of the agreed insurance limits is of fundamental importance, which in the case of cyber security insurance will usually be negotiated individually, because apart from household insurance, these insurances will often be taken out according to the individual needs of the policyholder.

References

1. OECD (2017) Enhancing the role of insurance in cyber risk management. OECD Publishing, Paris, pp 19–56. https://www.oecd-ilibrary.org/finance-and-investment/enhancing-the-role-of-insurance-in-cyber-risk-management_9789264282148-en;jsessionid=E2O8Q9EkZJyYjW0-vPCsZTkB.ip-10-240-5-104. Accessed 30 Oct 2020
2. Kolouch J (2016) Cybercrime, CZ.NIC, Prague. ISBN 978-80-88168-15-7
3. Šulc V (2018) Cybersecurity, Aleš Čeněk, Plzeň. ISBN 978-80-7380-737-5
4. ENISA (2019) Press Release of 31. 10. 2019, Cyleex: Inside a simulated cross-border cyber-attack on critical infrastructure. https://www.europol.europa.eu/newsroom/news/cyleex19-inside-simulated-cross-border-cyber-attack-critical-infrastructure. Accessed 30 Oct 2020
5. Romanosky S, Ablon N, Kuehn A, Jones T (2019) Content analysis of cyber insurance policies: how do carriers write policies and price cyber risk? J Cybersecur 5(1):7
6. Antonucci D (ed) (2017) The cyber risk handbook, creating an measuring effective cybersecurity capabilities. Wiley, New Jersey. ISBN 9781119308805
7. OECD (2016) OECD project on cyber risk insurance. http://www.oecd.org/daf/fin/insurance/OECD-Project-Cyber-Risk-Insurance.pdf. Accessed 30 Oct 2020
8. Pavlík L (2018) Possibilities of moddeling the impact of cyber threats in cyber risk insurance. In: MATEC web of conferences, 22nd international conference on circuits, systems, communications and computers (CSCC 2018). https://www.matec-conferences.org/articles/matecconf/abs/2018/69/matecconf_cscc2018_04032/matecconf_cscc2018_04032.html. Accessed 30 Oct 2020
9. Berdykulova GMK (2019) CyberRisk management in digital environment: case of Kazakhstani Bank. Int J Eng Adv Technol 8(5):777–782
10. ČSOB (2020) General insurance conditions, Insurance of the internet risks VPP PIR 2020, entry into force. https://www.csob.cz/portal/documents/10710/1599957/vpp-pojisteni-internetovych-rizik.pdf. Accessed 6 Nov 2020
11. CYBERPLUS (2019) Insurance Company Colonnade Insurance S. A. (organizační složka), Insurance Conditions Colonnade, verze CP 01–05/2019, member of Fairfax financial group, entry into force. https://www.colonnade.cz/UserFiles/pojistne-podminky/VPP%20CyberPlus.pdf. Accessed 31 Oct 2020
12. Kooperativa (2014) Kooperativa Insurance Company, plc Vienna Insurance Group, Specific liability insurance conditions for infliction of loss within the provision of professional services (P-610-14). https://www.koop.cz/dokumenty/podnikatele-prumysl/file-1031-general-pdf/file_1031_GENERAL.pdf. Accessed 6 Nov 2020

13. Kooperativa (2017) Insurance conditions for supplementary insurance of the electronic devices and machinery in immovable assets, version M-190/17. https://www.koop.cz/dokumenty/poj isteni-majetku/pojistne-podminky-a-informace-proklienta/dokumenty-k-pojisteni-majetku-a-odpovednosti-obcanu-102018/Dokumenty%20k%20pojištění%20majetku%20a%20odpověd nosti%20občanu°%20102018.pdf. Accessed 31 Oct 2020

14. Generali (2020) Generali Allgemeine Versicherungsbedingungen (AVB) für die kombinierte Haushaltsversicherung PRISMA Flex, Ausgabe 2020, version 2. https://www.generali.ch/pri vatkunden/wohnen-bauen/cyberversicherung. Accessed 31 Oct 2020

15. AXA (2015) AXA Insurance Company plc, Insurance Conditions—Insurance of the On-line Identity PP-PPA-Id-MB, entry into force

16. ČSOB (2018) General insurance conditions—special part, cybersecurity insurance VPP CRC 2018, entry into force. https://www.csobpoj.cz/documents/10332/32946/10N9059+VPP_CRC_2018_10-2018.pdf/dc55ba8d-b5e3-17c8-0954-63591b631e67?t=1576162606907. Accessed 31 Oct 2020

17. Chubb European Group (2016) Insurance conditions cyber enterprise risk management, version ERM 1-2016. https://www.chubb.com/content/dam/chubb-sites/chubb-com/cz-cz/for-business/financial-risk-professional-liability-insurance/documents/pdf/chubb_pp-cyber-risk-management.pdf. Accessed 31 Oct 2020

18. OECD (2019) Measuring digital security risk management practices. OECD Digital Economy Papers, No 283. https://doi.org/10.1787/20716826. ISSN: 20716826. https://www.oecd-ili brary.org/science-and-technology/measuring-digital-security-risk-management-practices-in-businesses_7b93c1f1-en. Accessed 30 Oct 2020

19. Blanco SM (2017) La Ciberseguridad y el uso de las tecnologías de la información y la comu-nicación (TIC) por el terrorismo, Revista Española de Derecho International 69(2):202. ISSN 0034-9380

20. Allianz (2018) Kundeninformationen und Allgemeine Versicherungsbedingungen Secure Cyber, Allianz Global Assistance. https://www.fhschweiz.ch/customer/files/2350/AVB_Cyber-security_Einzel_13-12-2018_DE.pdf. Accessed 31 Oct 2020

21. Lloyd (2019) Lloyd's market bulletin, Ref: Y5258. https://www.lloyds.com/market-resources/market-communications/market-bulletins/market-bulletins?page=3. Accessed 30 Oct 2020

22. Royal&Sun Alliance Insurance plc (2018) Cyber risk insurance, policy wording, version UK 05268, 21. https://www.rsabroker.com/system/files/Cyber%20Risk%20Insurance%20P olicy%20Wording%20-%20UKC05268A.PDF. Accessed 31 Oct 2020

23. Sardy S, Fleck M (2016) Cyberversicherung "plus"—Dienstleister im Schadenfall, Allianz. http://docplayer.org/75227375-Cyberversicherung-plus-dienstleister-im-schadenfall.html.

24. ENISA (2017) Commonality of risk assessment language in cyber insurance—recommenda-tions on Cyber Insurance. https://www.enisa.europa.eu/publications/commonality-of-risk-ass essment-language-in-cyber-insurance. Accessed 6 Nov 2020

25. ENISA (2016) Cyber-insurance a look at recent advances good practices and challenges. https://www.enisa.europa.eu/news/enisa-news/cyber-insurance-a-look-at-recent-advances-good-practices-and-challenges-by-enisa. Accessed 6 Nov 2020

26. EIOPA (2018) Understanding cyber insurance—a structured dialogue with insurance compa-nies, EIOPA. https://www.eiopa.europa.eu/content/understanding-cyber-insurance-structured-dialogue-insurance-companies_en. Accessed 30 Oct 2020

27. FERMA, Insurance Europe, BIPAR (2018) Preparing for cyber insurance, Report. https://www.ferma.eu/publication/preparing-for-cyber-insurance-report/. Accessed 6 Nov 2020

Creative Cyber Risk Management

Miroslav Čermák

Abstract The main aim of this paper is to analyse creative cyber risk management phenomena and identify the most frequently used techniques by management in the private and public sector. The used creative techniques and motivation of managers was identified through in-depth interviews with security experts responsible for identifying and analysing cyber risks in many organizations. The main motive for the using such techniques of creative cyber risk management by managers is the deliberate reduction of the value of risk to an acceptable level.

Keywords Creative risk management · Risk management · Cyber risk management

1 Introduction

The concept of creative risk management is completely new and originated during the writing of this paper. In essence, it best captures the practices that occur in many organizations in the field of cyber risk management.

Creative cyber risk management can also be described as a phenomenon that many employees, managers and owners do not pay attention to, nor are they aware of, that it is happening, and not at all that they themselves are committing creative risk management, whether either consciously or unconsciously.

At the beginning of this research was formulated hypothesis that *management under some circumstances intentionally manage risk creative way and uses creative methods and techniques*. The primary goal of the research was to verify this hypothesis and, if it was confirmed, to analyse the possible causes of this phenomenon and to propose appropriate measures to prevent this phenomenon.

This phenomenon is extremely undesirable mainly due to the fact that if the organization manages risks in an inappropriate manner, then in the longer term it can

M. Čermák (✉)
Department of Management and Informatics, Police Academy of the Czech Republic in Prague, Lhotecká 559/7, 143 01 Praha 4, Czech Republic
e-mail: cermak.miroslav@pacr.eu

© The Author(s), under exclusive license to Springer Nature Switzerland AG 2022
I. Tušer and Š. Hošková-Mayerová (eds.), *Trends and Future Directions in Security and Emergency Management*, Lecture Notes in Networks and Systems 257,
https://doi.org/10.1007/978-3-030-88907-4_20

lead to a significant deterioration of economic results and, in the extreme case, to the termination of the organization on the market.

The risk retention can be conscious or unconscious, and it is obvious that creative risk management can occur at any stage of risk management process and especially in case of serious cyber security risks, because management does not see added value in recommended cyber security measures leading to risk reduction.

2 Methodology

The research methodology is based on detailed knowledge of the risk management process and its phases. When we look at how risk management in organizations takes place and what stages the whole risk management process consists of, it is relatively easy to identify areas where some creativity can be developed to a greater or lesser extent, which can manifest itself and usually manifests itself in:

- **vague risk management**, the individual process steps, inputs, outputs and responsibilities for risk identification, analysis and management are intentionally not clearly stated;
- **inadequate risk management methodology**, which provides metrics but does not define who should determine the resulting value of assets, threats, vulnerabilities and the resulting risk;
- **the deliberate reduction of the value of the risk**, inherent risk through the value of the impact or probability of the threat, and residual risk primarily through the choice of ineffective security measures.

The above symptoms may not always be obvious at first glance, and even when they are noticed, this does not necessarily mean that we are dealing with any form of creative risk management. The fact that the risk management process is not implemented in the organization at all, or is implemented only at a low degree of maturity, may be due to insufficient security awareness of management, which does not have sufficient experience with risk management and:

- **has no idea how risk management could help** him in his business and how to implement the risk management process in his organization;
- **does not know how to approach risk analysis correctly** and acquires a tool and tries to use it in good faith;
- **makes unnecessary mistakes in choosing appropriate security measures** because they simply do not understand security.

We see that this is a completely different situation. In order for risk management to be described as truly creative, that risk management must take place in a completely conscious manner.

The expert team can easily get the opinion that this is not creative management and just start to doubt the competencies of the manager. Likewise, a manager may also doubt the competence of an expert team. But beware, this doubt will not appear anywhere, nor will it be expressed directly.

In this way, the manager can elegantly transfer his responsibility to experts, who are usually not insured against damage, and who will then bear the consequences and be able to be easily replaced. This is one of the many reasons why external experts are often used, and why their remuneration is also much higher.

Based on the above facts, a set of questions was compiled, which were asked in a **semi-structured interview** to experts who deal with cyber risk management in selected organizations. In collaboration with these experts, risk management policies and related standards, procedures and guidelines used by the organization were thoroughly analyzed.

3 Research

Within the research itself, it was verified whether there is no vague risk management, whether there is a quality risk management methodology and, last but not least, whether there is no reduction in the value of risk and, if so, how. In next chapters is presented result synthesis of information obtained through the interview with cyber security risk experts.

3.1 *Vague Risk Management*

Vague risk management enables conscious or even unconscious risk retention, consisting in not entering the risk into the risk register or its subsequent deletion. In common practice, **this technique is usually not used at all, as it is too conspicuous, and is usually identified relatively soon** as a finding that must be promptly removed, by the audit fulfilling the role of the 3rd level of defence.

And most importantly, a responsible manager could not claim that he is involved in risk management and that risk management is an integral part of his work and business processes. By issuing a methodology that defines the various stages of the entire process, specific responsibilities and required outputs, the manager demonstrates his clear intention to manage risk and can thus easily avert suspicion that he would commit any creative risk management at all.

3.2 *Insufficient Methodology*

The problem with most methodologies is that although they relatively precisely define individual metrics for assessing assets, threats, vulnerabilities and resulting risk, they do not define the necessary steps in risk identification process at all, and **do**

not specify by whom, how and on what basis the value of individual components and resulting risk is determined. It is not clear whether the individual risks and values of its components are to be determined by one expert or to be determined within an expert team. They also do not define who should be a member of this team, who should appoint him and how the final value should be determined.

It should be noted that each member of the expert team may come to a completely different conclusion, as he has different experiences. In such a case, is it appropriate to exclude extreme values, calculate the arithmetic mean, and use the median or does the expert team have to reach a consensus?

These questions are usually not answered by any methodology or the ISO standard referred to in the methodology. In other words, the reference to the standard is given rather only to create a false impression of quality.

3.3 Risk Reduction

Reducing the inherent value of risk is **the most popular technique**, as it allows low inherent risk to be accepted in accordance with the methodology. If we take into account the way the risk (R) is defined, usually as a threat (T), which exploits some vulnerability (V) and causes some impact (I), then we can write:

$$R = I \times T \times V$$

where $\times$ means that there is a certain relationship between the individual components of the risk. It is usually a product, but it can also be a sum. It is important that as soon as one value changes, the resulting risk also changes.

If the inherent risk cannot be reduced, it is necessary to reduce the current or residual risk by means of an effective safety measure (C), which is not really effective at all, only this effectiveness is attributed to it with a certain amount of creativity. It can be expressed as:

$$R = I \times T \times V/C$$

We see that **we can reduce the value of risk in several different ways**. By reducing the value of the impact, the likelihood of a threat or the degree of vulnerability [1]. Alternatively, by reducing the values of any two or even all three components. Alternatively, it is possible to design a whole set of measures, where the individual measures are completely ineffective, but soon they become effective measures thanks to synergy, so at least this can be said. We see that there are definitely no limits to creativity.

3.4 Risk Communication

In the field of risk management, there is one essential component that is wrongly neglected, and that is risk communication. In previous risk management models, this component was always depicted only in the last place, as it was intended to serve exclusively to inform management about risks.

But times have changed significantly and the risk communication component is now portrayed as the centre of the action and has an irreplaceable place in the modern risk management process, as it enables two-way communication between all stakeholders. Which certainly sounds good.

On the other hand, this method allows the management to have absolutely control over the whole process and influence it with their comments, and also to form the opinions of experts regarding risks and **come up with pseudo-mitigation security measures**, etc.

Either way, management can create ideal conditions for creative risk management in the organization by acting or not. At the same time, it can try to **significantly influence the expert team in the risk assessment phase**. After all, he has the management tools to do it.

Sometimes the **creative risk management can be quite obvious, and sometimes rather latent**. That creative management can also take place explicitly or implicitly. Latent implicit creative management is particularly dangerous because it is retrospectively perceived as the inexperience of the analytical team, but in reality, of course, this is not the case.

De facto, there is no explicit interference, but **latent manipulation of a person lower in the hierarchy**. And this person sometimes fulfils an unspoken request with a certain dose of servility, and often he or she becomes convinced that he should continue to do so. (In other words, there is a request from the top manager that something should be done about the risk.) Let's not forget that there are also nonverbal communication, NLP and argumentative fouls, all of which can be used.

Management can significantly influence the result using its managerial competencies, even if the risk analysis is performed by an independent expert team, its **complete independence can be questioned**. Especially in the case of an external expert team that does not know the situation in the organization and has to interview and obtain information from responsible employees of the organization, and the validity of this information may not always be able to verify.

In addition, the **analysis must always be paid by someone**. And in that case, he needs the risks to be assessed the way he needs to. With some exaggeration, one could talk about **risks on demand** in the context of creative risk management. Which can certainly be considered a completely extreme case, but it cannot be completely ruled out.

In such a case, **the expert team may be advised whether it does not want to reconsider its conclusions**, whether it has taken this or that into account, or whether it has simply misunderstood something and not provided true and complete information. After all, what he presented is certainly not the final report, but only DRAFT, and that he was deciding on the risks and uncertainties.

Of course, I am far from concluding that changing this paradigm has accelerated the development of creative risk management, but at least helped it. And I would also like to draw attention again to the fact that it is not always a matter of creative management and that erroneous evaluation can occur not because the experts are influenced by the manager, but because the team of experts or the expert himself:

- **attaches too much importance to the measure**, when in fact it does not reduce the vulnerability, threat or impact;
- **wants to avoid conflict** and so, where he has some doubts, he may tend, as the survey shows, to **choose medium values** or be influenced by the opinion of another expert;
- is **influenced by news** in the media and various security reports arising outside the organization, and according to them, he assesses the probabilities of a certain threat.

The question is whether it is possible to assess who actually contributed to influencing the experts, because when, for example, **risks are assessed by an expert team composed of experts from the organization**, they may be influenced by training, which could easily be prepared for them by management. Or they may be influenced by an undesirable **toxic corporate culture**, the attitude of a top manager taking on the role of an expert on everything and therefore also on risk management, and he can downplay risks, etc.

Experts must also analyze how ordinary employees and management perceive the risk, and they must adapt their communication accordingly. After all, the way in which risk is perceived is described in the SARF framework. We can evaluate how risk perception is strengthened and weakened [2]. In any case, the **risks are perceived differently by experts and otherwise by ordinary employees, including managers**, and they, locked in their own social bubble, tend to accept opinions from their surroundings uncritically.

3.5 Why Creative Management Occurs

The line between creative risk management and objective risk management is very thin, and can often be very easily crossed. Without a detailed analysis, it is very difficult to determine whether it was really creative risk management, or just the negligence, ignorance of management or inexperience of the expert team that performed the risk analysis and proposed a suitable way to manage the risk.

In addition, we may encounter actions that cannot be considered creative risk management, as it can only be an **arrogance of power**, where the manager decides to accept the risk only to show that he can do so and that no one will dictate what he should or should not do.

In publications devoted to risk management, we can read that the **ultimate responsibility for risk management always lies with the owner of the company**, who should, with the care of a proper manager, identify, analyse, evaluate and respond appropriately to all risks.

It is important to note that the owner usually delegates his responsibilities to top managers, who in turn delegate their responsibility for risk analysis to middle and lower management. And lower management then delegates this responsibility to ordinary employees.

Decisions on how to manage risks often become a collective matter and are decided by a committee. Personal responsibility for the consequences of a wrong decision is completely lost.

The question arises as to whether a professional manager is objective enough to decide on the appropriate way of managing risks, the materialisation of which does not have to take place immediately. Why? Because the **decision of the manager may not always be entirely in the interest of the real owner**. Especially if the manager comes to the conclusion that the event with a negative impact on the financial results can occur only after the end of his engagement in the organization.

It should be noted that if the manager is in the position of an executive or a member of the Board of Directors or the Supervisory Board, he is liable for the damage caused indefinitely in accordance with the Commercial Corporations Act. However, if the manager performs his function according to Act No. 262/2006 Coll. Labor Code §257, paragraph 2, and this also applies to the expert, so he is liable for any damage only up to 4.5 times his average monthly salary, unless he causes it intentionally [3].

At first sight, it seems that a member of a supreme body should be afraid of the consequences of his actions. But it would have to prove beyond a shadow of a doubt that he did or did not do so with the intention of causing damage, or that he could at least know that damage could occur, and did so at least carelessly.

In addition, each manager has the opportunity to take out liability insurance for damage caused by members of the statutory bodies known as Directors & Officers Liability, even in such a form that it is protected even at the time when the actual damage occurs after his departure. The manager does not have to worry too much that he could bear any consequences for his past decisions, and this is a **classic moral hazard**.

And if the **manager is evaluated for the results achieved**, such as increasing profit, market share, share price, etc., which he can achieve only by increasing revenue or reducing costs, or both, then he can go the way of reducing costs (which is the easiest way) and accepting risks that the manager should not accept.

Such an approach to risks can be described as creative. We can meet him practically everywhere, somewhere only sporadically, and somewhere very often. In principle, the larger the organization, the greater the turnover, the inappropriate motivation system, the more likely we are to encounter creative management.

While there may be no creative risk management for low-risk risk analyses, in the case of critical strategic projects, those pressures for creative risk management can be enormous. Especially if there is some high risk in the risk register and its elimination proves to be relatively time-consuming and costly and would significantly jeopardize the delivery date of the project or increase its costs and thus affect the expected ROI.

3.6 Recommendation

First and foremost, you need to get **support from top management**. Management must strive to ensure that risk management becomes an integral part of all business processes and that all risks appear in the risk register.

Subsequently, **the owner of the risk management process must be formally appointed**, the risk management process in the organization must be described and issued in the form of a policy document in order to become an integral part of the organization's regulatory system.

The policy **must describe each phase of risk management process**, determine the individual roles in the whole process, as well as the way of filling these roles and, last but not least, determine the responsibilities of individual roles, ideally in the form of RAM matrix.

Top management must define their **attitude to risk**, what risk and under what conditions it is willing to take. And he must formulate his position in the form of a clear statement, the so-called risk statement, which states his taste and tolerance for risk.

The risk management policy must specify the different risk categories, e.g. that the risks can be quantified as low, medium, high or critical and what level of management is then responsible for managing them, as well as what method of treatment is appropriate.

A separate document must then state why the risk is classified as low, medium, high or critical. The methodology must define a specific method of risk assessment and define intervals that will allow a clear classification of the risk into the relevant category.

Metrics for impacts, threats and vulnerabilities must also be defined, it must always be determined **who is responsible for updating these metrics**, on what basis the individual metrics are determined and under what conditions they can be changed, and who must approve the change.

Once the metrics are set, **the process for identifying, analyzing and assessing risks must be defined**. First of all, it is necessary to define who should perform these steps. It is obvious that if it is an individual, it will be easier to influence him than if it is an expert team.

For the above reason, **individual members of the expert team should not be subordinate to the manager who is responsible for managing the risk**.

It is necessary to clearly **define who should be part of the expert team, who should set up and manage this team**, in order to really ensure its independence, objectivity and transparency.

The method of determining the resulting value of individual risk components must be described and issued as a standard.

Ideally, even smaller subteams are created within the team's risk analysis as part of the risk analysis, with one subteam responsible for determining the value of the asset and the impact, the other for determining the size of the threat, and the third for determining the degree of vulnerability. It must always be stated who determined the value, when and what. The advantage of this approach is that the resulting value of the risk can be less influenced.

There should also always be evidence of the risk identification and analysis process, as this is the only way to ensure that the risk identified and analysed by the expert team is subsequently entered in the risk register in an unchanged form. Audit as the 3rd level of defence we will be able to compare whether the risks recorded in the register correspond to the risks assessed by the expert team.

If there is a risk of influencing the level of risk through measures, it is offered that one team assesses the inherent risk and the other team proposes a suitable way to manage the risk and in case of reduction then safety measures to reduce the risk.

4 Conclusion

The research confirmed the hypothesis that creative cyber risks management can occur in whatever organizations and in any phase of risk management process as is context setting, risk identification, risk analysis, risk evaluation, and risk treatment. The most common techniques that are used for this purpose by several managers were also identified, as well as the main motive that leads managers to use those unfair techniques. Further a number of relevant recommendations have been proposed which, if implemented, should prevent or make creative risk management much more difficult. The business has been, is and will always be associated with the risks that managers have to face, identify them, analyse them and decide how to deal with them. And since no one is infallible, there can sometimes be an incorrect risk assessment conducted. However, it is very difficult to prove whether this was just a misjudgement or a clear intention, and it can therefore be assumed that we will still struggle with creative risk management.

References

1. Čermák M (2012) Does the security measure reduce the threat, vulnerability or impact? (In Czech). https://www.cleverandsmart.cz/snizuje-bezpecnostni-opatreni-hrozbu-zranitelnost-nebo-dopad/. Accessed 20 Nov 2020
2. Pidgeon N, Henwood K (2010) The social ampli-fication of risk framework (SARF): theory, critiques, andpolicy implications. In: Risk communication and public health, pp 53–69. https://doi.org/10.1093/acprof:oso/9780199562848.003.04
3. Ožana J, Forejt J (2020) New rules of liability of members of the statutory body according to the big amendment to the Business Corporations Act or will the performance of the function of a member of the statutory body be only for the brave? (In Czech) EPRAVO.CZ. https://www.

epravo.cz/top/clanky/nova-pravidla-odpovednosti-clenu-statutarniho-organu-podle-velke-nov
ely-zok-aneb-bude-vykon-funkce-clena-statutarniho-organu-jen-pro-odvazne-112140.html.
Accessed 20 Nov 2020

Extraordinary Events, Preparation and Solutions

Use of Crisis Communication in Crisis Management

Pavel Otřísal and Dana Rebeka Ralbovská

Abstract This chapter deals with the use of crisis communication and crisis negotiation at the scene of an emergency by members of the integrated rescue system. An integral part is the issue of providing first psychological assistance to victims of emergencies and crisis situations at the scene. There are recommendations for the practical use of elements of crisis communication and crisis intervention in interaction with an aggressive individual and an individual demonstrating the intention of suicide to alleviate mental tension and de-escalate the overall situation. The process of calming the aggressor is also described in detail. Attention is also paid to the use of the plan of typical activities of the parts of the integrated rescue system in the joint intervention in the case of demonstrating the intention of suicide and in the provision of psychosocial assistance.

Keywords Crisis communication · Crisis negotiation · Crisis · Integrated rescue system

1 Introduction

In the exercise of his profession, a member of the integrated rescue system (IRS) often finds himself in situations where he has to manage professionally to provide assistance, but at the same time apply crisis communication and elements of crisis negotiation. These are emotionally tense situations (for example, coping with aggression by victims of emergency and crisis situations, individuals affected by a psychiatric diagnosis or intoxication, when communicating with an individual demonstrating

P. Otřísal (✉)
Faculty of Physical Culture, Palacký University Olomouc, Třída Míru 117, 771 11 Olomouc, Czech Republic
e-mail: pavel.otrisal@upol.cz

D. R. Ralbovská
Faculty of Biomedical Engineering, Czech Technical University in Prague, nám. Sítná 3105, 272 01 Kladno, Czech Republic
e-mail: rebeka.ralbovska@fbmi.cvut.cz

the intent to commit a suicidal act, etc.), for which these workers are prepared during their studies and subsequently in the process of lifelong learning [1, 2].

Due to the frequent occurrence of extraordinary events and crisis situations, tactical and verification exercises are carried out under the responsibility of individual IRS units, with the help of which the staff is not only practically prepared for managing extraordinary events and crisis situations, but also have the practical feedback [3]. During these exercises, attention is also paid to the provision of first psychological assistance, psychosocial assistance, and posttraumatic care to victims, as well as intervening professionals. An integral part is also the evaluation of communication skills in the field of crisis communication or in the field of crisis negotiation.

2 Part of Crisis Communication

Crisis communication is a part of crisis management and its goal is to provide correct, full, and reliable information regarding the procedures for dealing with an emergency or crisis. In the process of crisis communication, it is necessary to actively apply elements of positive social thinking, empathy, and the art of humility expressing. Even if the rescuers themselves are under the influence of acute stress, they must act professionally and decisively. Its approach must help those affected to regain a sense of security and safety. They provide information in a clear, concise, comprehensible manner and follow a slower speech pace. They avoid verbal aggression and are aware of the so-called spiralling of aggression, where everyone supports the aggressiveness of their counterpart.

Patience and the ability to actively listen are also an essential part of crisis communication, as victims of an emergency or crisis are very often in the acute phase of a stress response and will take considerable effort to understand their current crisis-specific needs. It is appropriate to show understanding, encourage the victim to communicate, and still be aware of their verbal and nonverbal expressions. For effective crisis communication, methods of active listening can be used, which include: maintaining eye contact, nonverbal expressions of listening (for example nodding), verbal expressions of listening, not talking too much (leaving space), asking open questions or questions that clarify the situation and paraphrasing. It is also important to use elements of assertiveness. An assertive person knows what he wants to do and how he will do it, he is fully aware of the responsibility for his behaviour and considers its possible consequences, which is essential for crisis communication.

In situations where it is necessary to obtain additional information from the victim or to verify that we have already understood the previously communicated information, it is effective to choose open-ended questions. Not questions that the individual only nods to our heads or answers yes or no. Open-ended questions encourage interaction and are a manifestation of our interest in the individual with whom we are communicating. By using open-ended questions, we prevent misunderstandings, clarify facts, and actively use feedback. The following sentences can be used:

"If I understood you correctly, then…".

"It follows from the context of our conversation…"

The victim has the full right to the maximum expression of all his emotions. It is undesirable for them to be displaced because the suppression of emotions can, on the contrary, cause a deepening of the experience, crying and wailing. It is, therefore, appropriate to confirm the victim's emotions and rights to them, even with the following sentences:

"I can't even imagine how difficult this situation must be for you. That is why I fully respect your right to mourning (or restlessness)…"

"I understand you have been very affected by this event and it is perfectly normal for you to react this way…"

"You don't have to be ashamed of your tears, they are a perfectly normal reaction in that situation."

When expressing emotional interest, get as close to the victim as possible (for example, when he is lying down, kneel) and offer physical support. "If it helps you and you don't mind, you can grab my hand or lean on me."

Try to anchor the victim using the following sentences, for example:

"I put a bottle of water next to you, so if you're thirsty you can have a drink."

"On the table next to which you are sitting, I will prepare the medication for applying."

"In which part of your body do you feel the most pain?"

"Don't you need a cover? Aren't you cold?"

Under no circumstances, use sentences of the type:

"I fully understand how you feel."

"Calm down here immediately and do not make a scene!"

"Maybe you are a man and you can handle an accident!"

"Cannot you react other than by crying or another hysterical reaction?"

"Death is a natural part of life, so take it that way. And now it has affected your child." Etc.

It is also important to avoid devaluation tendencies in the crisis communication process (shouting or reprimanding the victim, instructing, providing hasty advice, etc.). Likewise, jumping into speech is very inappropriate, which can arouse strong feelings of distrust.

At the scene of an emergency and in a crisis, each intervening member of the IRS is obliged to provide the first psychological assistance within the framework of crisis communication with the affected person (see Fig. 1). First, psychological assistance according to the State Type Activity 12/IRS Provision of psychosocial assistance from 2015 represents a set of simple procedures, the aim of which is to stabilize the mental state so that the situation for the affected person does not worsen. It includes the provision of basic human needs, including the promotion of a sense of security and the transfer to further care.

The steps of the first psychological help according to State Type Activity 12/IRS are: make contact—find the courage to address the affected person,

– determine the state of health,

Fig. 1 Simplified model of
providing first psychological
assistance

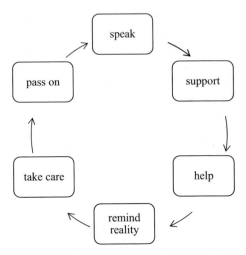

- protect privacy—pay attention to dignity. Protect the affected person from prying eyes or media,
- let me know what is going on and listen. Provide real information,
- identify and provide basic needs: heat, silence, fluids, transport, sedatives,
- watch out for safety,
- solve the situation here and now, or the immediate next steps, do not look for long-term solutions,
- hand over to the care of loved ones or other services.

When providing immediate psychosocial assistance at the scene of an emergency, the intervening member of the IRS is based on the following rules, which will create the abbreviation IMPRESS [4].

IM—immediately, Proximity—nearby, Expectancy—with expected ability, Security—safe, Simplicity—simply.

Simply, we can say, psychosocial assistance should be provided immediately at the scene of an emergency, with sensitive regard to the affected persons and their crisis needs.

Crisis negotiation is a specific way of communicating with an individual who is under the influence of a crisis and threatens to use some form of violence. It can deal with an auto-aggressive tendency (for example self-harm, suicidal behaviour, etc.) or a demonstration that he wants to hurt someone else. Through crisis negotiation, the negotiator strives for a calm and safe solution to crisis situations. It uses establishing and maintaining verbal contact and reaching a certain agreement by applying selected elements of crisis communication. It is a guide to a real solution to the situation that is acceptable to all. The possible results of the negotiations are shown in Fig. 2 [5].

Crisis communication can lead to conflicts in the provision of assistance. The reason may be, for example, that the intervener does not respond adequately to the findings. This can be seen by the victim as a sign of a lack of understanding. Consequently, the victim tends to repeat the information, which is unnecessary. Wasting

Fig. 2 Possible results of negotiations (modification according to Grada [5])

time is undesirable in an emergency or crisis. Other mistakes that can harm crisis communication are generalizations, insincerity, escaping from the topic, or communicating more information at once, without the space to absorb the communicated information.

To minimize crisis communication errors, it is recommended to follow the below stated procedures at the scene of an emergency or crisis:

- calming our own tension,
- awareness of our role at the scene of an emergency or crisis,
- avoiding well-known communication clichés,
- adequate reflection of the victim's feelings,
- meeting the basic needs of victims, supporting the return to reality and the state before an emergency or crisis,
- a summary of the whole event and an outline of a possible starting point.

3 The Use of Crisis Communication Within Demonstrating the Intent to Commit a Suicide

Suicide can be defined as a conscious and deliberate behaviour that is conducted to end one's life. The World Health Organization describes suicide by the following definition: "A suicide is an attack on a person with varying degrees of intent to die. Suicide is then a self-destructive act with a fatal result."

Dividing suicides:

- balanced suicide—is committed on the basis of an analysis of a life situation, where an individual is evaluated that suicide is the only logical option (for example, after the loss of a long-term life partner, feelings of loneliness, etc.),
- demonstrative suicide—in this type of suicide there is no intention to die, it is rather a form of calling for help or warning of a difficult life situation,
- impulsive (abbreviated) suicide, which occurs in the form of a quick and hasty reaction to a current event (for example, a bad report card, disagreement in love, etc.),

- extended suicide—in this type of suicidal behaviour, the suicide decides to end not only his own life, but also the life of family members or loved ones. An individual often suffers from the persistent notion that ending life is the best solution to a given crisis (for example, a mother experiencing a series of life's failures takes her child in her arms and jumps under a passing train). By this act, the suicide also becomes a murderer,
- about self-killing—death occurs not based on the intention to die, but accidentally (for example, in a fall by a drunk individual, accidental ingestion of a toxic substance, etc.),
- about parasuicide—there is no intention to die, it is more about self-harm, but even so it can end in death.

Self-harm can generally be characterized as intentional and long-term self-harm, which is associated with an effort to cope with negative mental states. The individual does not have suicidal thoughts, but a violation of bodily integrity occurs. These are usually incisional injuries, especially in the hands (for example, wrist, back of the hands, forearm)—cut wrist syndrome. It is very common to carve various characters and names into the skin, as well as burn the skin with a lighter or cigarette, etc. Often an individual uses a razor, scissors, or needle for self-harm.

Intentional self-harm syndrome occurs primarily in individuals with personality disorders, in individuals addicted to drugs or other addictive substances, as well as in individuals with eating disorders. With the help of perceived physical pain, he tries to suppress the experienced chronic mental pain.

We know a specific form of inducing disease and self-harm to gain benefits from those around us, called Münchhausen's syndrome. According to the International Classification of Diseases, it is the intentional induction or stimulation of somatic or mental disorders. An individual repeatedly pretends to have symptoms for no apparent reason and may also use elements of self-harm to induce individual disorders (for example, deliberately ingesting a substance to be hospitalized in a medical facility and receive appropriate attention, etc.).

A highly dangerous form is Münchhausen's syndrome by proxy or also Münchhausen's syndrome in deputies. These are situations where one of the parents tries to attract attention and gain the sympathy of his surroundings or medical staff by pretending to have various diseases in his child. Unfortunately, it does not protect its child from systematically and repeatedly harming it when it causes these diseases (for example, it suffocates its child in an attempt to mimic states of apnoea pause resulting in sudden death syndrome, gives high doses of table salt to cause dehydration, etc.).

As part of suicidal behaviour, we recognize:

- suicidal thoughts and ideas are without stronger intensity, which lack a tendency to commit suicide itself. The individual devotes a substantial part of the day to these ideas, and these ideas are difficult to confuse with other topics. They are manifested in everyday life by nonverbal and verbal expressions. (for example, reflections on what would happen if he committed suicide, how the family would function after his death, etc.),

- suicidal tendencies—they are used in the period of preparation for the implementation of suicide (for example, determining the method of carrying out
  suicide, obtaining information on the effects and lethal doses of drugs, purchasing
  weapons, etc.),
- suicide attempt—a life-threatening act that an individual performs with the intention of self-killing, but which does not have to be fatal. Suicidal attempts occur
  based on uncontrollable crisis situations (for example, in family or life tragedies,
  etc.), but also in the form of call for help (for example, in a demonstration suicide),
- completed suicide.

Soft methods are used within the implementation of suicide. These methods are
likely to save lives (for example, intoxication with drugs, etc.), or hard methods,
which cause almost immediate death (for example use of a firearm, jumping under
an express train, etc.).
Suicides often occur in connection with:

- mental disorders (for example, depression, borderline personality disorders,
  schizophrenia, etc.),
- serious changes in health status,
- abuse of alcohol or other narcotics, as well as gambling,
- emotional issues (for example, rejection by a loved one or separation, etc.),
- about feelings of loneliness,
- significant personal problems (for example, inability to repay debts, partnerships,
  divorce, unemployment, etc.).

A specific type of suicide that has occurred recently is suicide attempts by children
and adolescents who have succumbed to dangerous online games.
Suicides often occur because of an uncontrollable burden during a difficult period
of life. Authors Hartl and Hartlová define workload using the following definition:
"Workload that results from difficult social relationships or situations; its indicators may be: loss of well-being, growing feelings of insecurity, growing feelings
of internal or external threats, deepening feelings of inadequacy in social roles and
tasks, inability to adapt to new situations, feelings of life-threatening life-balance"
[6, p. 699].
At the scene of an emergency, members of the IRS often meet with an individual
who demonstrates the intention to commit suicide. In these cases, the procedure is
in accordance with the State Type Activity 12/IRS Demonstration of the Intent of
Suicide from 2005, in which this issue is comprehensively elaborated. During the
reported demonstration of the intention to commit suicide, the presence of individual
IRS units is necessary at the site of the emergency. The commander of the intervention
at the emergency site calls the negotiators from the Police of the Czech Republic
(CR). The negotiator then carries out a suicidal intervention or is the coordinator of
the negotiating team if the situation requires the presence of the whole team.
If a person demonstrating the intention to commit suicide carries a firearm or
pyrotechnic material, the intervention commander shall determine the boundary of

the danger zone and determine the area of the participating IRS units. The solution to this extraordinary event is the responsibility of the Police of the CR.

The tasks of ambulance crews include:

- the provision of professional medical assistance, not only to a person demonstrating the intention to commit suicide, but also to persons who might be injured in rescuing that person,
- providing urgent therapeutic intervention and physical restraint to a person, including the transport of a calmed person who has demonstrated suicidal intent to a destination medical facility,
- a statement of the person's death in the event of a completed suicide attempt and operations associated with the examination (examination) of the body of the deceased.

The following principles of dealing with a person with suicidal intentions can be selected from the above-mentioned State Type Activity 12/IRS:

- in a calm voice, address the person clearly, slowly, and simply and ask about his/her intentions.
- let the person talk about what they want. Do not convince her of anything. Speak as little as possible. It is important to listen. Negotiations should not worsen the situation,
- follow your feelings and reason. Get time until the arrival of a psychologist, crisis intervention, peer, or police negotiator.

In addition to the above, it is important when dealing with a suicidal individual:

- establish, in addition to verbal communication, emotional contact and show genuine interest (for example, pay attention to the harmony of verbal and nonverbal expressions in the crisis intervention, because the suicidal individual is very sensitive to this discrepancy, etc.),
- give space for verbal ventilation of aggression (for example, give the opportunity to describe to the suicidal individual all accumulated negative emotions or to describe the negative circumstances determining the origin of this crisis),
- briefly evaluate current circumstances with respect to: suicidal individual (for example is he/she oriented in person, time, and space? Is this the first suicide attempt or already has experience in these situations? etc.), the potential risk of the situation (for example, carry a weapon, do not stand too close to the edge of a bridge or window?
- try to work with a suicidal individual to form a way out of the situation so that the suicidal individual abandons the initial intention to end his/her life.

Procedures to avoid when interacting with a suicidal subject are:

- not to trivialize (for example, "Why do you make such a tragedy out of a given situation? The situation you describe does not seem as important to me as you perceive it.", etc.),
- not to condemn (for example, "You cannot talk, think, or feel like that!" etc.),

- not to blame (for example, "When I listen to you like that, you can actually blame yourself for the situation!", etc.),
- not to moralize (for example, "Suicide is a sin! Do you want to impose the burden of sin on your loved ones?", etc.),
- do not persuade (for example, "Please do not do this".… etc.),
- not to promise (for example, "If you give up your actions now, I will arrange for everything to be good.", etc.),
- not arguing with a suicidal individual,
- not provoke (for example, "I don't know why you do this play here, when you still don't have the courage to jump down the bridge".)
- not to accept the conditions imposed by the suicidal individual (for example, if we find out that he was betrayed and abandoned by his wife and he wants to take revenge on her by suicide, then in this case it is not reasonable to ensure the wife's presence at the scene of the emergency). suicide and to fulfil the vision, take revenge by the fact that the wife will not only see the execution of the suicide, but also others present will hear that she is responsible for his life) [7].

Suicidal issues, including communication with a suicidal individual, are given attention not only in the undergraduate education of IRS members but also in the context of lifelong learning. This comprehensive preparation reflects the requirements of practice where intervening rescuers encounter a variety of suicidal behaviours [1].

4 The Use of Crisis Communication in Managing Aggressive Behaviour

Aggressive manifestations of emergencies, victims and crisis victims or their relatives, acquaintances or only passers-by are often encountered by IRS members during their profession.

Tendencies to aggression generally arise from various factors: heredity, biological predisposition (for example, testosterone levels), sociocultural aspects, learning process (for example, imitation of the behaviour of parents or individuals in their immediate vicinity), group membership and the adoption of a group's culture, etc.

Under the conditions of providing professional help, aggression in individuals can occur in connection with various causes: pain, acute stress reaction, the presence of psychiatric illness, subjective feeling of danger to oneself, feeling helpless, feelings of injustice and injustice, drug effects, in chronically ill patients this may be in metabolic decompensation, manifestation of organic brain disease, etc.

Nakonečný states that: "aggression as an innate reaction to frustration was the cause of the social revolution. Adaptive behaviour has a selective advantage, but it can also be misleading, and aggression can compensate for the failure of adaptation" [6, p. 97].

Closely related to aggression is the term of hostility, by which we denote an individual's negatively or highly hostile attitude towards another individual or towards other individuals in general. "Hostility is defined less precisely, as a term referring to aggression, tendency to irritability, suspicion, non-cooperation, or jealousy" [8, p. 11]. For the sake of completeness, we state that the term "agitation" expresses psychomotor restlessness, which arises because of strong excitement.

Violent behaviour (violence) can be characterized as a type of aggressive behaviour aimed at physically harming another individual. This includes the following aggressive manifestations, such as beatings, kicks, scratches, throwing objects, demonstrating the intention to use weapons or even its use, etc.

For a better explanation and the interconnection of individual concepts, we will create a real situation. An individual who has a natural tendency to aggression becomes aggressive after attacking a drug and attacks his surroundings (violence). To the act of aggression comes on the basis of the interaction of the aggressor (the individual who initiates the conflict and is, therefore, the source of aggression) and the protector (the individual who is exposed to the attack of the aggressor, thus the protector of himself or another individual).

We recognize the following forms of aggression:

- auto aggression—focused on one's own person (for example, self-harm),
- hetero aggression—focused on other individuals,
- angry—this is a reaction to a previous situation. It can take the form of a gesture of resentment (impulsive nature of the reaction) or a manifestation of retribution (escalation of feelings of anger or hatred, and after an aggressive act, the individual has feelings of inner satisfaction),
- instrumental—represents a means to achieve the set goal,
- spontaneous—an act of aggression in which causing pain leads to emotional satisfaction of the individual,
- predatory aggression—represents a type of aggression in which an individual manifests as a predator and the other party represents its prey,
- fear-induced aggression—inadequate response of an individual to a felt stressful situation associated with an excessive feeling of fear,
- parental—a type of aggression that serves to protect offspring from a potential or real threat,
- irritant—arises based on various stimuli such as pain, hunger, fatigue, etc.
- under the influence of hallucinations, delusions, and other psychotic symptoms.

Different types of aggression according to their intensity are shown in Fig. 3.

It is important to clarify the sources and forms of potential aggression on the part of an aggressive individual, thus how aggression manifests itself in relation to intervening professionals, whether there are possibilities or ways in which he could mitigate negative impacts, etc. As part of effective procedures to calm an aggressive individual, it is therefore necessary for the intervening professional to have theoretical knowledge and practical skills regarding aggression and the possibility of influencing it.

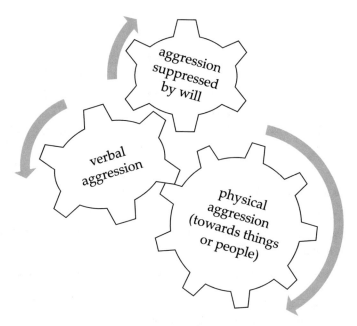

Fig. 3 Division of aggression according to the intensity of aggression

During contact with the victim of an emergency or crisis, in addition to the perception and assessment of the health condition, the intervening professional must also consider the risk of his potential aggression. This can be detected by detecting the following so-called warning signals before the conflict itself:

– physical signals—increased motor activity, psychomotor restlessness, increased muscle tone, stamping, tapping fingers on the table, tension in posture, clasped hands, excited gestures, threatening gestures, aggression against objects, etc.,
– signals based on the current mental state—speech becomes shorter, more concise, louder, verbalization of aggressive thoughts, reluctance to communicate, complaints, impaired ability to concentrate, perceive and process information, confused thoughts, the occurrence of qualitative disorders of consciousness: for example, hallucinations, etc.,
– signals from the perception of borders—the feeling of disturbing one's own or another's personal space,
– signals from previous experience with the patient—knowledge that the patient has previously had aggression, drug use, etc.

The scheme in Fig. 4 represents escalation of aggressive behaviour.

The process of calming the aggressor—de-escalation is a specific method of communication that can be effectively used to reduce tension. It consists of the following components:

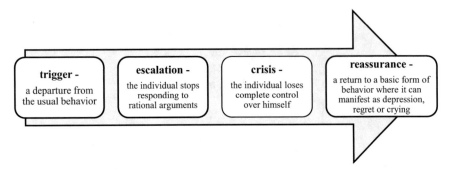

Fig.4 Phases of the attack cycle (modification according to [9])

- evaluation of the situation—collection of information and subsequent rapid analysis of the following factors:
  - the place and space in which the aggressive individual is located, including any escape routes,
  - the presence of objects indicating the possible use of psychedelic substances,
  - causing causes and sources of aggression, etc.

- crisis communication—adequate use of elements of crisis communication (for example, calm tone of speech, maintaining eye contact, slow entry into the personal to the intimate zone of the individual, etc.) can induce a sense of security and safety and thus eliminate or completely prevent other manifestations of aggression, for example, by the intervening professional will defer personal protective equipment in the event of an intervention involving the release of toxic substances [10, 11]. After the initial contact, it is advisable to remove the height advantage (for example, when the healthcare professional kneels to the lying patient). It is advisable to use sentences of the type:
  - "I am trying to understand what you are going through…".
  - "I understand that you have a right to be angry in this situation", etc.

It is also very convenient to use paraphrasing. Aggressive tendencies in an individual can be "disarmed" to a certain extent through kind and moderate behaviour.

Further escalation of aggression can be mitigated by the fact that the victim of an emergency or crisis intervening professional patiently explains all necessary information, continuously describing the procedures for providing professional assistance, including informing him in advance of any painful sensation associated with this process. Another recommended practice is for the victim to be involved in codecision on the next steps to address his or her health problems.

- negotiation tactics. The provision of professional assistance can be significantly complicated by aggression on the part of the victim, and thus the intervening professional gets into time constraints, among other things. He must manage to alleviate aggression through crisis communication in a short period of time and then carry out rescue procedures. To successfully master all above procedures, they must be able to handle effective forms of negotiation.

If the escalation of aggression through crisis communication or negotiation cannot be managed, the following procedures are recommended:

– use of pharmacotherapy - drugs are used that are expected to have a rapid onset of action: reassuring the patient and the associated possibility of shortening physical limitations,
– demonstration of strength,
– physical restraints (for example, manual fixation, bed straps, etc.).

The use of physical restraint must be accompanied by regular checking of the patient's state of health and, of course, entry in the medical records. The patient must be clearly sufficiently and in accordance with his current state of health, informed of the reasons for his physical limitations by healthcare professionals.

5 Conclusion

A large proportion of emergencies that escalate to a physical attack on an intervening professional are caused by a combination of the following factors: incorrect analysis and evaluation of the crisis, inability to call other intervening professionals, ignorance of the attack cycle, inadequate negotiated crisis communication and inadequate or unprofessional professionals.

Due to the above facts, members of the IRS are repeatedly theoretically educated and practically trained in the field of aggression as well as coping with aggressive behaviour. In addition, standardized procedures for dealing with violent incidents arising in the provision of professional care have been developed. The aim of these preventive measures is to minimize the impact on the mental and physical health of all parties involved.

If an IRS member is attacked as a victim of an emergency or crisis, it is advisable to offer him the possibility of professional assistance from either a peer crisis intervention agent or psychologist. This professional assistance is intended to prevent the short-term or long-term consequences of psychological trauma.

References

1. Tušer I, Bekešienė S, Navrátil J (2020) Emergency management and internal audit of emergency preparedness of pre-hospital emergency care. Qual Quant. https://doi.org/10.1007/s11135-020-01039-w
2. Tušer I, Jánský J, Petráš A (2021) Assessment of military preparedness for naturogenic threat: the COVID-19 pandemic in the Czech Republic. Heliyon 7(4). https://doi.org/10.1016/j.heliyon.2021.e06817

3. Potůček R (2020) Life cycle of the crisis situation threat and its various models. Stud Syst Decis Control 208:443–461. https://doi.org/10.1007/978-3-030-18593-0_32

4. Stodola P (2013) The real-time control algorithm and control curves for servomotors. Int J Circuits Syst Signal Process 7(2):118–125

5. Plamínek J (2012) In Czech *Konflikty a vyjednávání: umění vyhrávat, aniž by někdo prohrál.* 3., upr. a dopl. vyd. Praha: Grada. Poradce pro praxi. ISBN 978-80-247-4485-8. Psychological dictionary

6. Nakonečný M (2009) In Czech *Sociální psychologie.* Vyd. 2., rozš. a přeprac. Praha: Academia, 2009. ISBN 978-80-200-1679-9. Social psychology

7. Špatenková N (2011) In Czech *Krizová intervence pro praxi.* Praha: Grada. Psyché (Grada). ISBN 978-80-247-2624-3. Crisis intervention for practice.

8. Látalová K (2013) In Czech *Agresivita v psychiatrii.* Praha: Grada. ISBN 978-80-247-4454-4. Aggression in psychiatry

9. Honzák R (2015) In Czech *Svépomocná příručka sestry: (psychothriller).* Praha: Galén. ISBN 978-80-7492-142-1. Nurse's self-help guide

10. Otřísal P, Florus S (2014) Present and perspectives of physical and collective protection against the effects of toxic substances (in Czech). Chem Listy 108(12):1168–1171

11. Otrisal P et al (2018) Preparation of filtration sorptive materials from nanofibers, bicofibers, and textile adsorbents without binders employment. Nanomaterials 8(8):564. https://doi.org/10.3390/nano8080564

The Posttraumatic Care and a Crisis Intervention System for Parts of the Integrated Rescue System in the Czech Republic

Dana Rebeka Ralbovská and Pavel Otřísal

Abstract This chapter deals with the problems of the impact of emergencies, crisis situations, and traumatic events on the mental state of the Integrated Rescue system's (IRS) members. Due to the performance of the profession in a specific environment, the need to deal with crisis situations, and participate in eliminating the consequences of emergencies, it is clear that IRS's members get into situations not only physically but also emotionally demanding. The topicality of this topic is therefore based on the complexity of the work, especially on the increased psychological burden of these professionals and the risk of acute stress response, posttraumatic stress disorder, or other mental disorders arising in connection with the trauma. Attention is paid to the system of posttraumatic care, crisis intervention, and peer support within individual parts of the IRS in the Czech Republic (CR). The individual methods of critical incident stress management and components of posttraumatic development are also discussed.

Keywords Integrated rescue system · Posttraumatic intervention care · Peer support · Crisis intervention · Posttraumatic growth

1 Introduction

Members of the integrated rescue system (IRS) find themselves in a number of situations in which their psyche is affected by the resolution of emergencies and crisis situations, problem behaviour by affected persons, their families, or bystanders within their profession. Almost daily, they encounter emotionally tense situations,

D. R. Ralbovská (✉)
Faculty of Biomedical Engineering, Czech Technical University in Prague, nám. Sítná 3105, 272 01 Kladno, Czech Republic
e-mail: rebeka.ralbovska@fbmi.cvut.cz

P. Otřísal
Faculty of Physical Culture, Palacky University Olomouc, Třída Míru 117, 771 11 Olomouc, Czech Republic
e-mail: pavel.otrisal@upol.cz

such as traffic accidents with serious injuries or deaths, reporting deaths to survivors, crackdowns on armed offenders, negotiating with people, demonstrating intent to commit suicide, resolving difficult conflicts with aggressive people, etc. All above-mentioned crisis situations and extraordinary events have a negative impact on the mental state of IRS members. Although they undergo a demanding psychological examination when entering employment (they must meet the condition of mental health and resistance to stress), long-term actions of factors that provoke an acute or chronic stress response can cause psychological trauma and gradually contribute to a possible disturbance of the individual's mental balance.

With regard to the above facts, a system of providing posttraumatic care was created in the individual components of the IRS, and the issue of providing this care is the subject of the following chapter.

2 The Influence of Crisis Situations and Traumatic Events on the Psyche

The crisis situation can be described as a situation that is accompanied by massive emotional pressure, excessive exposure to stressors and fundamentally deviates from the ordinary experience of the individual [1]. Crisis situations related to the performance of a profession in IRS include: sudden death of a child, failure of cardiopulmonary resuscitation, devastating injury, injury or death of a colleague in the profession, meeting a victim of extreme damage to the integrity of the human body due to a crime, especially cruel execution of a suicidal act, mass disability with a large number of wounded and dead and, last but not least, heavily mediated (presence of indomitable media and family members of victims at the scene) emergency or disaster, terrorist attack, mass disaster, etc. [2, 3].

The individual is overwhelmed with feelings of intense fear, danger and restlessness in a traumatic event. The traumatic event often causes a violation of physical and mental integrity.

In the case of catastrophic events caused by natural disasters, mental integrity is violated and negative emotions arise, which are, however, partially offset by a sense of belonging and mutual help (for example neighbourhood assistance in floods, fires). On the other hand, the human psyche is more difficult to cope with traumatic events intentionally caused by another person, especially in cases of brutal violence. Especially if the victim was a small child, or it was a particularly brutal act.

Psychological trauma "is caused by an extremely stressful experience or a long-lasting stressful situation that has the following characteristics: the cause comes from outside, is extremely frightening and induces the experience of endangering life, physical or mental integrity and induces feelings of helplessness" [4]. A state of helplessness can also be achieved in the case of interventions associated with the release of toxic substances, in which the intervening specialists are forced to use personal protective equipment [5, 6].

It is therefore a significant violation of mental integrity, which manifests itself especially in the emotional area.

As Ralbovská [7] states, based on the experience of a highly traumatizing event, within individual victims they develop psychological traumatization, which is divided into three types according to the way they encounter trauma:

– primary traumatization—individuals are directly affected by the negative effects of a traumatic event (for example, in the case of victims of domestic violence, these are all forms of harm by the aggressor),
– secondary trauma—the individual encounters trauma to another, often close person,
– tertiary trauma—a person is in direct contact with a person primarily or secondarily traumatized and experiences negative emotions through it.

It is necessary to keep in mind that members of the IRS are endangered by secondary and tertiary trauma.

The concept of vicarious traumatization was introduced into the literature by [8]. Using their theory, they tried to clarify the cause of traumatic symptoms occurring in a number of professionals (for example paramedics, etc.), who themselves did not experience primary trauma, but based on the performance of their profession worked with victims of emergencies, crises and traumatic events. As a result of secondary trauma in these individuals, there are long-term changes in cognitive patterns, but also intrusive memories (flashbacks) of traumatic events that are typical of posttraumatic stress disorder. It is therefore a mediated trauma.

Crisis definitions usually cover three basic components of a crisis:

– a specific trigger event occurs,
– the individual perceives it as threatening, dangerous,
– the usual ways of coping with it fail, if the situation cannot be managed, a crisis will ensue [9].
– the usual ways of coping with it fail, if the situation cannot be managed, a crisis will ensue [9].

Each individual has a different degree of resistance to the effects of disaster, and each also deals in different ways with the intensity of the situation. In general, a psychic reaction is a response to an event. Victims' responses are highly individual and show high intra-individual variability over time. The primary reaction of every person is to protect their own life, the lives of loved ones and their property. At this point, the affected person was abandoned by any emotion. During the initial impact phase, we may encounter a wide range of behaviours and emotional expressions.

As a result of the negative traumatic effect, affected victims are created, including those who directly "saw, heard, touched, felt", their loved ones (or survivors)—family, friends, colleagues, or neighbours and residents of the place where the disaster occurred, and IRS members who intervened at the scene of the emergency are also affected. In the literature we can also meet the term survivor—the one who survived [7].

Symptoms that usually appear within one hour of an emergency, crisis or traumatic event include: anxiety, helplessness and indifference, limited perception of oneself and one's surroundings, detachment from reality, automatic and impersonal behaviour, feeling affected the person is not moving to outside his body, etc. There are also the following physical symptoms such as sweating or palpitations, increased blood pressure, stomach pressure, limb tremor, dizziness, headache, hot flushes or cold, feelings of fatigue and exhaustion.

As a result of the above-mentioned negative events, an acute stress reaction, posttraumatic stress disorder, reactive depression, reaction to loss or persistent personality changes may occur.

Among typical symptoms of the above disorders (except the acute stress response) can be included: repeated feelings of fear and anxiety, difficulty falling asleep and sleep disorders, psychotic reminiscences—episodes of recurrent trauma that take the form of lingering memories or occur in dreams, emotional numbness, persistent feeling alienated, avoiding contact with people, inadequate response to environmental stimuli, exaggerated startle reactions, avoiding activities and situations reminiscent of trauma, depression, suicidal thoughts, etc. [10].

3 Posttraumatic Care for Parts of IRS

Post-traumatic care at the CR Fire and Rescue Service (FRS) was established based on the knowledge that the profession of firefighter is one of the most endangered professions, with extreme physical and mental strain. In 2002, the concept of the psychological service of the CR FRS was approved and psychological workplaces were established in individual regions. To ensure the provision of posttraumatic care, a Post Traumatic Care Team (PCT) was established in each region. The team coordinator is a psychologist of the regional fire brigade. The members are trained firefighters with personal preconditions and motivation to help others who are appointed director of the regional fire brigade.

Membership in the PCT is voluntary and a member or employee who completes professional training in the form of a Posttraumatic Care course, according to the curriculum approved by the Ministry of the Interior—General Directorate of the CR FRS can become a member. Members of the PCT then provide posttraumatic care to victims of crisis situations and emergencies, members of the CR FRS, as well as their families. The Concept of the Psychological Service of the CR FRS for the years 2017–2025 is currently valid.

For the Police of the CR, posttraumatic intervention care (PIC)—including the procedure for requesting—it was regulated by a Binding instruction of the Police President No. 21/2009, which was further amended by Binding instruction of the Police President No. 79/2010. The methodological management of the PIC system was the responsibility of the Chief Psychologist of the Police of the CR, who proposed the coordinators of the individual PIC teams. The guarantor of the ethical and professional quality of the PIC system was the psychological workplace of the Ministry of the Interior (Department of Psychology, Department of Personnel, Ministry of the Interior of the CR). This provided the training of crisis interventions and co-participated with the chief psychologist of the Police of the CR in the conception and development of the PIC system. Within the Police of the CR, there were regional PIC Teams, which have been gradually replaced by the Peer Support System since 2016.

Within the Police of the CR, members and employees can currently use the following options for posttraumatic care:

- services of a police psychologist,
- crisis intervention,
- peer support,
- anonymous helpline in crisis.

The police psychologist offers police officers psychological care, consultations, psychological counselling, crisis intervention and possibly also psychotherapy. Psychological care at the Police of the CR uses knowledge of the police environment. Police psychologists undergo accredited training and further continuous education as a part of lifelong learning.

According to the Binding Instruction of the President of the Police No. 231 from the 26th of September 2016 on psychological services, crisis intervention means "short-term specialized assistance provided for the purpose of reducing adverse psychological consequences caused by a traumatic event and restoring mental balance; Crisis intervention is provided, for example, in the form of first psychological assistance, a crisis intervention interview or by arranging contacts to obtain further professional assistance" [11].

The system of peer support was established by the Binding Instruction of the President of the Police No. 231 from the 26th of September 2016 on psychological services In the CR. The peer support system is defined as follows:

(1) the system of peer support means the provision of psychological support to police officers and employees, or their relatives who are in a complicated, psychologically demanding life situation via providers of peer support, so-called peers.

(2) the system of peer support consists mainly in offering an interview, sharing feelings and problems, specific help, or information (for example, about appropriate procedures, institutions that can contribute to solving a current problem), or in offering to provide psychological help to other professionals.

(3) The aim of the peer support system is to prevent the development of psychological difficulties of police officers and employees and to expand the possibilities of psychological support [11].

The system of peer support emphasizes the so-called advisory interview, in which the recipient of the service can share their experiences and emotions with the peer. The peer then (at his own discretion) offers specific professional assistance or provides information (for example about appropriate coping strategies, about institutions that can contribute to solving the current problem, etc.). It can also mediate the help of a psychologist or an expert in the issue under whose professional responsibility the intervention was carried out. In the case of interventions involving the release of toxic substances, the interview may, for example, focus on building confidence in the individual protection equipment with which the intervening specialists are equipped [12].

Another variant of professional assistance for IRS members is the operation of an anonymous Crisis Assistance Line. Employees of the anonymous Crisis Assistance Line provide round-the-clock telephone psychological support to members and employees of the Police of the CR, the CR FRS, as well as their family members, or close people.

As part of the emergency medical service, the System of Psychosocial Intervention Service (SPIS) has been enshrined in Act No. 374/2011 Coll., on the emergency medical service, since 2012. This professional service is focused on supporting health care workers, especially those who have been and are most exposed to acute stress and posttraumatic influences. These are employees of the rescue service (pre-hospital emergency care), emergency income (hospital care), but also other fields. However, the service can be used by healthcare professionals if necessary, regardless of the department where they work.

The objectives of the SPIS are: prevention, to create resistance to psychological stress, to lead in direction to the understanding of the response to a crisis event, to normalize the stress response, to restore the normal level of functioning of an individual, to teach appropriate adaptation to stress and psychological stress, etc.

Within the activities of the SPIS, attention is paid to the education of individual members. These are interventions that provide so-called PEER support to their colleagues within the organization. As well as medical interventions that provide psychosocial intervention care to those affected at the scene of an emergency. The National Centre for Nursing and Non-Medical Health Professions participates in the education of these professionals.

From the 30rd of March 2020, in connection with the occurrence of COVID 19, a Peer Support Line for Healthcare Workers was established. As stated by [13]: "The purpose of the peer support line is to provide health professionals with a safe space for supportive conversation, basic recommendations for managing their stress, anxiety, frustration, emotions, meeting basic needs."

4   Critical Incident Stress Management

Emergencies, crisis situations and traumatic events have certain similar features: they arise suddenly, unexpectedly, their further development is often unpredictable, they are accompanied by strong emotional pressure, etc. Preparation for their effective management therefore requires a specific procedure.

Parks [14], based on knowledge from practice in the education of employees of security and rescue services, recommends educating these employees in lifelong learning in the field of Critical Incident Stress Management (CISM). He justifies his opinion by the fact that members of the IRS are obliged to provide the first psychological assistance to the affected persons, but also to their colleagues, at the scene of emergencies and crisis situations.

CISM methods can be characterized as "a prevention-oriented process that has the nature of discussion, support, structured encounters and education about stress. It is not about treatment (therapy, psychotherapy) or advisory" [15].

The goal of Critical incident stress management is to enable the affected persons to return to normal life more quickly and to reduce their likelihood of developing a mental disorder after a traumatic event. It includes preliminary preparation for managing an acute crisis situation up to techniques that eliminate its consequences.

Demobilization, defusing and debriefing are crisis intervention techniques that effectively serve to cope with traumatic events.

Demobilization is used in the event of a large-scale emergency. It should be performed immediately after moving from the intervention, for ten minutes, followed by a 20-min relaxation. The aim of demobilization is to eliminate unpleasant mental states at the end of the shift with an above-limit load, reduce the stress that is associated with the load, and start the forces' recovery.

Defusing takes place immediately or within 8 h after a traumatic event. It most often takes the form of a relaxing conversation, which aims to eliminate accumulated emotions and alleviate the impact of acute stress. The main goal is to alleviate cognitive, emotional, and physiological symptoms, so it takes place in the near future after the event. The group participating in the session should be small, homogeneous, composed of people who have experienced a traumatic event together.

An important therapeutic element is the collective grasp of the event and the finding that all negative emotions and experiences are essentially a normal response to an abnormal event. The defusing participant is also helped by the finding that other feelings of his colleagues appear similar to his own (for example doubts, fears, feelings of guilt, etc.). Defusing may not always have professional guidance. It is possible for the commander to create an atmosphere of well-being and support after a demanding intervention. It can often occur spontaneously, by jointly caring for those who need it most from the team.

Debriefing is a discussion associated with education. It works with memories of a traumatic event and with the thoughts that the memories evoked. It is an intervention that focuses on reducing stress, stabilizing the situation, and mobilizing forces so that the intervening professional can function normally as soon as possible [16]. It is performed in small groups from one to ten days after the event. It facilitates the mental closure of the event. The interview is usually conducted by a psychologist or peer at the debriefing venue. Debriefing should be carried out as part of a crisis intervention program, should be led by a trained expert and should take place at a certain distance from the event.

The ways in which an individual successfully handles stressful and traumatic events are called coping strategies. This strategy means how an individual adapts and conforms to the demands of life. Individual procedures are determined by basic responses to stress. The basic goal of these strategies is to ensure one's own survival.

In addition to the innate tendencies of an individual, the coping style is also influenced by the personality characteristics of the given individual and can use the feedback to shape the personality characteristics.

Considerable attention is also paid to the development of resilience, which can be understood as the general ability of an individual to develop with the intentions of normal or healthy development, despite the presence of crisis situations and traumatic circumstances. For members of the IRS it is important to understand that it is possible to find meaning in what happened to the individual and to take away the positives from this traumatic event in the future. According to the literature, the central goal of resilience can be considered the individual's ability to effectively solve problems and cope with stress [17].

According to the authors Lepore and Rovenson [18], we distinguish the following forms of resistance to traumatic events:

– recovery—after the end of the negative effects of trauma, the individual is able to completely eliminate the negative consequences of the event,
– resistance—can also be characterized as a specific way of processing the negative effect of the experienced event, in which the effect of the event on the individual's behaviour cannot be observed,
– reconfiguration—due to the negative effect of trauma there is a change in the personality of the affected individual, either in a positive direction (based on the traumatized event and its subsequent processing will be ready to successfully manage other traumatic events in the future) or in a negative direction, when there is temporary or permanent adverse personality change. Symptoms of an unwanted personality change can then take the form of feelings of hopelessness, threat, and alienation.

Defensive reactions that occur because of an emergency, crisis or traumatic event can be divided into adaptive (effective) and maladaptive defensive reactions. Maladaptive defensive reactions include those conditions where the individual is unable to cope with the effects of a traumatic event and is, for example, excessive aggression, social isolation, problem solving with addictive substances, etc.

Members of the IRS may feel guilty in connection with the performance of their profession, in the form of strong emotional stress, which is then intensively reflected in the work area. In the professional literature, you can find the concept of performance-related guilt, which can be observed within members of the IRS, who provide professional assistance at the scene of the emergency and their goal is to acutely mitigate the effects of the emergency on the affected persons. The basic symptoms of feelings of guilt in connection with the performance of the profession include:

– retroactive feelings of failure,
– reproaches for incorrect or slow decisions,
– slow dexterity in providing professional care, etc. [17].

5 Posttraumatic Development

Posttraumatic development or growth, which is related to the experienced trauma, can be characterized as a positive change in the individual's cognitive, educational and emotional life. It arises as a result of an individual's struggle with a life crisis or traumatic event that is important to him. By development in this case, we mean a change in which the individual gets above his current level of adaptation, psychological functioning and understanding of life. The long-term legacy of trauma includes both loss and gain. The survivor may concentrate on one or the other, but both are present—loss of illusions and expectations, loss of the future and readiness, sensitivity and strength [19].

Initially, after experiencing traumatic events, cognitive processes tend to be rather compulsive and appear in the form of intrusive thoughts, where the individual is constantly engaged in experienced events. The individual must endure the enormous distress that manifests itself because of trauma. Only after going through this stage does another phase of rumination begin, which is no longer significantly overwhelmed by emotions and takes place rather in the form of contemplation, thinking about the consequences of a traumatic event from various perspectives. As a result, the individual tries to find a certain meaning of the traumatic event and its effects. Thus, in addition to the negative aspects of the impact of the event, for the first time, some positive aspects may also begin to manifest themselves. The individual suddenly realizes that his world has changed and that under the pressure of these changes, he himself has changed internally.

As a result of a traumatic event, posttraumatic development can occur precisely due to traumatic experience and the need to cope with distress. It is not just a reestablishment of the balance that has been lost because of the crisis, but a process in which the individual must make a significant effort to bring about change. It is not a gain from the traumatic event itself, but rather a way of reacting to cope with the consequences of the traumatic event. Posttraumatic development is both an interpretive process that individuals perform in order to reinterpret experience as manageable, and the result of adaptation [17].

Based on the experienced traumatic event, the individual acquires life wisdom within the posttraumatic development, which can have the following dimensions:

- recognizing and managing uncertainty—openness to change and experience is an essential feature here,
- integration of emotions and cognition—means understanding one's own emotional experiences,
- recognition and acceptance of one's own possibilities—this dimension has above all an existential dimension, it includes the perception of finiteness and at the same time the real values of life.

Posttraumatic development can manifest itself in the following areas:

- the first area is relationships with other people—development can manifest itself in, for example, greater involvement of other people in their own lives, internal or external, greater compassion for their needs, a sense of greater belonging and closeness, better expression of feelings towards others, more attention to relationships in general,

- the second area is new life opportunities—there the development manifests itself, for example, in the discovery of new areas of interest and opportunities, improved management of time, directing the will to things that need to be changed,
- The third area is personal development—here the development is manifested, for example, by a feeling of increased self-confidence, better management of life difficulties, and increased acceptance of the life course,
- the fourth area is spiritual change—development is manifested, for example, by understanding the spiritual needs of oneself or others, by greater interest in the spiritual realm, sometimes by strengthening faith or religious needs,
- the last, fifth area is the appreciation of life—development is manifested, for example, by thinking about the essential things in life, by recognizing the possibilities that life gives us all.

The model in Fig. 1 illustrates the course of posttraumatic development.

Posttraumatic development represents a part of the theory of well-being, especially in the dimensions of self-acceptance, autonomy, environmental management, and the meaning of life. From the character traits, extroversion, openness of experience, friendliness, and conscientiousness associate positively with posttraumatic development [17].

Posttraumatic development can also be understood as the opposite of posttraumatic stress disorder, which leads to the emphasis on the fact that a certain positive benefit in the form of development is manifested even in the most serious traumatic circumstances and despite the stress that simultaneously causes.

Posttraumatic development is not just about emergency victims, it is also about intervening professionals. Although these professionals are trained in Critical Stress Management and Posttraumatic Care, they are exposed and threatened to psychological trauma by providing assistance to emergency victims contact with primarily traumatized (emergency victims) but also secondary traumatized victims (family members, acquaintances and friends of emergency victims). On the one hand, they provide professional assistance, posttraumatic care to victims of emergencies, but they are also often exposed to high-risk, life-threatening situations. Thus, they are negatively affected by the stress they get from helping victims (for instance, the ability to make quick decisions, often in the absence of information, coping with emotionally negative states of victims, interactions with individuals who are under the influence of psychedelic substances, the need to report bad news and death reports, etc.). The stress they experience in the direct interaction and assistance provided to victims who are in danger to their health or life and within which they have their specific crisis needs also plays a role.

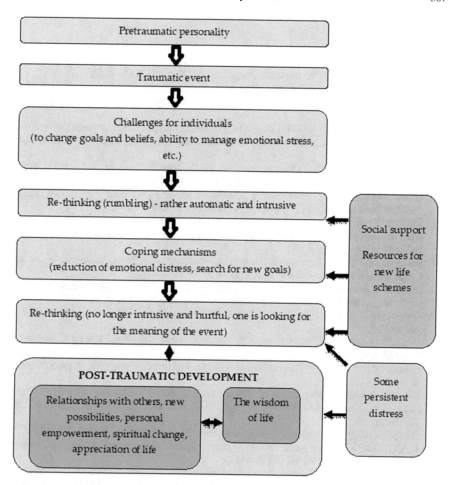

Fig. 1 Posttraumatic development (modification according to Lawrence Erlbaum Associates [19])

6 Conclusion

If a member of the IRS subsequently experiences a significant impact on his/her current way of life after experiencing a highly traumatic event, it is up to him/her to process his/her experience. It is also up to him/her whether he/she seeks professional help (contact or noncontact methods of posttraumatic care) or uses coping strategies that are used to manage emergencies, crisis situations or traumatic events. If he is trying to find a way that will help him/her to achieve a positive conclusion of a traumatic event, including its processing, then it can be said that it is a posttraumatic development.

References

1. Potůček R (2020) Life cycle of the crisis situation threat and its various models. Stud Syst Decis Control 208:443–461. https://doi.org/10.1007/978-3-030-18593-0_32
2. Tušer I, Bekešienė S, Navrátil J (2020) Emergency management and internal audit of emergency preparedness of pre-hospital emergency care. Qual Quant. https://doi.org/10.1007/s11135-020-01039-w
3. Tušer I, Jánský J, Petráš A (2021) Assessment of military preparedness for naturogenic threat: the COVID-19 pandemic in the Czech Republic. Heliyon 7(4). https://doi.org/10.1016/j.hel iyon.2021.e06817
4. Špatenková N et al. (2011) Crisis intervention for practice. In Czech Krizová intervence pro praxi. Praha: Grada. Psyché (Grada). ISBN 978-80-247-2624-3
5. Otřísal P, Florus S (2014) Present and perspectives of physical and collective protection against the effects of toxic substances (in Czech). Chem Listy 108(12):1168–1171
6. Štěpánek B, Otřísal P (2012) The development and establishment process of centres of excellence in North Atlantic organization. Croatian Journal of Education-Hrvatski Casopis za Odgoj i obrazovanje 14(1):169–174
7. Ralbovská DR (2017) Psychological aspects of emergencies. In Šín R Disaster medicine. In: Czech Medicína katastrof. Praha: Galén. ISBN 978-80-7492-295-4
8. McCann IL, Pearlman LA (1990) Vicarious traumatization: a framework for understanding the psychological effects of working with victims. J Trauma Stress 3(1):131–144
9. Špatenková N et al. (2017) Krize a krizová intervence. Praha: Grada. Psyché. ISBN 978-80-247-5327-0
10. Mitchell JT (2003) Major misconceptions in crisis intervention. Int J Emerg Ment Health 5(4):185–197
11. Police of the Czech Republic's President Binding instruction No. 231/2016, on psychological services
12. Otrisal P et al (2018) Preparation of filtration sorptive materials from nanofibers, bicofibers, and textile adsorbents without binders employment. Nanomaterials 8(8):564. https://doi.org/10.3390/nano8080564
13. Humpl L (2020) Peer support line (in Czech). https://1url.cz/fzh0p. Accessed 27 Sep 2020
14. Parks E (2015) Building a foundation for crisis intervention in Eastern Europe. Int J Emerg Ment Health Human Resil 17(1):352–355
15. Baštecká B (2005) Field crisis work: psychosocial intervention teams. In: Czech Terénní krizová práce: psychosociální intervenční týmy. Praha: Grada. Psyché (Grada). ISBN 80-247-0708-x
16. Stodola P (2013) The real-time control algorithm and control curves for servomotors. Int J Circuits Syst Signal Process 7(2):118–125
17. Mareš J (2012) Posttraumatic human development. In: Czech Posttraumatický rozvoj člověka. Praha: Grada. Psyché (Grada). ISBN 978-80-247-3007-3
18. Lepore SJ, Revenson TA (2006) Resilience and posttraumatic growth: recovery, resistance, and reconfiguration. In: Calhoun, Tedeshi (eds) Handbook of posttraumatic growth. Research and practice. ISBN 0-8058-5196-8
19. Lawrence G, Tedeshi RG (2006) Handbook of posttraumatic growth: Research and practice. sRoutledge Member of the Taylor and Francis Group. ISBN 978-0805857672

Scenarios of Supply Chain Security and Resilience Under the Conditions of Uncertainty and Sudden Demand Shifts

Pavel Foltin⬤ and Petr Tulach

Abstract Supply chain security and resilience represent the key aspects of the long-term sustainability and well-being of current societies. The ability of global supply chains to be resilient and flexible has to face a wide range of threats, uncertainties, and sudden shifts in case the time and necessary resources for the adequate reaction are limited. Challenges affecting supply chains, which companies have to deal with, result from a changeable environment and threats in the form of sudden demand shifts as well as unavailability of necessary resources caused by security threats. A conceptual model based on a modified CPM method was created on the basis of the scientific literature review research, focusing on available open sources and data, and subsequently applied to the further development of global distribution chains, logistics infrastructure, and availability of logistics capabilities. Possible approaches and methods of mitigation of the constraints to supply chains were discussed.

Keywords Supply chain security · Supply chain resilience · Logistics infrastructure · Scenarios development

1 Introduction

The contemporary world is becoming increasingly more complex, interdependent, interconnected, and mutually influenceable. This applies also to the life of the society, the range of influencing factors, and their variability over time. The interconnectedness and complexity can also be seen in the distribution of material stocks and the overall distribution chain implementation [1]. In the civilian concept, global supply

P. Foltin (✉)
Department of Logistics, Faculty of Military Leadership, University of Defence, Kounicova 65, 662 10 Brno, Czech Republic
e-mail: pavel.foltin@unob.cz

P. Tulach
Logio, s.r.o., Evropská 2588/33a, 160 00 Praha 6, Czech Republic
e-mail: tulach@logio.cz

© The Author(s), under exclusive license to Springer Nature Switzerland AG 2022
I. Tušer and Š. Hošková-Mayerová (eds.), *Trends and Future Directions in Security and Emergency Management*, Lecture Notes in Networks and Systems 257,
https://doi.org/10.1007/978-3-030-88907-4_23

chains provide for the operation of national economies, primarily through the distribution of material stocks across continents [2]. The concept of the armed forces involved in multinational military operations includes transport of entire units with military equipment, supplies, and personnel to crisis areas outside the territory of the despatching states. The subsequent setting up of the military logistics chain in relation to the civilian logistics infrastructure availability in the given regions is related to it.

Civilian and military entities are dependent on the logistics infrastructure elements due to the distances covered and have to face a whole range of influencing factors. One of the sources of potential logistics infrastructure disruption is natural factors, however, the impact of the human factor, whether intentional or unintentional, also plays a significant role [3]. For this reason, it is appropriate to focus on the identification of the key influencing factors and their impacts on the implementation of distribution chains [4].

2 Influencing Factors of Supply Chain Security and Sustainability

Technological progress, further supplemented by information and communication aspects, can be considered a key aspect of changes during the 20th and early twenty-first centuries. The data availability and its sharing at real time affect almost all aspects of society, from livelihood to trading, and the possibilities of material stocks distribution [5, p. 2]. It is possible to perceive the ever-increasing dynamics of social change in the context of technological development together with a growing spectrum of threats.

Rising trade levels can affect international politics in a very subversive way, while the trade growth within global economic systems will neither reduce the likelihood of international tensions nor provide greater international stability [6 p. 65]. Analyses of factors influencing the security environment are reactive in nature, at the expense of preventiveness [7, 8]. Moreover, current threats cannot be clearly defined or their boundaries defined [9]. In this context, David Simchi-Levi defines the paradigms of current supply chains as follows [10, p. 3]:

- The supply chain takes the form of a network with a global overlap;
- Individual elements of supply chains often have different, conflicting goals, e.g., suppliers seek to deliver large volumes of goods and materials in constant times and thus minimize unit costs which is in clear contrast to the view of customers who prefer just-in-time deliveries according to current preferences and needs;
- Supply chains form dynamic systems that are constantly evolving, both customer demand and supply offer change and at the same time the relationships between the various actors of the logistics chain also evolve and change over time;
- Supply chains change over time, e.g., due to the seasonal nature of demand (e.g., pre-Christmas capacity utilization of Asia-Europe capacity supply chains in weeks

48–50 of the year, similarly in weeks 3–5 of the year preceding the Chinese New Year celebrations), but also due to advertising and product promotion.

The complexity of the influencing factors depends on the change trends, which are reflected in the logistics capabilities and sustainability of distribution chains [11 p. 350]. The extent of this risk is enhanced if threats are combined [12 p. 256] or if other negative security factors act in parallel [13, pp. 13–14]. The practical implications are:

- Limited usability of logistics nodes capacities;
- Time unavailability of transport performance;
- Lack of experienced staff.

3   Goals, Methodological Approach, and Research Limitations

The aim of the research was to identify the theoretical links and interactions between the decisive factors determining the efficiency of distribution chains and the possibility of their subsequent optimization in conditions of uncertainty and indeterminacy. The following research question was formulated to fulfil the set goal:

Q   Is it possible to determine a conceptual model usable for predicting the development of the intensity of the negative factor activity affecting distribution chains using existing standard analytical tools?

An analysis of available scientific literature sources was used, from a methodological point of view, to process the initial preconditions of conceptual model creation. The results of the research into the initial assessment of attributes and criteria for the Critical Path Method (CPM) modification and its subsequent use in the conceptual model creation were elaborated within the habilitation thesis *Selected Aspects of Logistics Chains Security of Armed Forces* [14], the study *Application of Modified CPM on Portfolio of Security Criterion on Logistics Chains* [15] and the study *Assessment of military preparedness for naturogenic threat* [16]. The approach of neuro-dynamic programming of prof. Bertsekas, especially his main idea of the interconnectedness of the weight of current decisions in terms of optimality of further future steps, was applied based on the identified decisive factors [17]. Following these findings, a proposal to extend the standard CPM and other evaluation criteria, which are easily predictable and allow for an extension by a time sequence of the future development of influencing factors affecting the distribution chain, was made.

The research carried out was limited to data available on 31 December 2020. The research itself was limited to the assessment of available scientific literature sources and the subsequent determination of a basic approach to the conceptual model creation. No classified information or data was used in the course of the research.

4 Analytical Consideration

The additional risks affecting distribution chains stem from the distance travelled when chains run across national borders of various states with different conditions and availability of logistics infrastructure. Other influencing factors with potential impacts on the implementation of distribution chains may also be price rate changes, on which the prices of transmission services are based, or risk factors in the country or region [18]. In terms of possible disruptions to the functionality of logistics systems, it is not possible to completely eliminate uncertainty [19 p. 322], however, it is possible to work with it and reduce its probability, either to act proactively in order to minimize the risks or increase the ability to react flexibly in case that an undesirable situation has already occurred. The following approaches can be used to manage fluctuations and possible disruptions to logistics systems [20]:

- *Risk avoidance*: e.g., by choosing suitable suppliers, changing the approach in logistics chain management;
- *Risk mitigation*: e.g., by using a larger number of resources, timely involvement of an additional number of suppliers, quality management of implemented processes, creation of insurance stocks;
- *Risk compensation*: e.g., by creating increased stocks to cover identified crisis scenarios;
- *Risk transfer*: e.g., transfer of risk to a third party.

It is possible to identify four main categories of dependence in terms of the interconnection of supply chains with the critical infrastructure of the state, namely physical, cybernetic, geographical, and logical [21, p. 12]. Disruption of some of these links brings negative consequences for the functionality and availability of logistics systems. When designing the distribution chain structure, the incorrect classification of the terms "unknown" and "unlikely" appears. Highly unlikely events, although identifiable in the design of the distribution chain structure, are usually not included in the decision-making process itself. As Nathan Freier notes: *"What looks strange is thought improbable, what is improbable need not be considered seriously"* [22, p. 18].

The effect of influencing factors can be from short-term, lasting for several minutes, up to permanent damage. At the same time, it is possible to identify the effect of disrupting the distribution chain functionality, which is located in one part of the chain or gradually passed on within the chain, thereby threatening the chain functionality as a whole. Due to the nature of a possible disruption to the distribution chain functionality, uncertainty in the chain can manifest itself in three basic forms [23, p. 206]:

- *Chain deviations*: a change in one or more control parameters, such as changes in costs, demand, or delivery times within the distribution chain, however, without a change in the logistics chain original structure;

- *Chain disruption*: i.e., a significant change in the logistics chain structure due to the unavailability of some logistics nodes or their edges, due to unexpected events caused by human or natural factors;
- *Disruption due to a disaster*: i.e., temporary irreversible closure of the logistics chain due to an unexpected disaster, which may cause total disruption of the functionality of logistics nodes and transport capabilities.

Two main sources can be identified in terms of possible disruption of logistics systems, namely disruption caused by intentional human behaviour [24] or caused by anthropogenic natural factors. The main groups of potentially negative factors of the logistics chains disruption in international distribution chains can be identified as follows [11, p. 55]:

- *Internal factors*, i.e., the sum of influencing factors of elements and bonds within the chain;
- *Factors outside the specific logistics chain*, but still in the immediate vicinity of the logistics chain under consideration, which may adversely affect the chain functionality;
- *Factors originating completely outside the considered distribution chain*, which are therefore difficult to predict and influence in any way.

In addition to the identified sources of possible disruption of the distribution chain functionality and usability, it is also possible to identify a variable level of their influence [25]. Unexpected events can disrupt the material flow, i.e., the supply function, on the way from the original suppliers to the end-users [12, p. 7]. The emergence of threats manifests itself in irregular, catastrophic, and hybrid threats resulting from possible disruptions of functionality, based on an unfavourable supply system internal structure (so-called *threats of purpose*) or threats arising from the chain surrounding (so-called *threats of context*) [22, p. 7].

The importance of the ability to predict potential trends in new threats, which have not yet been properly taken into account, in order to provide the creation of suitable distribution chain variants of design and implementation, has been growing [26, p. 3]. Applications of localization models, according to the number of localized objects, number of available places, number of deployed objects, and interdependence of objects placed represent a possible approach to the optimization of created and evaluated variants of the distribution chain structure design. The number of chain objects has an analogy in looking for the number of distribution warehouses in the supply chain distribution network, their characteristics, and geographical delimitation [27]. When considering localization, it is necessary to take into account future developments, i.e. the perspective of the chosen location during the individual life cycle phases of the considered distribution chain [28, p. 9]. This creates the preconditions for the sustainability of logistics nodes and their capacities.

Two basic types of indicators, namely static and dynamic, can form the basis for evaluating the elements, elements functionality, and supply chain links. Static indicators can be considered to be the outputs of a more complex evaluation, which is usually time-consuming or resource-intensive, e.g., expert evaluation is time-consuming and at the same time demanding in terms of expert capacity. These static data changes 1–2 times a year. In addition to these static indicators, it is also possible to identify dynamic indicators, which are based on ongoing evaluations, where the outputs of these evaluations may in extreme cases be available in real, or almost real time (e.g., delays in minutes).

The use of the neuro-dynamic programming approach by Prof. Bertsekas [17, p. 1], where the initial premise is the sequence of decisions, i.e., individual decisions are made in individual steps, which corresponds to the standard military decision-making process when planning an operation, can represent a possible approach how to include dynamics into the decision-making process. The assessment of the current variant of the decision in relation to other subsequent variants of future decisions can be formulated as follows:

$$i_V \approx j \cdot p_{ij}(V)$$

where

V Adopted decision variant
i Status (situation, conditions) under which the variant of decision V has been adopted
j New situation (situation, conditions), after decision V has been adopted
$p_{ij}(V)$ Probability of reaching a new j state.

It is necessary to include and respect the cost factor $g\,(i,\,V,\,j)$ into the assessment of variants for the V_i decision implementation. Despite the inclusion of a cost factor, the decision itself can be considered incomplete as it takes into account the number of costs, while not taking into account other aspects of the decision, e.g. how much the future j state is desirable. The contribution of prof. Bertsekas's amplification of the view on the assessment of individual variants by evaluating the future state j after the current decision V from the point of view of optimal costs $J^*(j)$ can be expressed as follows [17], p. 2):

$$J^*(i) = \min_V E\big\{g(i, V, j) + J^*(j) \mid i, V\big\}, \quad \text{for all } i,$$

where

j State following the state i
$E\{i, V\}$ Indicates the expected value of j according to the selected variant V in the initial state i.

In the general definition of Bertsekas's neuro-dynamic programming, individual decisions are evaluated on the basis of the sum of the expected costs of the current designs and the optimal expected costs of all future decisions. However, this approach is suitable for a limited number of decisions to be made, e.g., in a game of chess, where the game has a clearly defined end of the game, i.e., a checkmate or a draw. However, in real conditions for complex issues such as planning and subsequent implementation of the proposed distribution chain, it is usually not possible to determine in advance the final number of decisions to resolve the decision issue. This approach is also not suitable for the known time interval of individual chain life cycle phases, especially with regard to the conditions of uncertainty and the degree of the influencing factors activity under which the chain is implemented.

The method of dynamic programming, therefore, focuses in the next step on reducing the requirement of optimal costs J^* of the current decision future consequences. This simplifies the overview of the decision-making problem solution and at the same time, it enables to reduce the number of decision steps. In the standard design, the optimal costs are replaced by an appropriate approximation $\tilde{J}(j, r)$, where r is the parameter vector and a suboptimal design $\tilde{\mu}(i)$ is used in the state i, which makes it possible to achieve approximately minimal costs:

$$\tilde{\mu}(i) = \arg\min_{V} E\{g(i, V, j) + J^*(j, r) \mid i, V\},$$

Subsequent steps already lead to determining such conditions that create the preconditions for the optimal variant adoption with regard to the next step of the decision-making process, based on current knowledge. The following steps of the neuro-dynamic approach can be considered inspiring, however, moving away from real conditions. For this reason, a prerequisite for the conceptual model creation is to focus on the principle of taking into account the influencing factors of both the current and the subsequent variants of the design concurrently.

5 Creating a Conceptual Model by CPM Amplification

The vast majority of methodological approaches fail mainly in partial evaluations since they do not sufficiently take into account the broader context [22, pp. 6–8], such as the impact of security threats and ongoing evaluation of its development trends. At the same time, risk factors themselves cannot be minimized by an improved level of information or knowledge. Moreover, the spectrum of minor events may significantly contribute to the materialization of some security threats or may act as an event catalyst enhancing their negative impact and spread. Examples include combinations of events that are currently taking place on their own, such as combinations of civil unrest due to the global financial crisis in the concurrent presence of avian influenza (H5N1) epidemic or the spread of Covid-19 pandemic [22, p. 17]. The overall sensitivity of distribution chains to safety factors then represents the functions of the negative impact significance of the identified risk events and their consequences [29, p. 77]. These consequences can be characterized by the probability of changes and events that may subsequently occur and cause disruption to the functionality, availability, and sustainability of logistics chains or their immediate surroundings [30, p. 1016].

It is appropriate to set priorities in accordance with the three logistics imperatives, namely the priorities of time, cost, and quality, within the design of distribution
chains. Delimitation of time priority is based on a summary of partial time calculations of individual logistics activities under the standard conditions. It is proposed
to assign appropriate probabilities to these time calculations, based on the possible
disruption to logistics chains, whether caused by natural or human factors. A similar
approach is proposed to delimit the priority of costs [31, p. 273], where it is possible
to calculate relatively accurately the costs of individual logistics activities from the
available price calculations and previous experience.

In terms of quality priority, however, the situation is different from the priorities of
time and costs, mainly due to the complexity and diversity of the concept of quality
in the implementation of logistics activities under the conditions of uncertainty and
indeterminacy. Due to the large range of influencing factors, the security and sustainability of logistics entities, i.e., logistics nodes and the links between them, can be
perceived as crucial. When examining the potential impacts of security factors on
the distribution chain, it is possible to make use of the principles of the graph theory
method, which examines the dependencies of key logistics entities in the supply,
resp. logistics chain. The standard CPM method provides only three basic items of
information about the nodes and edges of a chain, namely [32, pp. 191–199]:

- Identification of the node in the system, usually by the node number, characterizing
  its position in the system;
- Identification of two key variables, usually the times of entry and exit of monitored
  items to and from the system node.

The interconnections between nodes (i.e. the edges of a chain) are usually defined
by the length of the connection between the two nodes considered. Figure 1 graphi-
cally shows the characteristics provided in the case of two nodes and the connections
between them, which the CPM method works by default with.

where

N_1, N_2 Node identification in the chain
$P_{t_1}^{N_1}$ $P_{t_1}^{N_1}$ represents the activity at the node entrance N_1 at entrance time t_1
$P_{t_2}^{N_1}$ Similarly, $P_{t_2}^{N_1}$ represents an activity at the node that exists at time t_2 in the
       same N_1 node

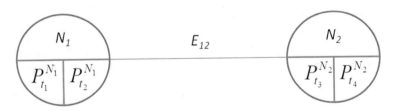

Fig. 1 Identification of two nodes in the system and connections between them. *Source* [14, p. 32]

E_{12} Edge identification in the chain (E_{12} represents the connection between N_1 and N_2 nodes).

When using CPM, the optimal path is formed by such nodes and their connections (edges), for which the sum of total times to overcome the shortest path is minimized.

5.1 Taking into Account the Portfolio of Security Aspects of the Logistics Chain Entity

However, the standard concept of the CPM method is not fully sufficient for the needs of the analysis of possible impacts of risk factors on the distribution chain. In terms of the complexity of the researched issue, it is proposed to further extend the standard CMP method to take into account security factors, i.e. to extend the characteristics of logistics entities (nodes and edges) by a multi-criteria approach by a group of attributes which shall provide additional information to perform critical path optimality analysis in case of disruption of some logistics entity, i.e. node, edge or implemented logistics activity. Taking into account the delimitation of the conceptual model, decisions on the use, resp. non-use of the given logistic entity (i.e. node, edge, or logistic activity) form part of the accepted variant of the decision V_i. It is proposed to extend the characteristics of the entity by:

- A group of attributes of the three logistics imperatives, i.e., the attribute of costs A_N, the attribute of time $A_Č$, and the attribute of quality A_K;
- A group of attributes according to the STEEP, i.e., groups of attributes from the following areas: social A_S, technological A_T, economic A_{EK}, environmental A_{EV}, and political A_P;
- A group of security attributes A_{H_i}, i.e., intentional A_{Int} and non-intentional A_{NonInt} threats;
- Attributes of information inputs AI_i.

Individual attributes can be viewed as separate characteristics of individual logistics entities or some attributes can be marked as a priority, e.g., attributes of the three logistics imperatives. In the case of joint evaluation of individual attributes, it is possible for each initial and final situation arising from the V_i decision to be characterized as the result of the activity of the attributes of the three logistics imperatives, which simultaneously maximize the achieved benefit $u(Vi)$ as follows:

$$V_i \approx \arg\max \sum_{i=1}^{n} \left(A_{trojim_i} + A_{STEEP_i} + A_H \right) \cdot \sum_{i=1}^{n} AI_i$$
$$\text{for } u(v_i) \rightarrow \max .$$

5.2 Selection of Logistic Entity Attributes

In terms of functionality and usability of each logistics entity, it is possible to divide attributes into two groups, namely the attributes of the entity, which are represented by logistics nodes, i.e., *Pi*, and the attributes of the entity, which are of external character and are related to the delimitation of the given entity as a connection (edge) of the supply chain, i.e. *Qi*. A selection of crucial attributes which provide the basic characteristics of the logistics entity-node and the logistics entity-edge of the logistics chain is performed within the characteristics of the functionality of the created conceptual model. To distinguish the total attributes of the logistics entities, delimited as *A*, the selected entity-node attributes are delimited as *X* and the selected entity-edge attributes are delimited as *Y*.

It is subsequently possible to assign security and sustainability attributes $P_{t_j}^{N_i}$ at time t_j to a logistics entity-node P_i as follows:

- *Time attribute X_1*, i.e. taking into account the time needed for a given node to fulfil its function (e.g. for a node representing a port, this is the time required for loading or unloading, material handling, etc.);
- *Location attribute X_2*, i.e., taking into account the local conditions in a given node from both a quantitative and qualitative point of view (e.g., for a node representing a port it takes into account the availability of qualified personnel, port infrastructure, i.e. handling capacity, storage capacity, the ability to unload/load several ships at the same time, equipment for fuel handling and storage, accommodation capacity for accompanying staff, etc.);
- *Cost attribute X_3*, i.e., taking into account the financial intensity of the use of a given node (e.g., for a node representing a port, these are the total financial costs of the given port use, e.g. handling costs, storage costs, administrative costs, customs duties, insurance, etc.);
- *Information attribute X_4*, i.e., taking into account the availability of quality and reliable information about a given node;
- *Flexibility and resilience attribute X_5*, i.e., taking into account the degree of note rigidity, unable to flexibly respond to changes in implementation conditions, and at the same time its inability to cope with possible disruption of the distribution chain.

It is also necessary to expand the spectrum of information about individual edges, i.e. the connections of two adjacent nodes, from the point of view of the formulated proposal completeness. The standard CPM for links between nodes, i.e., the edges of a chain, only works solely with the criterion of the distance between two adjacent nodes. This criterion can be considered decisive in the case of edge analysis. However, when examining the influence of safety factors on the edge functionality and its potential usability, it is suggested to take into account further criteria. The portfolio of edge security criteria $Q_{t_j}^{E_{rs}}$ of the edge E_{rs} at time t_j can be subsequently formulated as follows:

- *Time attribute Y_1*, i.e., taking into account the times needed to cover the distance between two nodes;
- *Route attribute Y_2*, i.e., taking into account the distance and geographical conditions through which the given connection between two nodes takes place;
- *Cost attribute Y_3*, i.e., taking into account the financial demandingness of overcoming the distance between two nodes;
- *Information attribute Y_4*, i.e., taking into account the availability of quality and reliable information about the given edge and the availability of information during its possible use;
- *Flexibility and resilience attribute Y_5*, i.e., taking into account the degree of flexibility of the edge use, e.g., ability to flexibly respond to condition changes in nodes implementation which it connects and within the edge itself), and at the same time, in the event of the edge functionality disruption, the ability to quickly restore its original functionality;
- *Quantity attribute Y_6*, i.e., taking into account the amount of transported material, resp. the range of services provided during transport on the route between the two nodes;
- *Quality attribute Y_7*, i.e., taking into account the possibility of quality disruption of the chosen method of overcoming the distance between two nodes.

The individual portfolios of the security criteria of nodes and edges express the relative security of a given element of the chain. In terms of further use, all security criteria of nodes and edges are formulated as minimizing and can take values $\langle 0; 1 \rangle$, i.e., in percentage from 0 to 100%. A value of 0 represents the zero influence of the safety criterion on the overall functionality of the node or edge, resp. a value of 1 represents the extreme influence of the given criterion on the safety of the node or edge and the real malfunction of the considered node or edge, e.g. the non-functioning of the airport due to the declared strike alert.

The principle of creating a portfolio of security criteria of nodes and edges allows to add other criteria or omit criteria that are not essential for the scenarios considered while maintaining the functionality of the approach under consideration. A graphical representation of these portfolios for two nodes and their edge is shown in Fig. 2.

where

N_1, N_2 Identification of the node in the chain

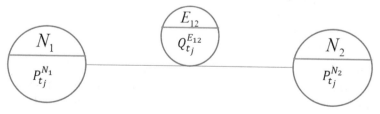

Fig. 2 Identification of two nodes in the system and link between them through a portfolio of criteria. *Source* Authors' own source

$P_{t_j}^{N_1}$, $P_{t_j}^{N_2}$ Portfolio of security nodes N_1 and N_2 at time t_j

E_{12} The edge between the nodes N_1 and N_2

$Q_{t_j}^{E_{12}}$ Portfolio of security criteria of the edge E_{12} at time t_j

j For $j = 0, 1, 2, \ldots, m$.

The different composition of the portfolios of security criteria of nodes and edges is based on a different perception of the influence of security factors on individual elements of the logistics chain and their interconnection. It also follows from the above list of criteria that the optimal path is such a sequence of logistics nodes and edges, which will show the optimal combination of individual security criteria for the considered crisis situation, resp. crisis scenario, at a given moment. The optimal variant is to find such a path that will pass through the nodes and edges, where the individual portfolios of criteria will acquire the total sum of the lowest values for the whole considered path.

5.3 Taking into Account the Security Criteria Weight in the Node and Edge

It is suggested to supplement the individual security attributes by their significance factors due to the fact that individual security criteria in nodes and edges may take on different meanings, depending on specific conditions and identified security threats. These relevance factors represent the weight of each monitored security attribute of a node or edge. For the needs of creating a simplified conceptual model, the weight of the security attribute X_k of a given node is marked with the symbol α_k, where $k = 1, \ldots, 5$, while $\alpha_k \in \langle 0; 1 \rangle$. Similarly, the weight of the security attribute Y_k of a given edge is marked by the symbol β_k, where $k = 1, \ldots, 7$, while $\beta_k \in \langle 0; 1 \rangle$.

It is necessary to monitor security attributes and their weights over time due to changing and further evolving conditions of the distribution chain implementation. Their reassessment or confirmation of their invariance must be carried out for each monitored period of time t_j. It is possible to divide the weights of security attributes of nodes and edges into four basic categories according to the probability of functionality disruption of the given node or edge to determine the weights of individual security attributes [33, pp. 100–105]:

- *Low*: range 0–30% (green);
- *Medium*: range 31–60% (yellow);
- *High*: range 61–90% (light red);
- *Very high*: range 91–100% (red).

The individual weights of the security attributes can, as a result, lead to the evaluation of the node, resp. the edge as unsuitable for possible use. At the same time, the weights of individual attributes make it possible to include the error factor in the evaluation of the safety criteria of nodes and edges. At a given time t_j, a formalized expression of the security criteria portfolio $P_{t_j}^{N_i}$ of the node N_i can be suggested for the considered node security criteria as follows [33, p. 37]:

$$P_{t_j}^{N_i} = \sum_{k=1}^{5} \alpha_k X_k$$

where

X_k Security criteria in node N_i and time t_j, $X_k \in \langle 0; 1 \rangle$, $k = 1, \ldots, 5$
t_j Expression of time, for $j = 0, 1, 2, \ldots, m$
N_i Node identification in the system, for $i = 1, 2, \ldots, n$
α_k Security criterion weight X_k in node N_i at time t_j, $\alpha_k \in \langle 0; 1 \rangle$.

At a given time t_j, a formalized expression of the portfolio of edge security criteria $Q_{t_j}^{E_{rs}}$ for the edge E_{rs} can be suggested for the considered security criteria $Y_1, \ldots, Y_7$ as follows [14], p. 38):

$$Q_{t_j}^{E_{rs}} = \sum_{k=1}^{7} \beta_k Y_k$$

where

Y_k Security criteria for the edge E_{rs} at time t_j, $Y_k \in \langle 0; 1 \rangle$, $k = 1, \ldots, 7$
t_j Expression of time, for $j = 0, 1, 2, \ldots, m$
N_i Node identification in the system, for $i = 1, 2, \ldots, n$
β_k Security criterion weight Y_k for the edge E_{rs}, at time t_j, $\beta_k \in \langle 0; 1 \rangle$.

When searching for the optimal variant of the critical path at a given time t_j for $j = 0, 1, 2, \ldots, m$, and at the same time taking into account the security criteria portfolio in individual nodes and edges of the system, it is necessary to find the optimal scenario $S_{t_j}^{opt}$ from the total number of possible scenarios U_{t_j} at given time t_j. All nodes $P_{t_j}^{N_i}$ and edges $Q_{t_j}^{E_{rs}}$ at time t_j are evaluated. The formalized expression of the search for the optimized combination of nodes and edges creating the optimized critical path at time t_j can be formulated as follows [14, p. 38]:

1. The security portfolio of all nodes N_i of the logistics chain $P_{t_j}^{N_i}$, for $i = 1, 2, \ldots, n$ is found.
2. The security portfolio of all edges E_{rs} of the logistics chain $Q_{t_j}^{E_{rs}}$, where N_r and N_s are the adjacent system nodes, while $Q_{t_j}^{E_{rs}} = Q_{t_j}^{E_{sr}}$ is found.
3. A set of all possible scenarios U_{t_j} at given time t_j and their safety evaluation S_{t_j} are found:

$$S_{t_j} = \sum_{N_i \in S_{t_j}} P_{t_j}^{N_i} + \sum_{E_{rs} \in S_{t_j}} Q_{t_j}^{E_{rs}}$$

4. The optimal scenario $S_{t_j}^{opt}$ from the total number of possible scenarios U_{t_j} at given time t_j is found as:

$$S_{t_j}^{opt} = \min_{S_{t_j} \in U_{t_j}} S_{t_j}$$

5. If the minimum occurs for more than one such scenario, one of them is chosen as optimal.

It is suggested to recalculate all the portfolios of security attributes of all nodes and edges of the logistics chain at time t_0. At time t_1, attention is focused only on possible changes in the security portfolios of nodes and edges. Ideally, no changes are identified and the optimal variant chosen at time t_0 is also the optimal variant at time t_1. From a practical point of view, however, the security environment is evolving and it is necessary to perform ongoing verifications of the originally adopted scenario. In case of identified differences, all security criteria of nodes and edges are recalculated. In the extreme case, a situation may arise that the originally optimal variant at time t_0 is no longer practically feasible at time t_1 and it will be necessary to go back to the previous, respectively default node.

6   Conclusion

The concept of a distribution chain design in conditions of uncertainty using a modified CPM method was created within the research based on the performed scientific literature research. The dynamics of the logistically relevant information inputs was included in the formulated proposal. The use of these indicators and indices makes it possible to analyse the availability and sustainability of the logistics infrastructure, including the potential impact of influencing factors and trends in their development. On this basis, the research question presented can be affirmatively answered.

In the case of the availability of longer periods of 20–25 years for the monitored indicators and indices, it is possible to make predictions of future development or to use them as early warning signals about expected changes (so-called *early warning signals*). However, this approach does not make it possible to capture potential "black swan" events, but only to help identify indications for a possible trend reversal.

The proposed use of a large range of freely available data sources by data mining methods (so-called data-mining) and obtaining indicators of potential changes is an effective and cost-effective way to increase the efficiency of funds spent. It can also lead to an increase in the quality of decisions taken, in the planning and subsequent implementation of the logistics of multinational operations, especially in relation to the regions with a lower level of available infrastructure.

Acknowledgements The work presented in this chapter has been supported by the institutional support of the Ministry of Defence of the Czech Republic, *Development of capabilities and sustainability of logistic support systems* (research project DZRO ROZVOLOG *Rozvoj schopností a udržitelnosti systémů logistické podpory*, 2015–2021).

References

1. Lovecek T et al (2017) Determining the resilience of transport critical infrastructure element: use case. Kaunas University of Technology, Juodkrante. ISSN 1822296X
2. Dujak D (2019) Causal analysis of retail distribution system change from direct store delivery to centralized distribution. IBIMA Bus Rev 2019:1–14
3. Zeman T, Urban R (2019) The negative impact of terrorism on tourism: not just a problem for developing countries? DETUROPE 11(2):75–91
4. Dinçer H, Hošková-Mayerová Š, Korsakienė R et al (2020) IT2-based multidimensional evaluation approach to the signaling: investors' priorities for the emerging industries. Soft Comput 24:13517–13534. https://doi.org/10.1007/s00500-019-04288-6
5. Tulach P et al (2020) Replacement possibilities of the medium-sice truck transport capability by UAVs in the disturbed logistics infrastructure. Springer International Publishing AG, Roma, pp 139–153. ISBN 978-3-030-70739-2
6. Huntington S (2001) Clash of civilizations (Střet civilizací). Rybka, Prague, p 65. ISBN 80-861-8249-5
7. Rehak D, Hromada M, Ristvej J (2017) Indication of critical infrastructure resilience failure. In: Safety and reliability—theory and applications—Proceedings of the 27th European safety and reliability conference, ESREL 2017. CRC Press/Balkema, Portorož, pp 963–970. ISBN 978-113862937-0
8. Korsakienė R, Bekesiene S, Hošková-Mayerová Š (2019) The effects of entrepreneurs' characteristics on internationalisation of gazelle firms: a case of Lithuania. Econ Res-Ekonomska Istraživanja 32(1):2864–2881. https://doi.org/10.1080/1331677X.2019.1655658
9. Strnádek J (2004) Aspects of world integration and globalization (Aspekty světové integrace a globalizace). Vojenské rozhledy 13(45):2
10. Simchi-Levi D, Kaminsky P, Simchi-Levi E (2002) Designing and managing the supply chain: concepts, strategies, and case studies, 2 edn. McGraw-Hill/Irwin, Boston, p 3. ISBN 978-0072845532
11. Zsidisin G, Ritchie B (2009) Supply chain risk. Springer, New Delphi, p 350. ISBN 978-0-387-79934-6
12. Waters D (2007) Supply chain risk management: vulnerability and resilience in logistics. Kogan Page, Philadelphia, p 256. ISBN 978-0-7494-4854-7
13. EU (2010) Internal security strategy for the European Union—towards an European security model. European Union, Brussels, pp 13–14. ISBN 978-92-82-824-2679
14. Foltin P (2012) Selected aspects of logistics chains security of armed forces. Brno Univ Defence Brno 2012:71. https://doi.org/10.13140/2.1.2759-1364
15. Sedlačík M et al (2014) Application of Modified CPM on Portfolio of Security Criterions on Logistics Chains (Aplikace modifikované metody CPM na portfolio bezpečnostních kritérií v logistických řetězcích) [Final report of the specific research project]]. University of Defence in Brno, Brno
16. Tušer I, Jánský J, Petráš A (2021) Assessment of military preparedness for naturogenic threat: the COVID-19 pandemic in the Czech Republic. Heliyon 7(4). https://doi.org/10.1016/j.heliyon.2021.e06817
17. Bertsekas D (1996) Neuro-dynamic programming: an overview [Online] 1996. Cited: 14 October 2020. http://web.mit.edu/people/dimitrib/NDP_Encycl.pdf.

18. National Research Council (1999) Reducing the logistics burden for the Army after next: doing more with less. The National Academies Press, Washington, DC. ISBN 978-0-309-06378-4
19. NATO Code of Best Practice for C2 Assessment. CCRP, 2002, p 322. ISBN 1-893723-09-7
20. Entchelmeier A, Hartman E, Henke M (2006) Supply risk assessment: a utility value based concept. IMP Group [Online] 2006. Cited: 14 March 2021. https://www.impgroup.org/uploads/papers/5682.pdf
21. Rinaldi S, Peerenboom J, Kelly T (2001) Critical infrastructure interdependencies. Control Syst Mag 21(6):12
22. Freier N (2008) Known unknowns: unconventional "strategic shocks" in defence strategy development, vol PKSOI Papers. U.S. Army War College, Carlisle, p 18. ISBN 978-158-4873-686
23. Tang C, Teo C-P, Wei K (2008) Supply chain analysis. Stanford University, Stanford, p 206. ISBN 978-0-387-75240-2
24. Ayling J (2009) Criminal organizations and resilience. Int J Law Crime Just 37:182–186
25. Chapman C (2015) Supply chain risk management—a comparative study of small to medium enterprises vs. large enterprises. University of Saskatchewan, Bracebridge. MBA 992 Research Project
26. Schwartz P (2004) Inevitable suprises—thinking ahead in a time of turbulence. Penguin Group Inc., New York, p 3. ISBN 1-592-40069-8
27. Foltin P et al (2018) Discrete event simulation in future military logistics applications and aspects. Springer International Publishing AG, Roma, pp 410–421. ISBN 978-3-319-76072-8
28. Vlkovský M (2020) Impact of vehicle type and road quality on cargo securing. Commun—Sci Lett Univ Zilina 22(1):9–14
29. Paulsson U (2007) On managing disruption risks in the supply chain—the DRISC model. Lund University, Lund, p 77
30. Vlkovský M, Veselík P (2019) Cargo securing—comparison of different quality roads. Acta Universitatis Agriculturae et Silviculturae Mendelianae Brunensis 67(4):1015–1023
31. Gros I, Barančík I, Čujan Z (2016) Big book of logistics (Velká kniha logistiky). University of Chemistry and Technology Prague, Prague, p 273. ISBN 978-80-7080-952-5
32. Jablonský J (2002) Operations research: quantitative models for economic decision making (Operační výzkum: kvantitativní modely pro ekonomické rozhodování). Professional Publishing, Brno, pp 1991–1199. ISBN 80-86419-23-1
33. Foltin P (2011) Security of logistics chains against terrorist threats. In: The 17th international conference "the knowledge-based organization". Nicolae Balcescu Land Forces Academy, Sibiu, pp 100–105. ISSN 1843-6722

Modern Approaches in Czech Prison Staff Education and Training Against a Background of Comenius' Thoughts

Miloslav Jůzl and František Vlach

Abstract This contribution is conceived as a reminiscence of the birth of a modern prison staff training in the Czech Republic since the beginning of 1960th. The authors discuss the origins of the epistemology foundations of philosophical and pedagogical theories of predecessors of the times, the process and their utilization within the theoretical part of the initial education of civilians and the practical part of the initial training of uniformed staff of the Prison Service of the Czech Republic. Special attention is paid to the last twenty years as a considerable development of education programmes—systematic initial training, vast long-life education opportunities, final examinations of officers after their probationary period, and implementation of good practices from abroad. The Czech Prison Service Academy has partners in France, Latvia, Lower Saxony and other countries. Not only mere adapting of foreign educational programmes but mutual exchange of experience is taking place today. The peak of the academic work is represented by the cooperation with high schools and by the scientific research of the Cabinet of Documentation and History dealing with the Czech penitentiary past. The Czech Prison Service Academy has a high professional potential to organize numerous international conferences. The most significant conference is the traditional Penology Days. The training centre of the Prison Service of the Czech Republic has become a prestige educational institution.

Keywords Prison Service Academy · Education · Penitentiary · Penology · Innovative programmes · International penitentiary

M. Jůzl
Comenius University Prague, Prague, Czech Republic
e-mail: juzl.miloslav@ujak.cz

F. Vlach (✉)
Prison Service Academy, Stráž pod Ralskem, Czech Republic
e-mail: fvlach@avs.justice.cz

© The Author(s), under exclusive license to Springer Nature Switzerland AG 2022
I. Tušer and Š. Hošková-Mayerová (eds.), *Trends and Future Directions in Security and Emergency Management*, Lecture Notes in Networks and Systems 257,
https://doi.org/10.1007/978-3-030-88907-4_24

1   Introduction—The Origin and Development of Pedagogy

Education as deliberate forming of personality has existed along with mankind from time immemorial. We can trace it at the beginning of history: to regulate the development of the growing generation, to form their cognition, confessions and behaviour in compliance with the needs and ideals of a specific society. This activity was gradually institutionalized and professionalized. Its problems were more and more reflected in the theoretical debates of philosophers, pedagogues, and other specialists. Gradually, the theory of pedagogy, as a generalized reflection of educative activities and as a tool helping the society to rationally frame, organize and ensure education of youth and adults, arouse.

Pedagogy is a very old science. Its basic issues were often discussed in connection with other sciences, mainly philosophy, for centuries. It was constitutionalised as a scientific field only in the new era. The term pedagogy appeared in the work of the German thinker Johann Friedrich Herbart and the Czech pedagogue Gustav Adolf Lindner in the first half of the nineteenth century. Its development has been recorded since then. However, the development became significant for pedagogy as a science in connection with other social sciences such as psychology and sociology in the twentieth century.

It was John Amos Comenius (Jan Amos Komenský) who played a crucial role in the rise of a new science—pedagogy. However, he called it didactics (see his work *The Great Didactic*). His contribution to the field of education is mentioned in another fundamental work of Comenius *General Consumption on an Improvement of All Things Human (in Czech: Všeobecná porada o nápravě věcí lidských)*.

1.1   The Science of Pedagogy and Its Subject Matter

The structure of all sciences, no matter the variety of topics and themes, is largely universal. To become a science, each science should meet certain criteria, i.e. it has to contain:

- **Theory**, by the means of which the theme is specified, described and clarified, and it systematically forms findings of the study.
- **Research activities** which produce detailed information, data, findings on the different specifics, functioning and implications, reasons and consequences of the respective theme.
- **Methodology**, which means a set of research methods and established processes and conventions that help carry out the research. It interrelates with previous research activities.
- **Basic informational organizational infrastructure**, which is collected, extended and offered to use thanks to the scientific research.

This structure also exists within pedagogy, penology and penitentiary science. Today, pedagogy is not perceived as a universal science, conceived unitarily. It has different variations according to individual authors' views (the same is true for example for psychology). For example German authors A. Kaiser and R. Kaiser in their *Textbook of Pedagogy (Studienbuch Pädagogik: Grund und Prüfungswissen)* begin with the explanation of the anthropological basics of education, because they find the phenomenon of education an issue that is present in all human societies, past and present, advanced and primitive—cultural anthropology.

American scientists Ornstein and Levine begin their introduction to pedagogy with an article on the teaching profession in compliance with a pragmatic conception stating that pedagogy is a concept of what happens in everyday interaction between the teacher and the students in the classroom.

Finally, a French specialist Franc Morandi begins his explanation of pedagogy with his debate on socialization of children and the learning processes occurring in the school environment [7].

Nowadays, pedagogy is a term used in two meanings. Vladimír Jůva conceptualizes it as a science about permanent education, lifelong education of children, juveniles and adults [1].

Other advocates the explanation of the term pedagogy stated above agree that it is a science on education of children and juveniles. To address education of adults, the term of andragogy has been used for a few decades (Greek andr-ós = man/adult). There is also one more term in pedagogy and that is gerontology (Mühlpachr) dealing with education of the elderly (see universities of the third age).

2 Philosophy of Education

2.1 Introduction

The contentual core of "philosophy of education" consists of the rich history of this field, which includes the research on pedagogy conceptions of big thinkers of the past as well as of forming of modern scientific and philosophical trends that have been significantly influencing basic concepts of education in various countries for a few centuries. If teachers of philosophy of education focus on explanations of what the apparent and implicit bases of individual educational concepts are, they usually pay their attention to influential thinkers and trends of ideas that have been changing and influencing pedagogical thinking.

Philosophy of education is formed and kept as reflexive of critical thinking on the everydayness of the educational process. In the field of theoretical research, a question has been raised on education and its fundamentals. All uncritically accepted and "natural" definitions and conceptions are considered problematic. Theoretical and empirical research undergoes analytical examinations. Philosophical thinking is also used when forming so-called educational goals. Research may be focused on basic anthropological patterns of thinking. A relatively wide stream of this issue is considered research on basic anthropological patterns of thinking.

Philosophy of education is linked to old traditions of philosophical research; it opens new questions and looks for substantiate answers. In this respect, a doubt emerges. Is it possible to reconstruct astonishment on disputableness of anything that appears laid, relatively stable or non-problematic at present? During lectures on philosophy, questions are asked if the birth of knowledge can be induced artfully and if an individual can be brought to the knowledge on the basis of institutionalized education.

If philosophy of education is considered a scientific discipline and it is taught mainly in the form of the history of ideas that have to a significant level lost live competency, everything that is substantial, that creates the core of philosophy, that means ability of questioning certain facts lively and timely, and asking new radical questions.

Teaching philosophy of education can be—just as teaching of some other disciplines—motivated by aspiration of giving the field the prestige of traditional academic fields.

In this case, the impact of philosophy of education presents "theoretical embellishment" with a little or insufficient influence on teaching. The effect is that practically conducted education is often totally isolated from philosophy and vice versa.

A typical situation is that on one hand, solitary academic groups of philosophers striving for their own prestige reinforcement, they hope to reach thanks to the recognition of their discipline as equal to other traditional sciences; on the other hand you can meet a lot of teachers making their decisions on the basis of their life experience but without deep and systematic reflection of basic philosophical starting points. From time to time, philosophy of education as a philosophical discipline critically deals with its own scope and achievements, examines the problems of its foundation and functioning, which guarantees an opportunity for it to pass radical reconstruction (as for other philosophical fields). If philosophy of education is framed and taught well, it cultivates critical thinking and mediates an overview of the world in a wide context and it reveals covered bases of our everyday activities. In that sense, it deserves to be a part of a live philosophical tradition [2].

2.2 Content of Philosophy of Education; Mainstreams in Philosophical Thinking

Philosophy of education examining historical development describes an inspirational concept and the theory of objectives and methods of education and it deals with the issues of normatively determined educational ideals. The origins of teaching for a fee are linked to speeches of sophists, the first professional educators in the Greek society. Their activities were criticized by their powerful opponents—founders of big philosophical schools and creators of their own pedagogical visions: Socrates, Plato and

Aristoteles. Sophism has undergone a number of innovations from scholastisticism till present and its traditional influence has been manifesting itself in the methods of disputation and oral dialogical examinations in a number of schools.

After the Antique decay, educational ideas evolved on the grounds of theoretical thinking by such thinkers as St. Augustin or Boëthius, religious thinker, philosopher and creator of classical textbooks of trivium and quadrivium used in the whole learning Europe for centuries.

It is important to mention one of the religious thinkers J. A. Comenius, author of a number of books on education and on its universal philosophical–theological implications. In the nineteenth century, J. F. Herbart laid the foundations of pedagogical science as a conceptually and systematically structured science. Forming of the nations in the eighteenth and nineteenth centuries initiated creation of a lot of pedagogical conceptions linking to the idea of national education. Until the nineteenth century, it had been possible to speak about big characters of European thinking who had been dealing with pedagogy within their study (Plato, Aristoteles, St. Augustin, J. A. Comenius, J. Locke, J. J. Rousseau, I. Kant, J. G. Herder, F. Schiller, G. W. F. Hegel, J. F. Herbart etc.). Later, antimetaphysically oriented movements arose. These had a significant influence on the basement of key modern pedagogical mainstreams of thinking (positivist roots of experimental pedagogy and pragmatism, Marxist pedagogy etc.) [3].

2.3 John Amos Comenius and General Consumption on an Improvement of All Things Human

John Amos Comenius (1592–1670), the last bishop of the Unity of Brethren, humanistic philosopher and pedagogue of the world importance, called the teacher of nations (Moravia, Bohemia, Germany, Poland, England, Sweden, Hungary, Holland) speaks in his uncompleted work *De rerum humanarum emendatione consultatio catholica (Obecná porada o nápravě věcí lidských)*, Amsterdam (1662), to all social institutions from the family to the state. His goal is the participation of all people on the improvement of the human society.

Individual volumes are set by the following titles:

- Pangresia (Všeobecné probuzení)
- Panaugia (Všeobecné osvícení).
- Pansophia (Všeobecná moudrost).
- Pampaedia (Všeobecná výchova).
- Panglottia (Všeobecný jazyk).
- Panorhosia (Všeobecná náprava).
- Pannuthesia (Všeobecné povzbuzení).

Considering the fact that one of the social pedagogy fields is the effort to improve an individual who has erred and he/she found him/herself on life slant due to the influence of various factors, we are going to try to apply Comenius' *General Consumption on an Improvement of All Things Human* on the persons who have ended up behind the bars (in prisons), on which the environment has a strong effect (negative process so called prisonisation) and so they need to socialize, reintegrate and come back to the society.

Even though John Amos Comenius did not deal directly with contemporary penitentiaries—his priority was always philosophy and didactics, his ideas however tackled penology and penitentiary science implicitly to that extend that it is necessary to dedicate him as a priest and teacher one chapter in this work. Let the inspiration be the paper by Aleš Kýr published in the magazine *Historická penologie III/2003* with the title *Odkaz Jana Amose Komenského nejen českému školství* (pp. 3–7) in which he masterly combines philosophy, sociology, penology and penitentiary sciences. Let us try to trace the ideas on penology and penitentiaristics enclosed in *General Consumption on an Improvement of All Things Human.*

Comenius, alarmed by the situation, calls for reformation of the ill world with his deeply human and at the same time clerical approach (social pathological concept). In his first work *Pangresia*, he asks himself the following question: "The reason why punishment has been demanded and imposed to those who commit crime is that without that it would not be possible to repress bad persons' self-will and no human society could exist in any nation. And even than it is not possible to eliminate not even supress all evil and violence, the evidence of which is given by jails, stakes, gallows and other means designated for controlling of criminals and crime, which are hardly ever at rest." (Comenius in Všetečka 1987, p. 54).

Further Comenius warns of the danger of misapplication of judicial instruments by judges with tendencies to submit others by the means of violence. That is why he asks himself a question if such use of violence is dignified: "And what are the ways? Universally violent: whips, shackles, jails, ropes, swards end so on. Does this correspond to the institute? Does it behove to treat a human being in this way?" (Comenius in Všetečka 1987, p. 42).

The implication for penitentiary practice on the basis of this work is that activation of all subjects of rehabilitation able to oppose to crime and its recidivism in the form of crime prevention, rehabilitation of imprisoned offenders as well as after their release. There is one more issue, Comenius suggests to think over—it is looking for new ways of punishment of offenders.

Panaugia is the only way to reform humankind by the means of universal attempt to gain wisdom. Getting wiser is the means to individual's redeem and mutual understanding without which the society reform is not possible.

Comenius indicates that the orientation of all subjects of remedy can also be applied to offenders and he recommends examination of causes, manifestations and consequences of delinquency and this is how he opens the space to the creation of future sciences (criminology and others).

According to Comenius, *Pansophia* emerges from the human nature, desire for good, appropriate ways of behaviour and means that makes it possible to distinguish good and evil and so to reach good and avoid evil. Except for human thinking and communication, free will of the human is the most important quality that distinguishes a human being from other beings and that is why it is necessary to pay attention to the development of conation.

Human will is influenced by intellectual cognition, but also by feelings and consciousness of moral accountability. Cultivation of emotions must aim to experience good, the truth and beauty. It can be concluded, that to get wiser, an individual needs to get knowledge along with volitional and emotional features that become also one of the basic principles of treatment of prisoners. The irreplaceable role plays in this process also Prison Chaplaincy Service and Prison Pastoral/Spiritual Care [6].

On the basis of *Pansomnia*, it is possible to apply universalisation of the impact of all subjects of remedy focused on purposeful and systematic preparation of imprisoned offenders for their return to the community—schemes for treatment programmes for convicts and their preparation in pre-release units (but also in admission units) and other continual post penitentiary care is inspiring—activities of curators, social workers and alas also of probation and mediation service and again of the church.

Pampaedia is defined by the accessibility, the scope and the profundity of education with the aim to ensure extensive preparation of an individual for his/her life in the society by gaining necessary knowledge, skills and by contracting moral principles. Voluntary approach of the breeding individual to education is a solid guarantee of the educational process effectivity. It is possible to gain this voluntariness by belief (motivation) that the person will learn something new and this will be done in the way of gentle schooling, weighty reasoning, developing of tenets, will and experience. It is possible to obtain the breeding person's interest in education if we listen to her/him, encourage them and reward them with approbation.

If there is a formula in Comenius' *Pansophia* how to proceed in the resocialisation of prisoners, than in *Pampaedi* there is a concrete instrumentalisation of treatment of imprisoned offenders by the means of a set of educational and therapeutic methods, forms and aids reflecting all components of education embodied into treatment programmes. This layout observes completion and extension of prisoners' education and facilitates their easier integration into the community.

Panglottia can be understood as an ideal tool to overcoming of communication barriers due to language diversities but also as a guide to unification of the meanings of the used words within a single language. It is important to firstly get to understand the other person and only then it is possible to educate them. Ineffectiveness of educational activities can be sequent on insufficiently comprehensible communication but also on interpretation. That is how it is possible to differently interpret such categories as the truth, law, justice and freedom, the proper comprehension of which has a big significance in rehabilitation of offenders.

Comenius suggests that sufficient communication between the subject of rehabilitation—and from the penitentiary point of view—offenders is considering mutual understanding and comprehension of the rehabilitation process extremely important for the whole process of resocialization (in contact with educators, guards and specialists) as well as for the connection with the outer world of family and friends.

Panorhosia concerns getting public affairs into a desirable state by the means of precedent approaches to the issues of the rehabilitation of the human society. Comenius results from the elementary postulate that an individual is basically good; otherwise rehabilitation would have no sense. The subject of rehabilitation can be either an individual, a group, a social institution, or even a social order. The agent of rehabilitation is the one who disposes of knowledge, possibilities and has a strong will to correct the undesirable state. At the same time, Comenius gives instructions what procedures is needed to reform the whole. Reform is according to him a purposeful and systematic activity, the effectiveness of which is depending on the level of disturbance of the subject of rehabilitation, on the level of proficiency, competencies and motivation of the agent of rehabilitation and on the use of the means of rehabilitation.

To reach the transformation of penitentiaries, the whole system must be changed, then rehabilitation of individuals can occur. It also includes adoption and observance of international standards, rules and obligations. Then, effective resocialization can seriously take place, as well as quantitatively higher levels of treatment of prisoners and integration of the society by the reduction of crime rate and reoffending thanks to successful process of rehabilitation. Again, the irreplaceable role plays here Prison Chaplaincy Service and Prison Pastoral/Spiritual Care.

Pannuthesia is dedicated to those who are able to take part in the accomplishment of the reform of the society on the basis of the Comenius' previous instruction. To encourage means to stimulate persons who have necessary knowledge to solve the problems as well as those who still hesitate to courage, activity and continual work.

Difficulties and uncertain results of correctional work can lead to the activity loss. That is why sound pedagogical optimism is necessary to cultivate and to approach the correction as the physician approaches the treatment. Encouraging of pedagogy staff is a motivational reminder of duties and missions, recognition of seriousness of their work and strengthening of their self-confidence, hope for success and hope for release from prison.

Motivation of agents of correction to revision or prevention of their resignation on offenders' rehabilitation is the reverse to encouragement of prisoners to their return to the society. They are also often exposed to feelings of waste of time and inconclusiveness of their pedagogical activities.

It is possible to deduce methodology principles from the conception of *General Consumption on an Improvement of All Things Human* that can help eliminate crime and reduce delinquency.

But first of all it serves all prison specialists as a methodology guide to treatment of prisoners. Also, penitentiary principles rising out of this work underlie success of resocialization processes of individuals sentenced to prison.

In this work of his, Comenius also follows (according his religious belief) his life educational goal: To prepare to eternal life that has three levels.

Get to know yourself and the world around you; gain control of yourself; lift yourself up to the Lord.

This implicates three elementary fields of education: education in sciences and in arts, in ethics and religion. The genius of the philosophical, pedagogical and religious works of John Amos Comenius reflexes explicitly in ideas that can be valued and implemented only at present and successfully use them not only in penology and penitentiary sciences as their philosophical essentials.

The Present Philosophy of Education—in accordance with Radim Palouš and social pedagogy, it is close to his idea that traditional ways of education are overcome and new ways to higher moral and ethical values should be found. Not only philosophical models of education based on forming, meaningful and purposeful impact on an individual etc., is concerned, but also the idea that education should teach an individual to freely orient themselves in the labyrinth and chaos of the present world and be able to find their way to higher values. Philosophy of values (the theory of values) is studied in so called axiology as a part of philosophy and it shifts determination of values and evaluation to a philosophical level. It means for a pedagogue, that it is him/her who has to be able to teach individuals to look for and to find the way to higher values. This is the interpretation of the philosophy of education—there must be changes in education.

And the same, even more amplified, holds true for penology, penitentiary science and penitentiary pedagogy.

From the economic point of view, money is the most valued. What will the pedagogic criteria be? Every discipline has a different hierarchy of values, but humankind has been adapting values for centuries according to the development of individual historical eras. They were different in primitive societies from those in medieval times and from present values. However, most values remain the same, others change. They can be divided into material and spiritual.

How are values seen by pedagogy? In principle, they are spiritual values and they can be viewed in the context of goals and components of education: Rational education seeks for the gaining broad and deep knowledge and skills in the field of natural, social and technical sciences. Moral education wants individuals to reach moral conviction. Labour education focuses on development of relationship towards work and related moral-volitional features, on gaining work skills and routines, aesthetic education is oriented towards cultivating refined taste, and physical training affects the individual in the terms of physical ability and his/her body development including hygiene, moral-volitional qualities and sport skills [7].

The final result is all-round harmonically developed and independent personalities who, on the basis of education, states and keeps his/her own values. Ordinarily, the values are health, job, family, education, culture, sport, hobbies and ethical values. As regards material values, they usually are living/housing, a car and as a final consequence money. And this all has to be considered also in professional preparation of prison staff that has to keep this in mind in the resocialisation process of prisoners.

3 Prison Service of the Czech Republic

Prison service is conceptualized as an indispensable security and social service to the community. Qualified and skilled prison staff that ensures both the dynamic and the static security at prison facilities followed by providing high-quality social services to individuals who have broken the law has a fundamental importance in the service performance. For over 25 years, extensive changes related to the treatment of prisoners have been taking place within the Prison Service. This implies increased demands on the professional level and personality preconditions of the applicants of professions within the Prison Service of the Czech Republic.

The importance of the professional training of employees of all organizational units of the Prison Service of the Czech Republic is constantly growing. Lifelong professional education is the base of all educational activities within the Prison Service of the Czech Republic, as the staff education and training have a significant impact on the improvement of educational activities provided for persons in custody, and consequently this leads to the reduction of the risk of the emergency situations occurrence. Lifelong education plays also an important role in the improvement of the workplace culture in prisons and remand prisons which is manifested in the decrease of the number of interpersonal conflicts and misunderstandings across all professional groups and organizational levels [5].

Uniformed and civilian prison staff cooperates to fulfil the tasks. High quality job and service performance of both the above mentioned Prison Service staff subcategories essentially depend on the system of the resort education, the prime guarantee of which is the Prison Service Academy.

3.1   The Czech Prison Service Academy

The Czech Prison Service is under the authority of the Ministry of Justice. The Prison Service cooperates with the Probation and Mediation Service of the Czech Republic, but they are separate organizations.

There are 35 prisons with the population of approximately 21,000 inmates. The tasks of the Prison Service are to ensure pre-trial detention and execution of sentence; to run two preventive detention facilities; to maintain law and order in the buildings of courts and the Ministry of Justice; and to provide escorts between prisons, to courts and to hospitals.

According to the law, the Czech prison staff is formed by two sub-groups. The ratio is about two thirds of uniformed staff members (directors and their first deputies, prison guards in direct contact with prisoners, perimeter security officers, prevention officers, escort officers, 9 unit members) and one third of civilians (psychologists, social workers, pedagogues, educators, medical staff, chaplains, administrative and support staff). The total number of staff is over 11,000, out of them about 1,000 Judicial Guards serve at judicial premises.

Fig. 1 Prison Service Academy in Stráž pod Ralskem

3.1.1 Basic Information About the Academy

The Academy is the only prison staff training centre in the Czech Republic. It is located at the border with Poland and with Germany in a little town of Stráž pod Ralskem (about 100 km from Prague) (Fig. 1).

3.1.2 History of Czech Prison Staff Training

Before 2000

The history dates back to 1960 when prison officers were trained in three-month courses. Since then, the school for prison staff has changed its residence; governing body; system of education; and its name several times.

The predecessor of the present institution is considered the vocational school founded by the then Minister of Justice of the former Czechoslovak Republic in Ostrov nad Ohří (close to Karlovy Vary) in 1970. The Standard Minimum Rules for the Treatment of Prisoners of 1957 were implemented in the professional training and the study programme was based on penology and criminology science. Initial course for cadets lasted seven weeks and in addition to physical and combat training it covered other subjects, such as education in corrections, psychology and pedagogy basics. A four-year study programme for future correctional educators was completed by a state leaving exam. The statistics show that between 1970 and 1980, over 4,700 new officers (out of them 350 women) passed the initial courses; and between 1970 and 1994, the number of the graduates from the secondary school programme reached 560.

In 1975, the school became a branch unit of the then Warrant Officers' School of the National Security Corps (Sbor nápravné výchovy) residing in Brno. Between 1982 and 1992 the branch occupied a six-floor accommodation and administration building and a school building. Over 2,800 new prison officers (out of them 270 women) passed the training between 1982 and 1989.

The Prison Service and Judicial Guard of the Czech Republic Act No. 555/1992 reformulated and newly stated the status of penitentiary personnel. In 1992, the then Minister of Justice renamed the school into the Institute of Education of the Prison Service. The base of the present system of education was laid. The length of initial courses varied according to the positions; courses for uniformed staff members took 16 weeks including a four-week practical training at selected prisons. The last state leaving exam at the secondary school programme was held in 1994.

In September 1994, a new Institute of Education branch was established in Stráž pod Ralskem and three classes for 90 new civilian staff members were opened. Another branch was opened in Kroměříž (Moravia) in 1996 and it finished its operation in 2006.

Since 1st November 1996, the Institute of Education was settled solely in Stráž pod Ralskem. However, the conditions were rather poor. Teachers' offices were in the students' hostel and theoretical lessons were held in a former nursery school. The Institute lacked the space for practical training.

Finally, two buildings opposite Stráž pod Ralskem Prison were purchased. They were interconnected by a central round part. The construction of a new school building began in autumn 1998.

2000–2019

Ceremonial inauguration of the facility took place on 31st August 2000. This year, it has been 20 years since the building opening. The Academy got its present name in 2013 and in 2016 it celebrated the 20th anniversary of the establishment of the prison staff training centre in Stráž pod Ralskem.

The efficiency of the Academy training is reflected in simple statistics:

- Between 1970 and 1994, nearly 5,000 staff members completed educational programmes.
- Between 2000 and 2019, over 50,000 persons attended the courses organized by the Academy.

The number of participants within the educational activities reached 3,776 in 2019. Out of them 623 new uniformed staff attended Type A Initial Course and nearly 500 new treatment specialists completed their initial training. It is not unusual that there are over 300 persons in the Academy on one day. The average number of lessons a teacher delivered in 2019 was 600.

3.1.3   The Academy in 2020

Main Tasks

The main tasks of the Academy do not differ from the tasks of any similar institution in Europe. It provides initial courses for new staff; it offers various specialized courses for prison practitioners; it organizes professional meetings, seminars and conferences; and it is also involved in scientific and research activities especially in the field of penal history. The Probation and Mediation Service, the Police and the Trade Unions use the learning spaces and the accommodation facilities for their activities [4].

Accommodation

To accommodate the attendees of initial courses, the Academy runs a hostel with 300 beds. A hotel is available to the participants of specialized courses, meetings, study visits and conferences.

Staff

The Academy is managed by the director and two deputies. Four departments— Education, Finance, Logistics and IT—are managed by the respective leaders. The total number of the Academy staff is 51. Two thirds of them are civilians. They are administrative and logistics staff and 6 civilian teachers. Uniformed managers and teachers make one third of the staff members.

In addition to the premises in Stráž pod Ralskem, the Academy operates a detached unit of History and Documentation of the Czech Penitentiary System in Prague. Apart from other tasks, the unit is responsible for the Czech penitentiary museum at Prague Pankrác Remand Prison. Three researchers are developing a very fruitful cooperation in the field of penal history with their French, Lower-Saxon and Slovak partners (Fig. 2).

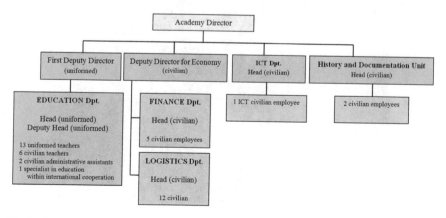

Fig. 2 Organization chart

Conditions for Education and Training

The school areas include an auditorium with 90 seats; 11 standard classrooms for theoretical lessons with the capacity of 24 students each; and one large and two small PC classrooms. Four classrooms for groups of 15 students are basically used for theoretical and practical final exams that succeed initial courses.

Practical training takes place in specialized areas. A multipurpose room with a capacity of 80 persons is suitable not only for lectures but also for specific training such as the use of the means of restraint. It is also appropriate for conference receptions. A simulation cell, a simulation prison/court entrance, a DO-JO for self-defence training, a shooting training classroom and language classrooms are very important. In 2016, a former kindergarten located opposite to the Academy was purchased and it is undergoing a thorough reconstruction. There will be a simulation centre for practical training in the future.

A permanent exhibition of illegal objects detected in Czech prisons is situated in the Academy as well. A library serves not only the prison staff but also the general public.

Sport activities, standard drill, marching or intervention techniques are carried out outdoor. A fitness room is open for students and teachers (Figs. 3, 4, 5, 6, 7 and 8).

Initial Training

Initial courses held at the Academy can be divided into three main types according to the recruits' future assignments.

New uniformed staff attend Type A Initial Course. The training takes 13 weeks.

Fig. 3 Aula

Fig. 4 Multipurpose room

Fig. 5 IT classroom

Type B Initial Course addresses the needs of new treatment specialists and other civilian employees. The initial course for psychologists, social workers, special pedagogues, educators and vocational training teachers lasts 7 weeks. One-day initial seminars for administrative and support staff are organized in prisons.

Successful students receive certificates at a ceremonial graduation meeting.

Fig. 6 Permanent exhibition of illegal objects

Fig. 7 Simulation prison/court entrance

At the end of their probationary (orientation) employment period that lasts three years, uniformed staff are invited to a short revision course and pass the professional exam. Only then they receive the unlimited contract.

Staff Enhancement Opportunities

To address prison staff's professional enhancement, the Academy opens a number of specialized courses for different groups of practitioners every year [9]. The contents

Fig. 8 Simulation cell

and duration of the courses vary according to the needs of the specific staff. Specialized courses for psychologists; social-psychological training courses; complementary courses on pedagogy; courses on mediation techniques for social workers; or courses focused on procedures of security checks at entrances to judicial buildings illustrate the fields the courses may focus on [8].

Conferences

A number of professional meetings and conferences take place at the Academy every year. The most significant conference is considered the Penology Days Conference. It is attended by a hundred of prison professionals as well as probation officers, scientists, representatives of Czech universities, NGOs and other guests. In September 2019, the 7th Penology Days Conference was held.

Innovative Programmes and Modern Handbooks

Teachers are encouraged to develop innovative educational programmes and modern manuals.

For example, English seminars for penitentiary practice and workshops on penitentiary Polish are unique staff training programmes in Europe. They combine progressive methods and specific approaches to effectively improve selected staff's professional communication in foreign languages. In a relatively short period with not too much effort, the participants are ready to take part in meetings with their partners from abroad. International professional English seminars are open to foreign prison professionals and are facilitated by distinguished experts like Doug Dretke, Executive Director of Correctional Management Institute of Texas, Sam Houston State University, Raluca Stuparu, International Relations Department Executive Member,

Romanian Prison Administration, or Jacek Matrejek, former Polish Prison Director, to name just a few.

Activities in the International Penitentiary Field

The Academy has concluded bilateral agreements with prison staff training centres in Ukraine, Slovakia, Poland, Lower Saxony, France, Romania and Latvia. A system of exchange study visits has been brought into practice. Every year, about 10 foreign delegations visit the Academy. Teachers and other staff members participate in about 10 study visits or international conferences abroad. The Academy joined EPTA (the European Penitentiary Training Association) in 2016.

4 Conclusion

This text is conceived as a reminiscence of the birth of a modern prison staff training in the Czech Republic since the beginning of 1960th. The authors discuss the origins of the epistemology foundations of philosophical and pedagogical theories of predecessors of the times, the process and their utilization within the theoretical part of the initial education of civilians and the practical part of the initial training of uniformed staff of the Prison Service of the Czech Republic. Special attention is paid to the last twenty years as a considerable development of education programmes—systematic initial training, vast long-life education opportunities, final examinations of officers after their probationary period, and implementation of good practices from abroad [11]. The Czech Prison Service Academy has partners in France, Latvia, Lower Saxony and other countries. Not only mere adapting of foreign educational programmes but mutual exchange of experience is taking place today. The peak of the academic work is represented by the cooperation with high schools and by the scientific research of the Cabinet of Documentation and History dealing with the Czech penitentiary past. The Czech Prison Service Academy has a high professional potential to organize numerous international conferences. The most significant conference is the traditional Penology Days. The training centre of the Prison Service of the Czech Republic has become a prestige educational institution [10].

References

1. Jůva V et al. (2001) Základy pedagogiky pro doplňující pedagogické studium. Paido, Brno. ISBN 80-85931-95-8
2. Jůzl M (2010) Základy pedagogiky. IMS Brno, Brno. ISBN 978-80-87182-02-4
3. Jůzl M (2017) Penitenciaristika jako věda žalářní. UJAK Praha, Praha. ISBN 978-80-7452-131-7
4. Kolitschova P, Kerbic P, Rak R (2018) Forensic and technical aspects of vehicle identification labels. In: 11th international scientific and technical conference on automative safety, Casta Papernicka, Slovakia, Apr 18–20, 2018. Proceedings paper IEEE Xplore, INSPEC

Accession Number: 17823676. https://doi.org/10.1109/AUTOSAFE.2018.8373340, UT WoS 000435296000046

5. Kopencova D (2020) Secondary education with security focus. INTED 2020 Proceedings. In: 14th international technology, education and technology and development conference. 2nd-4th March, 2020, Valencia, Spain, pp 2477–2481. ISBN: 978-84-09-17939-8, ISSN: 2340-1079. https://doi.org/10.21125/inted.2020.0755. https://library.iated.org/view/KOPENCOVA 2020CZE

6. Průcha J (2002) Moderní pedagogika. Portál, Praha. ISBN 80-7367-047-X

7. Průcha J (2006) Multikulturní výchova. Triton, Praha. ISBN 80-7254-866-2

8. Rak R (2019) Vehicle identification number—anatomy of error occurrence, MEIE 2019. In: 2nd international conference on mechanical, electric and industrial engineering, conference proceedings, 25–27 May 2019, Hangzhou, China, published on line, 2 September 2019. Journal of Physics: Conference Series (IOP Publishing: JPCS), vol 1303 (2019) 012146, Online ISSN 1742–6596. Print ISSN 1742-658. https://doi.org/10.1088/1742-6596/1303/1/012146. https://iopscience.iop.org/issue/1742-6596/1303/1

9. Rak R, Kopencova D (2020) Actual issues of modern digital vehicle forensic. Int Things Cloud Comput 8(1):12–16. ISSN: 2376-7715 (Print); ISSN: 2376-7731 (Online). https://doi.org/10.11648/j.iotcc.20200801.13 http://www.sciencepublishinggroup.com/journal/paperinfo?journalid=238&paperId=10048878

10. The Czech Prison Service Academy. https://www.vscr.cz/

11. Vlach F (2019) Analyse of the project results in the Czech Prison Service Staff in Public awareness about the importance of wellness for human life. College of PE and Sport Palestra, Prague, pp 85–98. ISBN 978-80-87723-52-4

Printed in the United States
by Baker & Taylor Publisher Services